Graduate Texts in Mathematics 100

# Graduate Texts in Mathematics

1 TAKEUTI/ZARING. Introduction to Axiomatic Set Theory. 2nd ed.
2 OXTOBY. Measure and Category. 2nd ed.
3 SCHAEFFER. Topological Vector Spaces.
4 HILTON/STAMMBACH. A Course in Homological Algebra.
5 MACLANE. Categories for the Working Mathematician.
6 HUGHES/PIPER. Projective Planes.
7 SERRE. A Course in Arithmetic.
8 TAKEUTI/ZARING. Axiometic Set Theory.
9 HUMPHREYS. Introduction to Lie Algebras and Representation Theory.
10 COHEN. A Course in Simple Homotopy Theory.
11 CONWAY. Functions of One Complex Variable. 2nd ed.
12 BEALS. Advanced Mathematical Analysis.
13 ANDERSON/FULLER. Rings and Categories of Modules.
14 GOLUBITSKY/GUILLEMIN. Stable Mappings and Their Singularities.
15 BERBERIAN. Lectures in Functional Analysis and Operator Theory.
16 WINTER. The Structure of Fields.
17 ROSENBLATT. Random Processes. 2nd ed.
18 HALMOS. Measure Theory.
19 HALMOS. A Hilbert Space Problem Book. 2nd ed., revised.
20 HUSEMOLLER. Fibre Bundles. 2nd ed.
21 HUMPHREYS. Linear Algebraic Groups.
22 BARNES/MACK. An Algebraic Introduction to Mathematical Logic.
23 GREUB. Linear Algebra. 4th ed.
24 HOLMES. Geometric Functional Analysis and its Applications.
25 HEWITT/STROMBERG. Real and Abstract Analysis.
26 MANES. Algebraic Theories.
27 KELLEY. General Topology.
28 ZARISKI/SAMUEL. Commutative Algebra. Vol. I.
29 ZARISKI/SAMUEL. Commutative Algebra. Vol. II.
30 JACOBSON. Lectures in Abstract Algebra I: Basic Concepts.
31 JACOBSON. Lectures in Abstract Algebra II: Linear Algebra.
32 JACOBSON. Lectures in Abstract Algebra III: Theory of Fields and Galois Theory.
33 HIRSCH. Differential Topology.
34 SPITZER. Principles of Random Walk. 2nd ed.
35 WERMER. Banach Algebras and Several Complex Variables. 2nd ed.
36 KELLEY/NAMIOKA et al. Linear Topological Spaces.
37 MONK. Mathematical Logic.
38 GRAUERT/FRITZSCHE. Several Complex Variables.
39 ARVESON. An Invitation to $C^*$-Algebras.
40 KEMENY/SNELL/KNAPP. Denumerable Markov Chains. 2nd ed.
41 APOSTOL. Modular Functions and Dirichlet Series in Number Theory.
42 SERRE. Linear Representations of Finite Groups.
43 GILLMAN/JERISON. Rings of Continuous Functions.
44 KENDIG. Elementary Algebraic Geometry.
45 LOÈVE. Probability Theory I. 4th ed.
46 LOÈVE. Probability Theory II. 4th ed.
47 MOISE. Geometric Topology in Dimensions 2 and 3.

*continued after Index*

Christian Berg
Jens Peter Reus Christensen
Paul Ressel

# Harmonic Analysis on Semigroups

## Theory of Positive Definite and Related Functions

Springer-Verlag
New York Berlin Heidelberg Tokyo

Christian Berg
Jens Peter Reus Christensen
Matematisk Institut
Københavns Universitet
Universitetsparken 5
DK-2100 København Ø
Denmark

Paul Ressel
Mathematisch-Geographische Fakultät
Katholische Universität Eichstätt
Residenzplatz 12
D-8078 Eichstätt
Federal Republic of Germany



---

AMS Classification (1980) *Primary*: 43-02, 43A35
*Secondary*: 20M14, 28C15, 43A05, 44A10, 44A60, 46A55, 52A07, 60E15

---

Library of Congress Cataloging in Publication Data
Berg, Christian
Harmonic analysis on semigroups.
(Graduate texts in mathematics; 100)
Bibliography: p.
Includes index.
1. Harmonic analysis. 2. Semigroups. I. Christensen, Jens Peter Reus. II. Ressel, Paul. III. Title. IV. Series.
QA403.B39 1984 515'.2433 83-20122

With 3 Illustrations.



Typeset by Composition House Ltd., Salisbury, England.
Printed and bound by R. R. Donnelley & Sons, Harrisonburg, Virginia.
Printed in the United States of America.

9 8 7 6 5 4 3 2 1

ISBN 0-387-90925-7 Springer-Verlag New York Berlin Heidelberg Tokyo
ISBN 3-540-90925-7 Springer-Verlag Berlin Heidelberg New York Tokyo

# Preface

The Fourier transform and the Laplace transform of a positive measure share, together with its moment sequence, a positive definiteness property which under certain regularity assumptions is characteristic for such expressions. This is formulated in exact terms in the famous theorems of Bochner, Bernstein–Widder and Hamburger. All three theorems can be viewed as special cases of a general theorem about functions $\varphi$ on abelian semigroups with involution $(S, +, *)$ which are positive definite in the sense that the matrix $(\varphi(s_j^* + s_k))$ is positive definite for all finite choices of elements $s_1, \ldots, s_n$ from $S$. The three basic results mentioned above correspond to $(\mathbb{R}, +, x^* = -x)$, $([0, \infty[, +, x^* = x)$ and $(\mathbb{N}_0, +, n^* = n)$.

The purpose of this book is to provide a treatment of these positive definite functions on abelian semigroups with involution. In doing so we also discuss related topics such as negative definite functions, completely monotone functions and Hoeffding-type inequalities. We view these subjects as important ingredients of harmonic analysis on semigroups. It has been our aim, simultaneously, to write a book which can serve as a textbook for an advanced graduate course, because we feel that the notion of positive definiteness is an important and basic notion which occurs in mathematics as often as the notion of a Hilbert space. The already mentioned Laplace and Fourier transformations, as well as the generating functions for integer-valued random variables, belong to the most important analytical tools in probability theory and its applications. Only recently it turned out that positive (resp. negative) definite functions allow a probabilistic characterization in terms of so-called Hoeffding-type inequalities.

As prerequisites for the reading of this book we assume the reader to be familiar with the fundamental principles of algebra, analysis and probability, including the basic notions from vector spaces, general topology and abstract measure theory and integration. On this basis we have included Chapter 1 about locally convex topological vector spaces with the main objective of proving the Hahn–Banach theorem in different versions which will be used later, in particular, in proving the Krein–Milman theorem. We also present a short introduction to the idea of integral representations in compact convex sets, mainly without proofs because the only version of Choquet's theorem which we use later is derived directly from the Krein–Milman theorem. For later use, however, we need an integration theory for measures on Hausdorff spaces, which are not necessarily locally compact. Chapter 2 contains a treatment of Radon measures, which are inner regular with respect to the family of compact sets on which they are assumed finite. The existence of Radon product measures is based on a general theorem about Radon bimeasures on a product of two Hausdorff spaces being induced by a Radon measure on the product space. Topics like the Riesz representation theorem, adapted spaces, and weak and vague convergence of measures are likewise treated.

Many results on positive and negative definite functions are not really dependent on the semigroup structure and are, in fact, true for general positive and negative definite matrices and kernels, and such results are placed in Chapter 3.

Chapters 4–8 contain the harmonic analysis on semigroups as well as a study of many concrete examples of semigroups. We will not go into detail with the content here but refer to the Contents for a quick survey. Much work is centered around the representation of positive definite functions on an abelian semigroup $(S, +, *)$ with involution as an integral of semicharacters with respect to a positive measure. It should be emphasized that most of the theory is developed without topology on the semigroup $S$. The reason for this is simply that a satisfactory general representation theorem for continuous positive definite functions on topological semigroups does not seem to be known. There is, of course, the classical theory of harmonic analysis on locally compact abelian groups, but we have decided not to include this in the exposition in order to keep it within reasonable bounds and because it can be found in many books.

As described we have tried to make the book essentially self-contained. However, we have broken this principle in a few places in order to obtain special results, but have never done it if the results were essential for later development. Most of the exercises should be easy to solve, a few are more involved and sometimes require consultations in the literature referred to. At the end of each chapter is a section called Notes and Remarks. Our aim has not been to write an encyclopedia but we hope that the historical comments are fair.

Within each chapter sections, propositions, lemmas, definitions, etc. are numbered consecutively as 1.1, 1.2, 1.3, . . . in §1, as 2.1, 2.2, 2.3, . . . in §2, and so on. When making a reference to another chapter we always add the number of that chapter, e.g. 3.1.1.

We have been fascinated by the present subject since our 1976 paper and have lectured on it on various occasions. Research projects in connection with the material presented have been supported by the Danish Natural Science Research Council, die Thyssen Stiftung, den Deutschen Akademischen Austauschdienst, det Danske Undervisningsministerium, as well as our home universities. Thanks are due to Flemming Topsøe for his advice on Chapter 2. We had the good fortune to have Bettina Mann type the manuscript and thank her for the superb typing.

*March* 1984

CHRISTIAN BERG
JENS PETER REUS CHRISTENSEN
PAUL RESSEL

# Contents

CHAPTER 1

# Introduction to Locally Convex Topological Vector Spaces and Dual Pairs

## §1. Locally Convex Vector Spaces

The purpose of this chapter is to provide a quick introduction to some of the basic aspects of the theory of topological vector spaces. Various versions of the Hahn–Banach theorem will be used later in the book and the exposition therefore centers around a fairly detailed treatment of these fundamental results. Other parts of the theory are only sketched, and we suggest that the reader consult one of the many books on the subject for further information, see e.g. Robertson and Robertson (1964), Rudin (1973) and Schaefer (1971).

**1.1.** We assume that the reader is familiar with the concept of a *vector space* $E$ over a field $\mathbb{K}$, which is always either $\mathbb{K} = \mathbb{R}$ or $\mathbb{K} = \mathbb{C}$, and of a *topology* $\mathcal{O}$ on a set $X$, where $\mathcal{O}$ means the system of open subsets of $X$.

Generally speaking, whenever a set is equipped with both an algebraic and a topological structure, we will require that the structures match in the sense that the algebraic operations become continuous mappings.

To be precise, a vector space $E$ equipped with a topology $\mathcal{O}$ is called a *topological vector space* if the mappings $(x, y) \mapsto x + y$ of $E \times E$ into $E$ and $(\lambda, x) \mapsto \lambda x$ of $\mathbb{K} \times E$ into $E$ are continuous. Here it is tacitly assumed that $E \times E$ and $\mathbb{K} \times E$ are equipped with the product topology and $\mathbb{K} = \mathbb{R}$ or $\mathbb{K} = \mathbb{C}$ with its usual topology. A topological vector space $E$ is, in particular, a *topological group* in the sense that the mappings $(x, y) \mapsto x + y$ of $E \times E$ into $E$ and $x \mapsto -x$ of $E$ into $E$ are continuous.

For each $u \in E$ the translation $\tau_u\colon x \mapsto x + u$ is a homeomorphism of $E$, so if $\mathcal{B}$ is a base for the filter $\mathcal{U}$ of neighbourhoods of zero, then $u + \mathcal{B}$ is a base for the filter of neighbourhoods of $u$. Therefore the whole topological structure of $E$ is determined by a base of neighbourhoods of the origin.

A subset $A$ of a vector space $E$ is called *absorbing* if for each $x \in E$ there exists some $M > 0$ such that $x \in \lambda A$ for all $\lambda \in \mathbb{K}$ with $|\lambda| \geqq M$; and it is called *balanced*, if $\lambda A \subseteq A$ for all $\lambda \in \mathbb{K}$ with $|\lambda| \leqq 1$. Finally, $A$ is called *absolutely convex*, if it is convex and balanced.

**1.2. Proposition.** *Let $E$ be a topological vector space and let $\mathscr{U}$ be the filter of neighbourhoods of zero. Then:*

(i) *every $U \in \mathscr{U}$ is absorbing;*
(ii) *for every $U \in \mathscr{U}$ there exists $V \in \mathscr{U}$ with $V + V \subseteq U$;*
(iii) *for every $U \in \mathscr{U}$, $b(U) = \bigcap_{|\mu| \geqq 1} \mu U$ is a balanced neighbourhood of zero contained in $U$.*

PROOF. For $a \in E$ the mapping $\lambda \mapsto \lambda a$ of $\mathbb{K}$ into $E$ is continuous at $\lambda = 0$ and this implies (i). Similarly the continuity at (0, 0) of the mapping $(x, y) \mapsto x + y$ implies (ii). Finally, by the continuity of the mapping $(\lambda, x) \mapsto \lambda x$ at $(0, 0) \in \mathbb{K} \times E$ we can associate with a given $U \in \mathscr{U}$ a number $\varepsilon > 0$ and $V \in \mathscr{U}$ such that $\lambda V \subseteq U$ for $|\lambda| \leqq \varepsilon$. Therefore

$$\varepsilon V \subseteq b(U) \subseteq U$$

so $U$ contains the balanced set $b(U)$ which is a neighbourhood of zero because $\varepsilon V$ is so, $x \mapsto \varepsilon x$ being a homeomorphism of $E$. □

From Proposition 1.2 it follows that in every topological vector space the filter $\mathscr{U}$ has a base of balanced neighbourhoods.

A topological vector space need not have a base for $\mathscr{U}$ consisting of convex sets, but the spaces we will discuss always have such a base.

**1.3. Definition.** A topological vector space $E$ over $\mathbb{K}$ is called *locally convex* if the filter of neighbourhoods of zero has a base of convex neighbourhoods.

**1.4. Proposition.** *In a locally convex topological vector space $E$ the filter of neighbourhoods of zero has a base $\mathscr{B}$ with the following properties:*

(i) *Every $U \in \mathscr{B}$ is absorbing and absolutely convex.*
(ii) *If $U \in \mathscr{B}$ and $\lambda \neq 0$, then $\lambda U \in \mathscr{B}$.*

*Conversely, given a base $\mathscr{B}$ for a filter on $E$ with the properties* (i) *and* (ii), *there is a unique topology on $E$ such that $E$ is a (locally convex) topological vector space with $\mathscr{B}$ as a base for the filter of neighbourhoods of zero.*

PROOF. If $U$ is a convex neighbourhood of zero then $b(U)$ is absolutely convex. If $\mathscr{B}_0$ is a base of convex neighbourhoods, then the family $\mathscr{B} = \{\lambda b(U) \mid U \in \mathscr{B}_0, \lambda \neq 0\}$ is a base satisfying (i) and (ii).

Conversely, suppose that $\mathscr{B}$ is a base for a filter $\mathscr{F}$ on $E$ and satisfies (i) and (ii). Then every set $U \in \mathscr{F}$ contains zero. The only possible topology on $E$ which makes $E$ to a topological vector space, and which has $\mathscr{F}$ as the filter of neighbourhoods of zero, has the filter $a + \mathscr{F}$ as filter of neighbourhoods of $a \in E$. Calling a nonempty subset $G \subseteq E$ "open" if for every $a \in G$ there exists $U \in \mathscr{B}$ such that $a + U \subseteq G$, it is easy to see that these "open" sets form a topology with $a + \mathscr{F}$ as the filter of neighbourhoods of $a$, and that $E$ is a topological vector space. □

In applications of the theory of locally convex vector spaces the topology on a given vector space $E$ is often defined by a family of seminorms.

**1.5. Definition.** A function $p: E \to [0, \infty[$ is called a *seminorm* if it has the following properties:

(i) *homogeneity*: $p(\lambda x) = |\lambda| p(x)$ for $\lambda \in \mathbb{K}$, $x \in E$;
(ii) *subadditivity*: $p(x + y) \leqq p(x) + p(y)$ for $x, y \in E$.

If, in addition, $p^{-1}(\{0\}) = \{0\}$, then $p$ is called a *norm.*

If $p$ is a seminorm and $\alpha > 0$ then the sets $\{x \in E \mid p(x) < \alpha\}$ are absolutely convex and absorbing.

For a nonempty set $A \subseteq E$, we define a mapping $p_A: E \to [0, \infty]$ by

$$p_A(x) = \inf\{\lambda > 0 \mid x \in \lambda A\}$$

(where $p_A(x) = \infty$, if the set in question is empty).

The following lemma is easy to prove.

**1.6. Lemma.** *If $A \subseteq E$ is*

(i) *absorbing, then $p_A(x) < \infty$ for $x \in E$;*
(ii) *convex, then $p_A$ is subadditive;*
(iii) *balanced, then $p_A$ is homogeneous, and*

$$\{x \in E \mid p_A(x) < 1\} \subseteq A \subseteq \{x \in E \mid p_A(x) \leqq 1\}.$$

If $A$ satisfies (i)–(iii) then $p_A$ is called the *seminorm determined by $A$.*

A seminorm $p$ satisfies $|p(x) - p(y)| \leqq p(x - y)$. In particular, if $E$ is a topological vector space then $p$ is continuous if and only if it is continuous at 0 and this is equivalent with $\{x \mid p(x) < \alpha\}$ being a neighbourhood of zero for one (and hence for all) $\alpha > 0$.

We will now see how a family $(p_i)_{i \in I}$ of seminorms on a vector space $E$ induces a topology on $E$.

**1.7. Proposition.** *There exists a coarsest topology on E with the properties that E is a topological vector space and each $p_i$ is continuous. Under this topology E is locally convex and the family of sets*

$$\{x \in E \mid p_{i_1}(x) < \varepsilon, \ldots, p_{i_n}(x) < \varepsilon\}, \qquad i_1, \ldots, i_n \in I, \quad n \in \mathbb{N}, \quad \varepsilon > 0,$$

*is a base for the filter of neighbourhoods of zero.*

PROOF. Let $\mathscr{B}$ denote the above family of sets. Then $\mathscr{B}$ is a base for a filter on $E$ having the properties (i) and (ii) of Proposition 1.4, and the unique topology asserted there is the coarsest topology on $E$ making $E$ to a topological vector space in which each $p_i$ is continuous. □

The above topology is called the *topology induced by the family* $(p_i)_{i \in I}$ *of seminorms.*

Note that in this topology a net $(x_\alpha)$ from $E$ converges to $x$ if and only if $\lim_\alpha p_i(x - x_\alpha) = 0$ for all $i \in I$.

The topology of an arbitrary locally convex topological vector space $E$ is always induced by a family of seminorms, e.g. by the family of all continuous seminorms as is easily seen by 1.4 and 1.6.

**1.8. Proposition.** *Let $E$ be a locally convex topological vector space, where the topology is induced by a family $(p_i)_{i \in I}$ of seminorms. Then $E$ is a Hausdorff space if and only if for every $x \in E \backslash \{0\}$ there exists $i \in I$ such that $p_i(x) \neq 0$.*

PROOF. Suppose $x \neq y$ and that $(p_i)_{i \in I}$ has the above separation property. Then there exist $i \in I$ and $\varepsilon > 0$ such that $p_i(x - y) = 2\varepsilon$. The sets

$$\{u \mid p_i(x - u) < \varepsilon\}, \{u \mid p_i(y - u) < \varepsilon\}$$

are open disjoint neighbourhoods of $x$ and $y$.

For the converse we prove the apparently stronger statement that the separation property of $(p_i)_{i \in I}$ is a consequence of $E$ being a $T_1$-space (i.e. the one point sets are closed). In fact, if $x \neq 0$ and $\{x\}$ is closed there exists a neighbourhood $U$ of zero such that $x \notin U$. By Proposition 1.7 there exist $\varepsilon > 0$ and finitely many indices $i_1, \ldots, i_n \in I$ such that

$$\{y \mid p_{i_1}(y) < \varepsilon, \ldots, p_{i_n}(y) < \varepsilon\} \subseteq U,$$

so for some $i \in \{i_1, \ldots, i_n\}$ we have $p_i(x) \geqq \varepsilon$. □

**1.9. Finest Locally Convex Topology.** Let $E$ be a vector space over $\mathbb{K}$. Among the topologies on $E$, which make $E$ into a locally convex topological vector space, there is a finest one, namely the topology induced by the family of all seminorms on $E$. This topology is called the *finest locally convex topology* on $E$. An alternative way of describing this topology is by saying that the system of all absorbing absolutely convex sets is a base for the filter of neighbourhoods of zero, cf. 1.4.

The finest locally convex topology is Hausdorff. In fact, let $e \in E \backslash \{0\}$ be given. We choose an algebraic basis for $E$ containing $e$ and let $\varphi$ be the linear functional determined by $\varphi(e) = 1$ and $\varphi$ being zero on the other vectors of the basis. Then $p = |\varphi|$ is a seminorm with $p(e) = 1$, and the result follows from 1.8.

Notice that every linear functional is continuous in the finest locally convex topology.

In Chapter 6 the finest locally convex topology will be used on the vector space of polynomials in one or more variables.

**1.10. Exercise.** Let $E$ be a topological vector space, and let $A, B, C, F \subseteq E$.

(a) Show that $A + B$ is open in $E$ if $A$ is open and $B$ is arbitrary.
(b) Show that $F + C$ is closed in $E$ if $F$ is closed and $C$ is compact.

**1.11. Exercise.** Let $E$ be a topological vector space. Show that the interior of a convex set is convex. Show that if $U$ is an absolutely convex neighbourhood of 0 in $E$ then its interior is absolutely convex. It follows that a locally convex topological vector space has a base for the filter of neighbourhoods of 0 consisting of open absolutely convex sets.

**1.12. Exercise.** Show that a Hausdorff topological vector space is a regular topological space. (It is actually completely regular, but that is more difficult to prove.)

**1.13. Exercise.** Let $E$ be a topological vector space and $A \subseteq E$ a nonempty and balanced subset. Then:

(i) if $A$ is open, $A = \{x \in E \mid p_A(x) < 1\}$;
(ii) if $A$ is closed, $A = \{x \in E \mid p_A(x) \leqq 1\}$.

**1.14. Exercise.** Let $p$, $q$ be two seminorms on a vector space $E$. Then if $\{x \in E \mid p(x) \leqq 1\} = \{x \in E \mid q(x) \leqq 1\}$ it follows that $p = q$.

**1.15. Exercise.** Let the topology of the locally convex vector space $E$ be induced by the family $(p_i)_{i \in I}$ of seminorms, and let $f$ be a linear functional on $E$. Then $f$ is continuous if and only if there exist $c \in ]0, \infty[$ and some finite subset $J \subseteq I$ such that $|f(x)| \leqq c \cdot \max\{p_i(x) \mid i \in J\}$ for all $x \in E$.

# §2. Hahn–Banach Theorems

One main result in the theory of locally convex topological vector spaces is the Hahn–Banach theorem about extensions of linear functionals. In the following we treat this and closely related results under the name of Hahn–Banach theorems.

We recall that a *hyperplane* $H$ in a vector space $E$ over $\mathbb{K}$ is a maximal proper linear subspace of $E$ or, equivalently, a linear subspace of codimension one (i.e. $\dim E/H = 1$). Another equivalent formulation is that a hyperplane is a set of the form $\varphi^{-1}(\{0\})$ for a linear functional $\varphi: E \to \mathbb{K}$ not identically zero.

Neither local convexity nor the Hausdorff separation property is needed in our first version of the Hahn–Banach theorem. However the existence of a nonempty open convex set $A \neq E$ is a strong implicit assumption on $E$.

**2.1. Theorem** (Geometric Version). *Let $E$ be a topological vector space over $\mathbb{K}$ and let $A$ be a nonempty open convex subset of $E$. If $M$ is a linear subspace of $E$ with $A \cap M = \emptyset$, there exists a closed hyperplane $H$ containing $M$ with $A \cap H = \emptyset$.*

PROOF. We first consider the case $\mathbb{K} = \mathbb{R}$. By Zorn's lemma there exists a maximal linear subspace $H$ of $E$ such that $M \subseteq H$ and $A \cap H = \emptyset$. Let $C = H + \bigcup_{\lambda > 0} \lambda A$.

The sum of an open set and an arbitrary set is open, hence $C$ is open, cf. Exercise 1.10. We now derive four properties of $C$ and $H$ by contradiction:

(a) $C \cap (-C) = \emptyset$.

In fact, if we assume $x \in C \cap (-C)$, we have $x = h_1 + \lambda_1 a_1 = h_2 - \lambda_2 a_2$ with $h_i \in H$, $a_i \in A$, $\lambda_i > 0$, $i = 1, 2$. By the convexity of $A$

$$\frac{\lambda_1}{\lambda_1 + \lambda_2} a_1 + \frac{\lambda_2}{\lambda_1 + \lambda_2} a_2 = \frac{1}{\lambda_1 + \lambda_2}(h_2 - h_1) \in A \cap H$$

which is impossible.

(b) $H \cup C \cup (-C) = E$.

In fact, if there exists $x \in E \setminus (H \cup C \cup (-C))$ we define $\tilde{H} = H + \mathbb{R}x$, so $H$ is a proper subspace of $\tilde{H}$. Furthermore $A \cap \tilde{H} = \emptyset$ because $y \in A \cap \tilde{H}$ can be written $y = h + \lambda x$ with $h \in H$ and $\lambda \neq 0$ ($A \cap H = \emptyset$), and then $x = (1/\lambda)y - (1/\lambda)h \in C \cup (-C)$, which is incompatible with the choice of $x$. Finally the existence of $\tilde{H}$ is inconsistent with the maximality of $H$ so (b) holds.

(c) $H \cap (C \cup (-C)) = \emptyset$.

In fact, if $x \in H \cap C$ then $x = h + \lambda a$ with $h \in H$, $a \in A$ and $\lambda > 0$, but then $a = (1/\lambda)(x - h) \in A \cap H$, which is a contradiction.

From (b) and (c) follows that $H$ is the complement of the open set $C \cup (-C)$, hence closed.

(d) $H$ is a hyperplane.

If $H$ is not a hyperplane there exists $x \in E \setminus H$ such that $\tilde{H} = H + \mathbb{R}x \neq E$. Without loss of generality we may assume $x \in C$ and we can choose $y \in (-C) \setminus \tilde{H}$. The function $f\colon [0, 1] \to E$ defined by $f(\lambda) = (1 - \lambda)x + \lambda y$ is continuous, so $f^{-1}(C)$ and $f^{-1}(-C)$ are disjoint open subsets of $[0, 1]$ containing, respectively, 0 and 1. Since $[0, 1]$ is connected there exists $\alpha \in ]0, 1[$ such that $f(\alpha) \in H$. But this implies $y = (1/\alpha)(f(\alpha) - (1 - \alpha)x) \in \tilde{H}$, which is a contradiction.

This finishes the proof of the real case.

A complex vector space can be considered as a real vector space, and if $H$ denotes a real closed hyperplane containing $M$ and such that $A \cap H = \emptyset$, then $H \cap (iH)$ is a complex hyperplane with the desired properties. □

The following important criterion for continuity of a linear functional will be used several times.

**2.2. Proposition.** *Let $E$ be a topological vector space over $\mathbb{K}$, let $\varphi: E \to \mathbb{K}$ be a nonzero linear functional and let $H = \varphi^{-1}(\{0\})$ be the corresponding hyperplane. Then precisely one of the following two statements is true:*

(i) *$\varphi$ is continuous and $H$ is closed;*
(ii) *$\varphi$ is discontinuous and $H$ is dense.*

PROOF. The closure $\bar{H}$ is a linear subspace of $E$. By the maximality of $H$ we therefore have either $H = \bar{H}$ or $\bar{H} = E$. If $\varphi$ is continuous then $H = \varphi^{-1}(\{0\})$ is closed. Suppose next that $H$ is closed. Let $a \in E \setminus H$ be chosen such that $\varphi(a) = 1$. By Proposition 1.2 there exists a balanced neighbourhood $V$ of zero such that $(a + V) \cap H = \varnothing$, and therefore $\varphi(V)$ is a balanced subset of $\mathbb{K}$ such that $0 \notin 1 + \varphi(V)$, hence $\varphi(V) \subseteq \{x \in \mathbb{K} \mid |x| < 1\}$. It follows that $|\varphi(x)| < \varepsilon$ for all $x \in \varepsilon V$, $\varepsilon > 0$, so $\varphi$ is continuous at zero, and hence continuous. □

**2.3. Theorem of Separation.** *Let $E$ be a locally convex topological vector space over $\mathbb{K}$. Suppose $F$ and $C$ are disjoint nonempty convex subsets of $E$ such that $F$ is closed and $C$ is compact. Then there exists a continuous linear functional $\varphi: E \to \mathbb{K}$ such that*

$$\sup_{x \in C} \operatorname{Re} \varphi(x) < \inf_{x \in F} \operatorname{Re} \varphi(x).$$

PROOF. Let us first suppose $\mathbb{K} = \mathbb{R}$, and consider the set $B = F - C$. Obviously $B$ is convex, and using the compactness of $C$ it may be seen that $B$ is closed, cf. Exercise 1.10. Since $F \cap C = \varnothing$ we have $0 \notin B$, so by 1.4 there exists an absolutely convex neighbourhood $U$ of 0 such that $U \cap B = \varnothing$. The interior $V$ of $U$ is an open absolutely convex neighbourhood (cf. Exercise 1.11) so $A = B + V = B - V$ is a nonempty open convex set (1.10) such that $0 \notin A$. Since $\{0\}$ is a linear subspace not intersecting $A$, there exists by Theorem 2.1 a closed hyperplane $H$ with $A \cap H = \varnothing$. Let $\varphi$ be a linear functional on $E$ with $H = \varphi^{-1}(\{0\})$. By 2.2, $\varphi$ is continuous. Now $\varphi(A)$ is a convex subset of $\mathbb{R}$, hence an interval, and since $0 \notin \varphi(A)$ we may assume $\varphi(A) \subseteq ]0, \infty[$. (If this is not the case we replace $\varphi$ by $-\varphi$). We next claim

$$\inf_{x \in B} \varphi(x) > 0,$$

which is equivalent to the assertion. If the contrary was true there exists a sequence $(x_n)$ from $B$ such that $\varphi(x_n) \to 0$. Since $V$ is absorbing there exists $u \in V$ with $\varphi(u) < 0$, but $x_n + u \in A$ so that $\varphi(x_n) + \varphi(u) > 0$ for all $n$, which is in contradiction with $\varphi(x_n) \to 0$.

In the case $\mathbb{K} = \mathbb{C}$ we consider $E$ as a real vector space and find a $\mathbb{R}$-linear functional $\varphi: E \to \mathbb{R}$ as above. To finish the proof we notice that there exists precisely one $\mathbb{C}$-linear functional $\psi: E \to \mathbb{C}$ with $\operatorname{Re} \psi = \varphi$ namely $\psi(x) = \varphi(x) - i\varphi(ix)$, which is continuous since $\varphi$ is so. □

Applying the theorem to two one-point sets we find

**2.4. Corollary.** *Let $E$ be a locally convex Hausdorff topological vector space. For $a, b \in E$, $a \neq b$, there exists a continuous linear functional $f$ on $E$ such that $f(a) \neq f(b)$.*

We shall now treat the versions of the Hahn–Banach theorem which are called extension theorems. Although they may be derived from the geometric version, we give a direct proof using Zorn's lemma.

The first extension theorem is purely algebraic and very useful in the theory of integral representations. It uses the following weakened form of the concept of a seminorm.

**2.5. Definition.** Let $E$ be a vector space. A function $p: E \to \mathbb{R}$ is called *sublinear* if it has the following properties:

(i) *positive homogeneity*: $p(\lambda x) = \lambda p(x)$ for $\lambda \geqq 0$, $x \in E$;
(ii) *subadditivity*: $p(x + y) \leqq p(x) + p(y)$ for $x, y \in E$.

A function $f: E \to \mathbb{R}$ is called *dominated* by $p$ if $f(x) \leqq p(x)$ for all $x \in E$.

**2.6. Theorem** (Extension Version). *Let $M$ be a linear subspace of a real vector space $E$ and let $p: E \to \mathbb{R}$ be a sublinear function. If $f: M \to \mathbb{R}$ is linear and dominated by $p$ on $M$, there exists a linear extension $\tilde{f}: E \to \mathbb{R}$ of $f$, which is dominated by $p$.*

PROOF. We first show that it is always possible to perform one-dimensional extensions assuming $M \neq E$.

Let $e \in E \setminus M$ and define $M' = \operatorname{span}(M \cup \{e\})$. Every element $x' \in M'$ has a unique representation as $x' = x + te$ with $x \in M$, $t \in \mathbb{R}$. For every $\alpha \in \mathbb{R}$ the functional $f'_\alpha: M' \to \mathbb{R}$ defined by $f'_\alpha(x + te) = f(x) + t\alpha$ is a linear extension of $f$. We shall see that $\alpha$ may be chosen such that $f'_\alpha$ is dominated by $p$.

By the subadditivity of $p$ we get for all $x, y \in M$

$$f(x) + f(y) = f(x + y) \leqq p(x + y) \leqq p(x - e) + p(e + y),$$

or

$$f(x) - p(x - e) \leqq p(e + y) - f(y).$$

Defining

$$k = \sup\{f(x) - p(x - e) \mid x \in M\},$$

$$K = \inf\{p(e + y) - f(y) \mid y \in M\},$$

we have

$$-\infty < k \leqq K < \infty.$$

It is easily seen that a necessary condition for $f'_\alpha$ to be dominated by $p$ on $M'$ is that $\alpha \in [k, K]$. This condition is also sufficient. In fact, for $\alpha \in [k, K]$, $x, y \in M$ and $t > 0$, we have

$$f(t^{-1}x) - p(t^{-1}x - e) \leqq \alpha \leqq p(e + t^{-1}y) - f(t^{-1}y).$$

Multiplying by $t > 0$ and rearranging yields

$$f(x) - t\alpha \leqq p(x - te), \qquad f(y) + t\alpha \leqq p(y + te)$$

and shows that $f'_\alpha$ is dominated by $p$ on $M'$.

We next consider the set $\mathscr{F}$ of pairs $(M', f')$, where $M' \supseteq M$ is a linear subspace of $E$ and $f'$ is a linear $p$-dominated extension of $f$ to $M'$. For $(M', f')$, $(M'', f'') \in \mathscr{F}$ we define $(M', f') \prec (M'', f'')$ if and only if $M' \subseteq M''$ and $f''$ is an extension of $f'$. Under this relation $\mathscr{F}$ is inductively ordered, so by Zorn's lemma there exists a maximal element $(\tilde{M}, \tilde{f})$. The preceding discussion shows that $\tilde{M} = E$, which finishes the proof. □

The following corollary was established by Choquet (1962) in his treatment of the moment problem.

**2.7. Corollary.** *Let $M$ be a linear subspace of a real vector space $E$, and let $P$ be a convex cone in $E$ such that $M + P = E$. Then every linear functional $f: M \to \mathbb{R}$, which is nonnegative on $M \cap P$, can be extended to a linear functional $\tilde{f}: E \to \mathbb{R}$ which is nonnegative on $P$.*

PROOF. On $E$ we define the order relation $x \leqq y$ by $y - x \in P$. For $x \in E$ there exist $y_1, y_2 \in M$ such that $y_1 \leqq x \leqq y_2$ because $x, -x \in M + P$. This implies that the expression

$$p(x) = \inf\{f(y) \mid y \in M, y \geqq x\}, \qquad x \in E$$

satisfies $-\infty < p(x) < \infty$, and it is clear that $p$ is sublinear and $f(x) = p(x)$ for $x \in M$. Let $\tilde{f}: E \to \mathbb{R}$ be a linear extension of $f$ which is dominated by $p$. We shall see that $\tilde{f}(x) \geqq 0$ for all $x \in P$. Indeed, for $x \in P$ we have $-x \leqq 0$ and hence $\tilde{f}(-x) \leqq p(-x) \leqq f(0) = 0$. □

**2.8. Theorem.** *Let $M$ be a linear subspace of a vector space $E$ over $\mathbb{K}$ and let $p: E \to [0, \infty[$ be a seminorm. If $f: M \to \mathbb{K}$ is linear and satisfies $|f(x)| \leqq p(x)$ for all $x \in M$, there exists a linear extension $\tilde{f}: E \to \mathbb{K}$ of $f$ which satisfies $|\tilde{f}(x)| \leqq p(x)$ for all $x \in E$.*

PROOF. The real case follows immediately from Theorem 2.6 since a seminorm $p$ is sublinear and satisfies $p(-x) = p(x)$.

In the complex case, we consider $E$ as a real vector space and extend $g = \operatorname{Re}(f)$ to a $\mathbb{R}$-linear functional $\tilde{g}: E \to \mathbb{R}$ satisfying $|\tilde{g}(x)| \leqq p(x)$ for $x \in E$. Let finally $\tilde{f}: E \to \mathbb{C}$ be the unique $\mathbb{C}$-linear functional with $\operatorname{Re}(\tilde{f}) = \tilde{g}$, i.e. $\tilde{f}(x) = \tilde{g}(x) - i\tilde{g}(ix)$ for $x \in E$. Since $\operatorname{Re}(\tilde{f}|M) = \tilde{g}|M = g = \operatorname{Re}(f)$ we necessarily have $\tilde{f}|M = f$. For $x \in E$ we choose $\alpha \in \mathbb{C}$ with $|\alpha| = 1$ such that $\alpha\tilde{f}(x) = |\tilde{f}(x)|$, and find

$$|\tilde{f}(x)| = \tilde{f}(\alpha x) = \operatorname{Re} \tilde{f}(\alpha x) = \tilde{g}(\alpha x) \leqq p(\alpha x) = |\alpha| p(x) = p(x). \qquad \square$$

**2.9. Corollary.** *Let E be a locally convex topological vector space and M a linear subspace. A continuous linear functional on M can be extended to a continuous linear functional on E.*

PROOF. There exists an absolutely convex neighbourhood $U$ of 0 in $E$ such that the linear functional $f$ on $M$ satisfies $|f(x)| \leqq 1$ for $x \in U \cap M$. Let $x \in M$ and let $\lambda > 0$ be such that $x \in \lambda U$. Then $\lambda^{-1}x \in U \cap M$ and hence $|f(x)| \leqq \lambda$. This shows that the seminorm $p_U$ determined by $U$ (cf. 1.6) satisfies $|f(x)| \leqq p_U(x)$ for $x \in M$. Let $\tilde{f}$ be a linear extension of $f$ satisfying $|\tilde{f}(x)| \leqq p_U(x)$ for $x \in E$. Then $|\tilde{f}(x)| \leqq \varepsilon$ for $x \in \varepsilon U$, which shows that $\tilde{f}$ is continuous. $\square$

**2.10.** If $E$ denotes a topological vector space we denote by $E'$ the vector space of continuous linear functionals on $E$, and $E'$ is called the *topological dual space*, which is a linear subspace of the *algebraic dual space* $E^*$ of all linear functionals on $E$.

**2.11. Exercise.** Let $E$ be a real vector space and $p$ a sublinear function on $E$. Show that

$$p(x) = \sup\{f(x) | f \in E^*, f \leqq p\}.$$

**2.12. Exercise.** Let $p_1, \ldots, p_n\colon E \to \mathbb{R}$ be sublinear functions on a real vector space $E$ and let $f\colon E \to \mathbb{R}$ be linear and satisfying $f(x) \leqq p_1(x) + \cdots + p_n(x)$ for $x \in E$. Show that there exist linear functions $f_1, \ldots, f_n\colon E \to \mathbb{R}$ such that $f = f_1 + \cdots + f_n$ and such that $f_i$ is dominated by $p_i$ for $i = 1, \ldots, n$. *Hint*: Consider the product space $E^n$.

**2.13. Exercise.** With the notation as in Theorem 2.6 we denote by $A(f, E)$ the set of linear extensions $\tilde{f}\colon E \to \mathbb{R}$ of $f$ which are dominated by $p$. Clearly $A(f, E)$ is convex. Show by a Zorn's lemma argument that $A(f, E)$ has extreme points. Let $x_0 \in E$. Show that there exists an extreme point $\tilde{f}_0$ in $A(f, E)$ such that

$$\tilde{f}_0(x_0) = \sup\{\tilde{f}(x_0) | \tilde{f} \in A(f, E)\}.$$

(For the notion of an extreme point see 2.5.1. The result of the exercise is due to Vincent-Smith (1966, private communication). For a generalization see Andenaes (1970).)

# §3. Dual Pairs

Let $\mathbb{N}_0 = \{0, 1, 2, \ldots\}$, let $E = \mathbb{R}^{\mathbb{N}_0}$ be the vector space of real sequences $s = (s_k)_{k \geqq 0}$ and let $F$ be the vector space of polynomials $p(x) = \sum_{k=0}^{n} c_k x^k$ with real coefficients. Note that $F$ can be identified with the subspace of sequences $s \in E$ with only finitely many nonzero terms. For $s \in E$ and $p \in F$ we can define

$$\langle s, p \rangle = \sum_{k=0}^{\infty} s_k c_k$$

and $\langle \cdot, \cdot \rangle$ is a bilinear mapping of $E \times F$ into $\mathbb{R}$, which clearly satisfies the axioms in the following definition, so $E$ and $F$ form a dual pair under $\langle \cdot, \cdot \rangle$.

**3.1. Definition.** Let $E$ and $F$ be vector spaces over $\mathbb{K}$ and $\langle \cdot, \cdot \rangle: E \times F \to \mathbb{K}$ a bilinear form, i.e. separately linear. We say that $E$ and $F$ form a *dual pair* under $\langle \cdot, \cdot \rangle$ if the following conditions hold:

(i) For every $e \in E \setminus \{0\}$ there exists $f \in F$ such that $\langle e, f \rangle \neq 0$.
(ii) For every $f \in F \setminus \{0\}$ there exists $e \in E$ such that $\langle e, f \rangle \neq 0$.

**3.2.** A locally convex Hausdorff topological vector space $E$ and its topological dual space $E'$ form a dual pair under the bilinear form $\langle x, \varphi \rangle = \varphi(x)$ for $x \in E$, $\varphi \in E'$. The condition (ii) is clearly true and (i) follows from Corollary 2.4.

A vector space $E$ and its algebraic dual space $E^*$ form a dual pair under the bilinear form $\langle x, \varphi \rangle = \varphi(x)$. This example is a special case of the above example if $E$ is equipped with the finest locally convex topology, cf. 1.9.

We see below that every dual pair $(E, F, \langle \cdot, \cdot \rangle)$ arises in the above way in the sense that there exist a topology $\eta$ on $E$, such that $E$ is a locally convex Hausdorff topological vector space, and an isomorphism $j: F \to E'$ such that $j(f)(e) = \langle e, f \rangle$ for $e \in E, f \in F$. Such a topology $\eta$ is called *compatible with the duality between E and F*. In general there exist many different topologies on $E$ of this kind, and we will now define one, which turns out to be the coarsest compatible with the duality and therefore is called the *weak topology*.

**3.3. Definition.** Let $E$ and $F$ be a dual pair under $\langle \cdot, \cdot \rangle$. The *weak topology* $\sigma(E, F)$ on $E$ is the topology induced by the family $(p_f)_{f \in F}$ of seminorms, where $p_f(e) = |\langle e, f \rangle|$.

Condition (i) of 3.1 implies that $\sigma(E, F)$ is Hausdorff, cf. 1.8. By reasons of symmetry there is also a weak topology $\sigma(F, E)$ on $F$.

**3.4. Proposition.** *The topology $\sigma(E, F)$ is the coarsest of the topologies compatible with the duality between E and F.*

Proof. If $\eta$ is a topology compatible with the duality then $e \mapsto \langle e, f \rangle$ is $\eta$-continuous for all $f \in F$, and so are the seminorms $(p_f)_{f \in F}$. By 1.7 it follows that $\sigma(E, F)$ is coarser than $\eta$. If $E$ is equipped with the weak topology then $e \mapsto \langle e, f \rangle$ is a continuous linear functional on $E$ for each $f \in F$, and the mapping $j\colon F \to E'$ given by $j(f)(e) = \langle e, f \rangle$ is linear and one-to-one (condition (ii) of 3.1). To see that $j$ is onto we consider a $\sigma(E, F)$-continuous linear functional $\varphi$ on $E$. By 1.7 there exists $\varepsilon > 0$ and $f_1, \ldots, f_n \in F$ such that $p_{f_i}(x) < \varepsilon$, $i = 1, \ldots, n$, implies $|\varphi(x)| \leqq 1$. This gives at once that

$$\{x \in E \mid \langle x, f_i \rangle = 0,\ i = 1, \ldots, n\} \subseteq \varphi^{-1}(\{0\}). \tag{1}$$

Let us consider the linear mapping $\psi\colon E \to \mathbb{K}^n$ defined by

$$\psi(x) = (\langle x, f_1 \rangle, \ldots, \langle x, f_n \rangle), \qquad x \in E.$$

The image $\psi(E)$ is a linear subspace of $\mathbb{K}^n$ and the inclusion (1) implies that $\tilde{\varphi}\colon \psi(E) \to \mathbb{K}$ is well defined by $\tilde{\varphi}(\psi(x)) = \varphi(x)$, $x \in E$. But a linear functional on a subspace of $\mathbb{K}^n$ may be written

$$\tilde{\varphi}(y) = \sum_{i=1}^{n} \lambda_i y_i, \qquad y \in \psi(E) \subseteq \mathbb{K}^n,$$

for a not necessarily unique vector $(\lambda_1, \ldots, \lambda_n) \in \mathbb{K}^n$, and this shows that $\varphi(x) = \langle x, f \rangle$ with $f = \sum_{i=1}^{n} \lambda_i f_i \in F$, hence $j(f) = \varphi$. □

It is only slightly more difficult to show that there is also a finest topology on $E$ compatible with the duality. This topology is called the *Mackey topology* and is denoted $\tau(E, F)$, cf. Exercise 3.13.

We now associate with each subset of one of the two vector spaces of a dual pair a subset of the other space of the pair, called the polar subset.

**3.5. Definition.** Let $E$ and $F$ be a dual pair under $\langle \cdot, \cdot \rangle$. For a subset $A \subseteq E$ the *polar* subset $A^\circ$ is given by

$$A^\circ = \{f \in F \mid \operatorname{Re}\langle e, f \rangle \leqq 1 \text{ for all } e \in A\}.$$

For $e \in E$ the set $\{e\}^\circ = \{f \in F \mid \operatorname{Re}\langle e, f \rangle \leqq 1\}$ is convex and closed in any topology $\xi$ on $F$ compatible with the duality. Therefore also

$$A^\circ = \bigcap_{e \in A} \{e\}^\circ$$

is $\xi$-closed and convex. Furthermore $0 \in A^\circ$.

**3.6. The Bipolar Theorem.** *Let $\eta$ be any topology on $E$ compatible with the duality between $E$ and $F$ and let $A \subseteq E$. The bipolar set $A^{\circ\circ} = (A^\circ)^\circ$ is the smallest $\eta$-closed and convex subset of $E$ containing $A \cup \{0\}$.*

Proof. From the above remark it follows that $A^{\circ\circ}$ is an $\eta$-closed and convex set containing $A \cup \{0\}$. To finish the proof we show that the existence of an $\eta$-closed convex set $B$ containing $A \cup \{0\}$ and a point $e \in A^{\circ\circ} \setminus B$ will lead to a contradiction. In fact, by the separation theorem (2.3) there exists an

$\eta$-continuous linear functional $\varphi\colon E \to \mathbb{K}$ and a number $\lambda \in \mathbb{R}$ such that

$$\operatorname{Re} \varphi(e) < \lambda < \inf_{b \in B} \operatorname{Re} \varphi(b).$$

Since $0 \in B$ we have $\lambda < 0$. If $f \in F$ is such that $\varphi(x) = \langle x, f \rangle$ for $x \in E$ we find

$$\sup_{b \in B} \operatorname{Re}\left\langle b, \frac{1}{\lambda} f \right\rangle < 1 < \operatorname{Re}\left\langle e, \frac{1}{\lambda} f \right\rangle.$$

The first inequality shows that $(1/\lambda) f \in A^\circ$ and the last inequality is then incompatible with $e \in A^{\circ\circ}$. □

**3.7. Remark.** If $A$ is balanced we have

$$A^\circ = \{f \in F \mid |\langle e, f \rangle| \leqq 1 \text{ for all } e \in A\}.$$

This is often used as a definition of the (absolute) polar set.

If $A$ is a cone (i.e. $\lambda A \subseteq A$ for all $\lambda \geqq 0$) we have

$$A^\circ = \{f \in F \mid \operatorname{Re}\langle e, f \rangle \leqq 0 \text{ for all } e \in A\},$$

which is a convex cone. With $A \subseteq E$ we also associate another convex cone $A^\perp \subseteq F$, which is closed in any topology on $F$ compatible with the duality between $E$ and $F$, namely

$$A^\perp = \{f \in F \mid \langle e, f \rangle \geqq 0 \text{ for all } e \in A\}.$$

Clearly $A^\perp \subseteq -A^\circ$ and if $E$ and $F$ are real vector spaces and if $A$ is a cone then $A^\perp = -A^\circ$.

For a set $A$ containing 0 the bipolar theorem states that $A^{\circ\circ}$ is the $\eta$-closed convex hull of $A$. Using translations we therefore have the following consequence of the bipolar theorem:

**3.8. Proposition.** *The closed convex hull of a subset of $E$ is the same for all topologies on $E$ compatible with a given duality.*

If $E$ is a finite dimensional vector space over $\mathbb{K}$, hence isomorphic with $\mathbb{K}^n$ where $n$ is the dimension of $E$, there is exactly one topology on $E$ compatible with the duality between $E$ and $E^*$. More generally there is exactly one Hausdorff topology on $E$ such that $E$ is a topological vector space. We will refer to this topology as the *canonical topology* of $E$. These assertions are contained in the following result.

**3.9. Proposition.** *Let $E$ be a finite dimensional subspace of a Hausdorff topological vector space $F$. Then $E$ is closed in $F$, and any algebraic isomorphism $\varphi\colon \mathbb{K}^n \to E$ ($n = \dim(E)$) is a homeomorphism, when $\mathbb{K}^n$ is equipped with the topology generated by the euclidean norm.*

PROOF. We first show by induction that any isomorphism $\varphi: \mathbb{K}^n \to E$ is a homeomorphism.

For $n = 1$ we put $\varphi(1) = e$. The continuity of scalar multiplication implies that $\varphi: \lambda \mapsto \lambda e$ is continuous from $\mathbb{K}$ to $E$. The inverse $\varphi^{-1}$ is a linear functional on $E$, and its kernel is the hyperplane $\{0\}$, which is closed since $E$ is Hausdorff. By 2.2 it follows that $\varphi^{-1}$ is continuous.

Let us assume that the above statement is true for all dimensions less than $n$ and let $\varphi: \mathbb{K}^n \to E$ be an algebraic isomorphism. As before the continuity of the algebraic operations shows that $\varphi$ is continuous. To see that $\varphi^{-1}: E \to \mathbb{K}^n$ is continuous it suffices to prove that each linear functional on $E$ is continuous. To get a contradiction let us assume that $\psi: E \to \mathbb{K}$ is a discontinuous linear functional and put $H = \psi^{-1}(\{0\})$. Then $H$ is a $(n-1)$-dimensional hyperplane, which is dense in $E$ by 2.2. Let $\|\cdot\|$ be the euclidean norm (or any norm) on $H$. By the induction hypothesis the norm topology on $H$ coincides with the topology induced from $E$, so there exists an open set $U$ in $E$ such that

$$U \cap H = \{x \in H \mid \|x\| < 1\}.$$

Since $H$ is dense in $E$ and $U$ is open, we have $\overline{U \cap H} = \overline{U}$, where the closures are in $E$. But the set $\{x \in H \mid \|x\| \leqq 1\}$ is compact in $H$, hence in $E$ and in particular closed in $E$, so we get

$$U \subseteq \overline{U} = \overline{U \cap H} \subseteq \{x \in H \mid \|x\| \leqq 1\}.$$

Since $U$ is absorbing in $E$ we get $E = H$. By this absurdity $\varphi$ is indeed a homeomorphism.

We finally show that $E$ is closed in $F$. If this is not true there exists $x \in \overline{E} \setminus E$. Then $\tilde{E} = \operatorname{span}(E \cup \{x\})$ is a $(n+1)$-dimensional space. If $e_1, \ldots, e_n$ is an algebraic basis for $E$, then $\varphi: \mathbb{K}^{n+1} \to \tilde{E}$ given by $\varphi(\lambda_1, \ldots, \lambda_n, \lambda) = \sum_{i=1}^n \lambda_i e_i + \lambda x$ is an algebraic isomorphism, hence a homeomorphism. It follows that $E$ is closed in $\tilde{E}$, hence $x \in \overline{E} \cap \tilde{E} = E$, which is a contradiction. □

**3.10. Exercise.** Let $E$ and $F$ be a dual pair under $\langle \cdot, \cdot \rangle$. Then the weak topology $\sigma(E, F)$ is the coarsest topology on $E$ for which the mappings $e \mapsto \langle e, f \rangle$ are continuous when $f$ ranges over $F$.

**3.11. Exercise** (Theorem of Alaoglu–Bourbaki). Let $E$ be a locally convex Hausdorff topological vector space with topological dual space $E'$ and let $U$ be a neighbourhood of zero in $E$. Show that $U^\circ$ is $\sigma(E', E)$-compact. *Hint*: Show that for $x \in E$ there exists $\lambda > 0$ such that $|\langle x, f \rangle| \leqq \lambda$ for all $f \in U^\circ$.

**3.12. Exercise.** Let $E, F$ be a dual pair under $\langle \cdot, \cdot \rangle$ and let $\eta$ be a topology on $E$ compatible with the duality. Let $U$ be a closed, absolutely convex neighbourhood of zero in $E$ and let $p_U$ be the seminorm determined by $U$, cf. 1.6. Show that

$$p_U(x) = \sup\{|\langle x, f \rangle| \mid f \in U^\circ\}, \qquad x \in E.$$

**3.13. Exercise** (Theorem of Mackey–Arens). Let $E$, $F$ be a dual pair under $\langle\cdot,\cdot\rangle$, and let $\mathscr{A}$ be the family of all absolutely convex and $\sigma(F, E)$-compact subsets of $F$. For $A \in \mathscr{A}$ we define

$$\|e\|_A = \sup\{|\langle e, f\rangle| \mid f \in A\}, \qquad e \in E.$$

Show that $\|\cdot\|_A$ is a seminorm on $E$. Use 3.11 and 3.12 to show that if $\eta$ is a topology on $E$ compatible with the duality then $\eta$ is induced by some subfamily of $(\|\cdot\|_A)_{A\in\mathscr{A}}$. Show finally that the topology induced by the family $(\|\cdot\|_A)_{A\in\mathscr{A}}$ is the Mackey topology, i.e. the finest topology on $E$ compatible with the duality.

## Notes and Remarks

In the period up to the 1940's most results in functional analysis were about normed spaces. The development of the theory of distributions of Schwartz was one main motivation for a study of general spaces, since the basic spaces of test functions and distributions are nonnormable in their natural topology. Today locally convex Hausdorff topological vector spaces are a natural frame for many theories and problems in functional analysis, e.g. the theory of integral representations, which we shall discuss in the next chapter. For historical information on the theory of topological vector spaces we refer the reader to the book by Dieudonné (1981).

CHAPTER 2
# Radon Measures and Integral Representations

## §1. Introduction to Radon Measures on Hausdorff Spaces

It is well known that the pure set-theoretical theory of measure and integration has its limitations, and many interesting results need a topological frame because measure spaces without an underlying "nice" topological structure may be very pathological. In classical analysis this difficulty was overcome by introducing the theory of Radon measures on locally compact spaces. On these spaces there is a particularly important one-to-one relationship between Radon measures and certain linear functionals (see below) which in many treatments on analysis leads to the definition, that a Radon measure *is* a linear functional with certain properties.

Another branch of mathematics with a need for a highly developed measure theory is probability theory. Here the class of locally compact spaces turned out to be far too narrow, partly due to the fact that an infinite dimensional topological vector space never can be locally compact. For example, it was found that the class of polish spaces (i.e. separable and completely metrizable spaces) was much more appropriate for probabilistic purposes.

Later on it became clear that a very satisfactory theory of Radon measures can be developed on arbitrary Hausdorff spaces. This has been done, for example, in L. Schwartz's monograph (1973). We shall follow an approach to Radon measure theory which has been initiated by Kisyński and developed by Topsøe. It deviates, for example, from the Schwartz–Bourbaki theory in working only with inner approximation, but we hope to show that it gives an easy and elegant access to the main results.

In the following let $X$ denote an arbitrary Hausdorff space. The natural $\sigma$-algebra on which the measures considered will be defined will always be the $\sigma$-algebra $\mathscr{B} = \mathscr{B}(X)$ of all *Borel subsets* of $X$, i.e. the $\sigma$-algebra generated by the open subsets of $X$. In our terminology a measure will always be non-negative; a measure defined on $\mathscr{B}(X)$ will be called a *Borel measure* on $X$. Later on we also need to consider $\sigma$-additive functions on $\mathscr{B}(X)$ which may assume negative values, these functions will be called *signed measures*.

**1.1. Definition.** A *Radon measure* $\mu$ on the Hausdorff space $X$ is a Borel measure on $X$ satisfying

(i) $\mu(C) < \infty$ for each compact subset $C \subseteq X$,
(ii) $\mu(B) = \sup\{\mu(C) \mid C \subseteq B, C \text{ compact}\}$ for each $B \in \mathscr{B}(X)$.

The set of all Radon measures on $X$ is denoted $M_+(X)$.

**Remark.** Many authors require a Radon measure to be *locally finite*, i.e. each point has an open neighbourhood with finite measure. There are good reasons for not having this condition as part of the definition, see Notes and Remarks at the end of this chapter. A finite Radon measure $\mu$ (i.e. $\mu(X) < \infty$) satisfies

$$\mu(B) = \inf\{\mu(G) \mid B \subseteq G, G \text{ open}\} \qquad \text{for} \quad B \in \mathscr{B}(X)$$

as is easily seen by considering the property (ii) for $B^c$. However for arbitrary Radon measures this need not be true as is shown by Exercise 1.30 below.

Let $\mathscr{K} = \mathscr{K}(X)$ denote the family of all compact subsets of $X$. Clearly the restriction to $\mathscr{K}$ of a Radon measure $\mu$ is a set function

$$\lambda\colon \mathscr{K} \to [0, \infty[$$

satisfying the axioms of a Radon content below.

**1.2. Definition.** A *Radon content* is a set function $\lambda\colon \mathscr{K} \to [0, \infty[$ which satisfies

$$\lambda(C_2) - \lambda(C_1) = \sup\{\lambda(C) \mid C \subseteq C_2 \setminus C_1, C \in \mathscr{K}\} \tag{1}$$

for all $C_1, C_2 \in \mathscr{K}$ with $C_1 \subseteq C_2$.

The key result in our approach to Radon measure theory is the extension theorem (1.4) below, the proof of which will need the following lemma.

**1.3. Lemma.** *A Radon content $\lambda$ has the following properties:*

(i) $\lambda(C_1) \leqq \lambda(C_2)$ *for all* $C_1, C_2 \in \mathscr{K}$, $C_1 \subseteq C_2$, *i.e. $\lambda$ is monotone.*
(ii) $\lambda(C_1 \cup C_2) + \lambda(C_1 \cap C_2) = \lambda(C_1) + \lambda(C_2)$, *i.e. $\lambda$ is modular.*
(iii) *If a net* $(C_\alpha)_{\alpha \in A}$ *in* $\mathscr{K}$ *is decreasing with* $C = \bigcap_\alpha C_\alpha$ *then* $\lambda(C) = \lim_\alpha \lambda(C_\alpha) = \inf_\alpha \lambda(C_\alpha)$. *In particular for a decreasing sequence* $C_1 \supseteq C_2 \supseteq \cdots$ *of compact sets we have* $\lim_{n\to\infty} \lambda(C_n) = \lambda(\bigcap_{n=1}^\infty C_n)$.

**Proof.** (i) as well as $\lambda(\emptyset) = 0$ is obvious.

(ii) We have $(C_1 \cup C_2)\backslash C_2 = C_1\backslash(C_1 \cap C_2)$ and therefore

$$\lambda(C_1 \cup C_2) - \lambda(C_2) = \lambda(C_1) - \lambda(C_1 \cap C_2)$$

as an immediate consequence of (1).

(iii) Assume that $\delta := \inf_\alpha(\lambda(C_\alpha) - \lambda(C)) > 0$. We choose a fixed set $C_{\alpha_0}$ and $C' \subseteq C_{\alpha_0}\backslash C$, $C' \in \mathscr{K}$ such that

$$\lambda(C_{\alpha_0}) - \lambda(C) - \lambda(C') < \delta.$$

Now $\bigcap_{\alpha \geqq \alpha_0}(C' \cap C_\alpha) = \emptyset$ and therefore $C' \cap C_{\alpha_1} = \emptyset$ for some $C_{\alpha_1} \subseteq C_{\alpha_0}$ since $C'$ is compact and $(C_\alpha)_{\alpha \in A}$ is decreasing. From (ii) we get

$$\begin{aligned}\lambda(C' \cup C_{\alpha_1}) &= \lambda(C') + \lambda(C_{\alpha_1}) \leqq \lambda(C_{\alpha_0}) \\ &< \lambda(C') + \lambda(C) + \delta\end{aligned}$$

implying $\lambda(C_{\alpha_1}) - \lambda(C) < \delta$, a contradiction. □

**1.4. Theorem.** *Any Radon content on a Hausdorff space has a unique extension to a Radon measure.*

**Proof.** Let $\lambda$ be a Radon content on $X$. We define for any subset $A \subseteq X$ the inner measure by

$$\lambda_*(A) := \sup\{\lambda(C) \mid C \subseteq A, C \in \mathscr{K}\}$$

and have to show that $\mu := \lambda_* \mid \mathscr{B}$ is a measure. Of course $\lambda_*$ is an extension of $\lambda$, but it may assume the value $+\infty$, if $\lambda$ is unbounded. In a certain analogy with Carathéodory's famous abstract measure extension theorem we consider the set system

$$\mathscr{A} := \{A \subseteq X \mid \lambda_*(C \cap A) + \lambda_*(C \cap A^c) = \lambda_*(C) \text{ for all } C \in \mathscr{K}\},$$

and we will show that $\mathscr{A}$ is a $\sigma$-algebra containing $\mathscr{B}$, on which the restriction of $\lambda_*$ is $\sigma$-additive.

From the very definition $\mathscr{A}$ is closed under complements and contains the empty set. The defining property (1) of a Radon content shows that $\mathscr{A}$ even contains all open subsets of $X$. Let $A_1, A_2 \in \mathscr{A}$ be disjoint and let $C_1 \subseteq A_1$, $C_2 \subseteq A_2$ be compact. Then the modularity of $\lambda$ gives

$$\lambda(C_1) + \lambda(C_2) = \lambda(C_1 \cup C_2) \leqq \lambda_*(A_1 \cup A_2)$$

and hence

$$\lambda_*(A_1) + \lambda_*(A_2) \leqq \lambda_*(A_1 \cup A_2),$$

i.e. $\lambda_*$ is "superadditive". As a consequence $\mathscr{A}$ may also be written as

$$\mathscr{A} = \{A \subseteq X \mid \lambda_*(C \cap A) + \lambda_*(C \cap A^c) \geqq \lambda_*(C) \text{ for all } C \in \mathscr{K}\}.$$

Now let a sequence $A_1, A_2, \ldots \in \mathscr{A}$ be given and fix $C \in \mathscr{K}$ as well as $\varepsilon > 0$. Then there exist compact sets $K_j \subseteq C \cap A_j$ and $L_j \subseteq C \cap A_j^c$ such that

$$\lambda(C) \leqq \lambda(K_j) + \lambda(L_j) + \frac{\varepsilon}{2^j}, \qquad j = 1, 2, \ldots. \tag{2}$$

From the modularity of $\lambda$ we get

$$\lambda\left(\bigcup_{j=1}^{n+1} K_j\right) + \lambda\left(K_{n+1} \cap \bigcup_{j=1}^{n} K_j\right) = \lambda\left(\bigcup_{j=1}^{n} K_j\right) + \lambda(K_{n+1}) \tag{3}$$

as well as

$$\lambda\left(\bigcap_{j=1}^{n+1} L_j\right) + \lambda\left(L_{n+1} \cup \bigcap_{j=1}^{n} L_j\right) = \lambda\left(\bigcap_{j=1}^{n} L_j\right) + \lambda(L_{n+1}). \tag{4}$$

We have

$$\tilde{K}_n := K_{n+1} \cap \bigcup_{j=1}^{n} K_j \subseteq C \cap \left(\bigcup_{j=1}^{n} A_j \cap A_{n+1}\right)$$

and

$$\tilde{L}_n := L_{n+1} \cup \bigcap_{j=1}^{n} L_j \subseteq C \cap \left(\bigcap_{j=1}^{n} A_j^c \cup A_{n+1}^c\right),$$

hence $\tilde{K}_n$ and $\tilde{L}_n$ are disjoint compact subsets of $C$, so that

$$\lambda(\tilde{K}_n) + \lambda(\tilde{L}_n) \leqq \lambda(C). \tag{5}$$

Adding the equalities (3) and (4) and inserting (2) and (5) give

$$\begin{aligned}\lambda\left(\bigcup_{j=1}^{n+1} K_j\right) + \lambda\left(\bigcap_{j=1}^{n+1} L_j\right) &= \lambda\left(\bigcup_{j=1}^{n} K_j\right) + \lambda\left(\bigcap_{j=1}^{n} L_j\right) + \lambda(K_{n+1}) + \lambda(L_{n+1}) \\ &\quad - \lambda(\tilde{K}_n) - \lambda(\tilde{L}_n) \\ &\geqq \lambda\left(\bigcup_{j=1}^{n} K_j\right) + \lambda\left(\bigcap_{j=1}^{n} L_j\right) - \frac{\varepsilon}{2^{n+1}}.\end{aligned} \tag{6}$$

If we add (6) over $n = 1, 2, \ldots, N-1$ and use (2) for $j = 1$ we get

$$\lambda\left(\bigcup_{j=1}^{N} K_j\right) + \lambda\left(\bigcap_{j=1}^{N} L_j\right) \geqq \lambda(C) - \varepsilon \cdot \sum_{i=1}^{N} \frac{1}{2^i}. \tag{7}$$

Put $A := \bigcup_{j=1}^{\infty} A_j$; then $\lambda(\bigcup_{j=1}^{N} K_j) \leqq \lambda_*(C \cap A)$ for all $N$ and

$$\lambda\left(\bigcap_{j=1}^{\infty} L_j\right) = \lim_{N\to\infty} \lambda\left(\bigcap_{j=1}^{N} L_j\right) \leqq \lambda_*(C \cap A^c) \tag{8}$$

by Lemma 1.3(iii), hence letting $N$ tend to infinity in (7) gives

$$\lambda_*(C \cap A) + \lambda_*(C \cap A^c) \geqq \lambda(C) - \varepsilon,$$

and since this holds for all $\varepsilon > 0$, we have in fact shown $A \in \mathscr{A}$, hence $\mathscr{A}$ is a $\sigma$-algebra containing the open sets and therefore the Borel sets.

Let us now furthermore assume that the sets $A_1, A_2, \ldots \in \mathscr{A}$ are pairwise disjoint and that $C \subseteq A$. Then $\lim_{N\to\infty} \lambda(\bigcap_{j=1}^{N} L_j) = 0$ by (8), and taking again the limit in (7) gives

$$\lambda(C) - \varepsilon \leqq \lim_{N\to\infty} \lambda\left(\bigcup_{j=1}^{N} K_j\right) = \lim_{N\to\infty} \sum_{j=1}^{N} \lambda(K_j) = \sum_{j=1}^{\infty} \lambda(K_j) \leqq \sum_{j=1}^{\infty} \lambda_*(A_j)$$

for all $\varepsilon > 0$. Letting $\varepsilon \to 0$ we find

$$\lambda_*(A) = \sup\{\lambda(C) | C \subseteq A, C \in \mathscr{K}\} \leqq \sum_{j=1}^{\infty} \lambda_*(A_j),$$

and since the reverse inequality is obvious by the superadditivity of $\lambda_*$ we have that $\lambda_* | \mathscr{A}$ is a measure, thus finishing our proof. □

The result we are now going to prove is a kind of monotone convergence theorem for Radon measures. The usual form of this theorem on general measure spaces deals with an increasing sequence of nonnegative measurable functions; however, if the underlying measure is a Radon measure and if the functions to be integrated are *lower semicontinuous* (i.e. $\{f > t\}$ is open for all $t \in \mathbb{R}$), then the sequence may be replaced by an arbitrary increasing net of functions, as we shall see.

In the sequel we shall make repeated use of the obvious inequality

$$0 \leqq f_n := \frac{1}{2^n} \sum_{i=1}^{\infty} 1_{\{f > i/2^n\}} \leqq f \tag{9}$$

being valid for arbitrary functions $f$ with values in $[0, \infty]$. If $f$ is finite the infinite series in (9) reduces to a finite sum (pointwise) and $f - f_n \leqq 1/2^n$. Note that $f_n$ increases to $f$ also if $f$ assumes the value $\infty$. Let us mention that the family of all lower semicontinuous functions is closed under finite sums, multiplication with a nonnegative constant, and that the supremum of an arbitrary subfamily of these functions is still lower semicontinuous. Noting finally that an indicator function $f = 1_G$ is lower semicontinuous if and only if $G$ is open, we see that the functions $f_n$ defined in (9) are lower semicontinuous if $f$ is.

**1.5. Theorem.** *Let $\mu$ be a Radon measure on the Hausdorff space $X$. Then the following holds:*

(i) *If a net $(G_\alpha)_{\alpha \in A}$ of open subsets of $X$ is increasing with $\bigcup_\alpha G_\alpha = G$ then*

$$\mu(G) = \sup_\alpha \mu(G_\alpha) = \lim_\alpha \mu(G_\alpha).$$

(ii) *If a net $(f_\alpha)_{\alpha \in A}$ of lower semicontinuous functions $X \to [0, \infty]$ is increasing with $\sup_\alpha f_\alpha = f$ then*

$$\int f \, d\mu = \sup_\alpha \int f_\alpha \, d\mu = \lim_\alpha \int f_\alpha \, d\mu.$$

PROOF. (i) Let $C \subseteq G$ be compact. Then finitely many $G_{\alpha_1}, \ldots, G_{\alpha_k}$ cover $C$ and by assumption there is some $\alpha_0$ such that $G_{\alpha_1} \cup \cdots \cup G_{\alpha_k} \subseteq G_{\alpha_0}$, implying $\mu(C) \leqq \mu(G_{\alpha_0}) \leqq \sup_\alpha \mu(G_\alpha)$ and therefore

$$\mu(G) = \sup\{\mu(C) | C \subseteq G, C \in \mathscr{K}\} \leqq \sup_\alpha \mu(G_\alpha).$$

The reverse inequality is trivial.

(ii) For every $t \in \mathbb{R}$ the open sets $\{f_\alpha > t\}$ increase to $\{f > t\}$. Using the functions $f_n$ and the corresponding $f_{\alpha,n}$ as defined in (9) we find

$$\int f_n \, d\mu = \frac{1}{2^n} \sum_i \mu\left(\left\{f > \frac{i}{2^n}\right\}\right) = \frac{1}{2^n} \sum_i \lim_\alpha \mu\left(\left\{f_\alpha > \frac{i}{2^n}\right\}\right)$$

$$= \lim_\alpha \frac{1}{2^n} \sum_i \mu\left(\left\{f_\alpha > \frac{i}{2^n}\right\}\right) = \lim_\alpha \int f_{\alpha,n} \, d\mu,$$

where the interchange of limits is justified, both limits being suprema, and using this device once more we get

$$\int f \, d\mu = \sup_n \int f_n \, d\mu = \sup_n \sup_\alpha \int f_{\alpha,n} \, d\mu$$

$$= \sup_\alpha \sup_n \int f_{\alpha,n} \, d\mu = \sup_\alpha \int f_\alpha \, d\mu,$$

applying, of course, the usual monotone convergence theorem. □

**1.6. Remark.** Theorem 1.5 can be applied to an upwards filtering family $A$ of sets or functions by defining an increasing net in the following way: The index set and the mapping of the net will be $A$ and the identical mapping.

A Borel measure satisfying property (i) of the above theorem is usually called a *τ-smooth* measure. The class of these measures is in general larger than the class of Radon measures, however, for finite Borel measures on locally compact spaces the two notions coincide. The generalized monotone convergence theorem expressed as property (ii) of the above theorem uses only the τ-smoothness of the underlying Radon measure and therefore remains valid for τ-smooth measures as well, see Topsøe (1970) and Varadarajan (1965).

We shall need in the following the notion of *restriction of a Radon measure* to a Borel subset. If $X$ is a Hausdorff space and $B \in \mathscr{B}(X)$, then $B$ is again a Hausdorff space with respect to the trace topology $\{B \cap G \mid G \text{ open in } X\}$ and it is easy to see that the Borel subsets of $B$ are given by

$$\mathscr{B}(B) = \{B \cap A \mid A \in \mathscr{B}(X)\} = \{D \in \mathscr{B}(X) \mid D \subseteq B\}$$

so that in fact $\mathscr{B}(B) \subseteq \mathscr{B}(X)$. For $\mu \in M_+(X)$ we now define

$$\mu \mid B: \mathscr{B}(B) \to [0, \infty]$$

as the restriction of $\mu$ to $\mathscr{B}(B)$, i.e. $(\mu \mid B)(A) := \mu(A)$ for $A \in \mathscr{B}(B)$. It is immediately seen that $\mu \mid B$ is again a Radon measure.

**1.7. Proposition.** *Let $\mu$ be a Radon measure on $X$. If the function $f: X \to [0, \infty]$ is Borel measurable, then*

$$\int f \, d\mu = \sup_{K \in \mathscr{K}} \int_K f \, d\mu, \tag{10}$$

*and if* $f: X \to [0, \infty[$ *is continuous then* $\nu: \mathscr{B}(X) \to [0, \infty]$ *defined by*

$$\nu(B) := \int_B f\, d\mu$$

*is again a Radon measure. The measure* $\nu$ *is often denoted* $f\mu$ *or* $f\,d\mu$.

PROOF. If $f = 1_B$ for some $B \in \mathscr{B}(X)$, then (10) follows from the definition of a Radon measure. It is obvious that (10) remains true if $f$ is an elementary measurable nonnegative function, i.e. $f = \sum_{i=1}^n \alpha_i 1_{B_i}$ with pairwise disjoint Borel sets $B_1, \ldots, B_n$ and $\alpha_1, \ldots, \alpha_n \geqq 0$. But it is well known that an arbitrary Borel measurable $f \geqq 0$ is the pointwise limit of some increasing sequence of elementary functions, so that the usual monotone convergence theorem and the possibility of interchanging two suprema give (10) in the general case also.

Let now $f: X \to \mathbb{R}_+$ be continuous and $\nu(B) = \int_B f\, d\mu$, $B \in \mathscr{B}(X)$. Obviously, $\nu$ is finite on compact sets. Applying (10) to the restrictions $\mu|B$ and $f|B$ we find

$$\int_B f\, d\mu = \sup\left\{\int_K f\, d\mu \,\middle|\, K \in \mathscr{K}, K \subseteq B\right\}. \qquad \square$$

**1.8.** Let $\mu$ be a Radon measure on $X$ and consider the family $\mathscr{G}$ of all open $\mu$-zero sets in $X$. The system of all finite unions of sets in $\mathscr{G}$ filters upwards to the union $G$ of all sets in $\mathscr{G}$ and $\mu(G) = 0$ by Theorem 1.5. The open set $G$ is therefore maximal in $\mathscr{G}$ and its complement is called the *support* of $\mu$ or abbreviated $\operatorname{supp}(\mu)$. It is immediate that $\operatorname{supp}(\mu)$ is closed and that

$$\operatorname{supp}(\mu) = \{x \in X \mid \mu(U) > 0 \text{ for each open set } U \text{ such that } x \in U\}.$$

Particularly simple examples of Radon measures are those with a finite support which we will call *molecular measures*, and among these are the *one-point* or *Dirac measures* $\varepsilon_x$ defined by $\varepsilon_x(\{x\}) = 1$ and $\varepsilon_x(\{x\}^c) = 0$. Of course $\operatorname{supp}(\varepsilon_x) = \{x\}$ and if $\mu = \sum_{i=1}^n \alpha_i \varepsilon_{x_i}$ is a molecular measure with $x_i \neq x_j$ for $i \neq j$, then $\operatorname{supp}(\mu) = \{x_i \mid \alpha_i > 0\}$. The set of molecular measures is denoted $\operatorname{Mol}_+(X)$.

In the usual set-theoretical measure theory, as well as in the theory of Radon measures, the notion of a product measure is of central importance. In the latter case we are immediately confronted with the following problem: Let $X$ and $Y$ be two Hausdorff spaces; then the product of the two $\sigma$-algebras of Borel sets, usually denoted $\mathscr{B}(X) \otimes \mathscr{B}(Y)$, is by definition the smallest $\sigma$-algebra on $X \times Y$ rendering the two canonical projections $\pi_X: X \times Y \to X$ and $\pi_Y: X \times Y \to Y$ measurable, i.e. $\mathscr{B}(X) \otimes \mathscr{B}(Y)$ is the $\sigma$-algebra generated by $\pi_X^{-1}(\mathscr{B}(X)) \cup \pi_Y^{-1}(\mathscr{B}(Y))$. By definition of the product topology these two projections are continuous on $X \times Y$ and therefore Borel measurable, so that always

$$\mathscr{B}(X) \otimes \mathscr{B}(Y) \subseteq \mathscr{B}(X \times Y).$$

On "nice" spaces we even have equality of these two $\sigma$-algebras on $X \times Y$, but this need not always hold, see the exercises below.

Our next goal will be to show existence and uniqueness of the product of two arbitrary Radon measures. This stands in some contrast to set-theoretical measure theory where usually $\sigma$-finiteness of the measures is required in order to guarantee a uniquely determined product measure. We begin with a lemma.

**1.9. Lemma.** *Let $Z$ be a Hausdorff space and let $\mathscr{A}$ be an algebra of subsets of $Z$ containing a base for the topology. If $\Lambda: \mathscr{A} \to [0, \infty[$ is finitely additive then $\lambda: \mathscr{K}(Z) \to [0, \infty[$ defined by*

$$\lambda(C) := \inf\{\Lambda(G) \mid C \subseteq G, G \text{ open}, G \in \mathscr{A}\}$$

*is a Radon content on $Z$.*

Proof. Let $C \subseteq Z$ be compact, then every point $x \in C$ has an open neighbourhood $G_x \in \mathscr{A}$. Finitely many of these neighbourhoods cover $C$ and their union is still in $\mathscr{A}$. Hence $\lambda(C)$ is certainly finite.

Now let two compact sets $C_1 \subseteq C_2$ be given. For $\varepsilon > 0$ there is an open set $G_1 \supseteq C_1$, $G_1 \in \mathscr{A}$ such that $\Lambda(G_1) - \lambda(C_1) < \varepsilon$. The set $C := C_2 \cap G_1^c$ is compact, too, allowing us to choose a further open set $G \in \mathscr{A}$, $G \supseteq C$ with $\Lambda(G) - \lambda(C) < \varepsilon$. Of course, $C_2 \subseteq G \cup G_1 \in \mathscr{A}$ so that $\lambda(C_2) \leqq \Lambda(G) + \Lambda(G_1)$ and therefore $\lambda(C_2) - \lambda(C_1) \leqq \Lambda(G) + \Lambda(G_1) + \varepsilon - \Lambda(G_1) < \lambda(C) + 2\varepsilon$. Hence

$$\lambda(C_2) - \lambda(C_1) \leqq \sup\{\lambda(C) \mid C \subseteq C_2 \backslash C_1, C \in \mathscr{K}\}.$$

The reverse inequality will follow immediately if we can show that $\lambda$ is additive on disjoint compact sets. Therefore let $K, L \in \mathscr{K}$ with $K \cap L = \varnothing$ be given. One direction, namely

$$\lambda(K \cup L) \leqq \lambda(K) + \lambda(L)$$

is obvious, so it remains to be shown that for arbitrary $\varepsilon > 0$

$$\lambda(K) + \lambda(L) \leqq \lambda(K \cup L) + \varepsilon.$$

By definition there is an open set $W \in \mathscr{A}$ containing $K \cup L$ such that $\Lambda(W) - \lambda(K \cup L) < \varepsilon$. The assumption made on the algebra $\mathscr{A}$ implies that $K$ and $L$ may be separated by open sets $G$, $H$ belonging to $\mathscr{A}$, i.e. we have

$$K \subseteq G, \qquad L \subseteq H, \qquad G \cap H = \varnothing.$$

Hence

$$\begin{aligned} \lambda(K) + \lambda(L) &\leqq \Lambda(G \cap W) + \Lambda(H \cap W) \\ &= \Lambda((G \cup H) \cap W) \\ &\leqq \Lambda(W) < \lambda(K \cup L) + \varepsilon, \end{aligned}$$

thus finishing the proof. □

Later on we shall need existence and unicity of certain Radon measures on the product of two Hausdorff spaces $X$ and $Y$ not only for the product of two measures, but also for so-called Radon bimeasures. If $(X, \mathscr{A})$ and $(Y, \mathscr{B})$ are just two measurable spaces (without an underlying topological structure) then a *bimeasure* $\Phi$ is by definition a function

$$\Phi: \mathscr{A} \times \mathscr{B} \to [0, \infty]$$

such that for fixed $A \in \mathscr{A}$ the partial function $B \mapsto \Phi(A, B)$ is a measure on $\mathscr{B}$ and for fixed $B \in \mathscr{B}$ the function $A \mapsto \Phi(A, B)$ is a measure on $\mathscr{A}$. Obviously, if $\kappa$ is a measure on $\mathscr{A} \otimes \mathscr{B}$, then $(A, B) \mapsto \kappa(A \times B)$ is a bimeasure, but in general not even a bounded bimeasure is induced in this way, cf. Exercise 1.31. Our next result will, however, show that for Radon bimeasures such pathologies do not exist, where by definition $\Phi$ is a *Radon bimeasure* if $\Phi$ is a bimeasure defined on $\mathscr{B}(X) \times \mathscr{B}(Y)$ such that $\Phi(K, L) < \infty$ for all compact sets $K$, $L$ and $\Phi(A, B) = \sup\{\Phi(K, L) \mid A \supseteq K \in \mathscr{K}(X),\ B \supseteq L \in \mathscr{K}(Y)\}$ for all Borel sets $A$ and $B$.

**1.10. Theorem.** *Let $X$ and $Y$ be two Hausdorff spaces and let $\Phi: \mathscr{B}(X) \times \mathscr{B}(Y) \to [0, \infty]$ denote a Radon bimeasure. Then there is a uniquely determined Radon measure $\kappa$ on $X \times Y$ with the property*

$$\Phi(K, L) = \kappa(K \times L) \qquad \textit{for all} \quad K \in \mathscr{K}(X), \quad L \in \mathscr{K}(Y).$$

*Furthermore, the equality*

$$\Phi(A, B) = \kappa(A \times B)$$

*holds for all Borel sets $A \in \mathscr{B}(X)$, $B \in \mathscr{B}(Y)$.*

PROOF. Denote $Z := X \times Y$ and let $\mathscr{A}$ be the algebra generated by the "measurable rectangles" $A \times B$, where $A \in \mathscr{B}(X)$ and $B \in \mathscr{B}(Y)$. This algebra contains, of course, the products of open sets in $X$ (resp. $Y$) and therefore a base for the topology on $Z$. It is easy to see that there is a uniquely determined finitely additive set function $\Lambda$ on $\mathscr{A}$ fulfilling

$$\Lambda(A \times B) = \Phi(A, B) \qquad \text{for all} \quad A \in \mathscr{B}(X) \quad \text{and} \quad B \in \mathscr{B}(Y).$$

Let us now first assume that $\Phi(X, Y) < \infty$. Then we may apply Lemma 1.9 which, combined with the extension theorem 1.4, shows the existence of a Radon measure $\kappa$ on $Z$ such that

$$\kappa(C) = \inf\{\Lambda(G) \mid C \subseteq G \in \mathscr{A}, G \text{ open}\}$$

for each compact set $C \subseteq Z$. If $C = K \times L$ is the product of two compact sets $K \subseteq X$, $L \subseteq Y$, then $C \in \mathscr{A}$ and

$$\kappa(K \times L) \geqq \Lambda(K \times L) = \Phi(K, L)$$

by monotonicity of $\Lambda$. On the other hand, we may use the two finite Radon measures $\mu(A) := \Phi(A, Y)$ on $X$ and $\nu(B) := \Phi(X, B)$ on $Y$ to provide us with open sets $G \supseteq K$, $H \supseteq L$ such that $\mu(G\backslash K) < \varepsilon$ and $\nu(H\backslash L) < \varepsilon$. Then

$$\Lambda(G \times H\backslash K \times L) \leqq \Lambda((G\backslash K) \times Y) + \Lambda(X \times (H\backslash L)) = \mu(G\backslash K) + \nu(H\backslash L) < 2\varepsilon,$$

and thus

$$\kappa(K \times L) \leqq \Lambda(K \times L),$$

i.e. we have the desired equality.

If $A \in \mathscr{B}(X)$, $B \in \mathscr{B}(Y)$ and $C$ is a compact subset of $A \times B$, then the projections $K := \pi_X(C)$ and $L := \pi_Y(C)$ are still compact and $C \subseteq K \times L \subseteq A \times B$, implying

$$\begin{aligned}\kappa(A \times B) &= \sup\{\kappa(C) \mid C \subseteq A \times B, C \in \mathscr{K}(Z)\} \\ &= \sup\{\kappa(K \times L) \mid K \subseteq A, L \subseteq B, K \in \mathscr{K}(X), L \in \mathscr{K}(Y)\} \\ &= \sup\{\Phi(K, L) \mid K \subseteq A, L \subseteq B, K \in \mathscr{K}(X), L \in \mathscr{K}(Y)\} \\ &= \Phi(A, B),\end{aligned}$$

using in the last equality once more that $\Phi$ is a Radon bimeasure. We also see from the preceding argument that $\kappa$ is indeed uniquely determined from its values on products of compact sets (still assuming $\Phi(X, Y) < \infty$).

In the second step we abandon the finiteness restriction on $\Phi$. For two compact sets $K \subseteq X$, $L \subseteq Y$ we know that there is a uniquely determined $\kappa_{K,L} \in M_+(K \times L)$ such that

$$\kappa_{K,L}(A \times B) = \Phi(A, B)$$

for all Borel sets $A \subseteq K$, $B \subseteq L$. Of course these measures $\kappa_{K,L}$ are compatible in the sense that $K_1 \subseteq K_2$, $L_1 \subseteq L_2$ implies

$$\kappa_{K_2,L_2} | K_1 \times L_1 = \kappa_{K_1,L_1}.$$

If now $C \subseteq Z$ is compact, then $C \subseteq K \times L$ for suitable compact sets $K \subseteq X$, $L \subseteq Y$, and irrespective of the choice of $K$ and $L$ the value

$$\kappa(C) := \kappa_{K,L}(C)$$

is well defined; furthermore, we see immediately that $\kappa$ is even a Radon content on $Z$ whose extension to a Radon measure on $Z$ we still denote by $\kappa$.

Repeating the argument already used we see that also in this case

$$\kappa(A \times B) = \Phi(A, B) \qquad \text{for all} \quad A \in \mathscr{B}(X), \quad B \in \mathscr{B}(Y).$$

Since the values $\kappa(C)$ for compact subsets $C \subseteq Z$ are uniquely determined by the values $\kappa(K \times L)$ for $K \in \mathscr{K}(X)$, $L \in \mathscr{K}(Y)$, so is finally $\kappa$ itself, thereby finishing the proof. □

A particularly important special case is the following: let $\mu \in M_+(X)$ and $\nu \in M_+(Y)$ denote two Radon measures, then $\Phi(A, B) := \mu(A) \cdot \nu(B)$ is of course a Radon bimeasure leading to

**1.11. Corollary.** *If $\mu$ and $\nu$ are two Radon measures on the Hausdorff spaces $X$ and $Y$, then there is a uniquely determined Radon measure on $X \times Y$, called the product of $\mu$ and $\nu$ and denoted $\mu \otimes \nu$, with the property*

$$\mu \otimes \nu(K \times L) = \mu(K) \cdot \nu(L) \qquad \textit{for all} \quad K \in \mathscr{K}(X), \quad L \in \mathscr{K}(Y).$$

*For all Borel sets $A \subseteq X$ and $B \subseteq Y$ we have*

$$\mu \otimes \nu(A \times B) = \mu(A) \cdot \nu(B),$$

*so that, in particular, the restriction of $\mu \otimes \nu$ to the product $\sigma$-algebra $\mathscr{B}(X) \otimes \mathscr{B}(Y)$ is a product measure of $\mu$ and $\nu$ in the usual sense.*

Later on we shall also need an amended version of the Fubini theorem, being more general in allowing the interchange of the order of integration for some Borel measurable functions on the product $X \times Y$ which are not necessarily measurable with respect to $\mathscr{B}(X) \otimes \mathscr{B}(Y)$. In particular, this interchange will be possible for all nonnegative continuous functions on $X \times Y$.

**1.12. Theorem.** *Let $\mu$ and $\nu$ be two Radon measures on the Hausdorff spaces $X$ and $Y$ and let $f: X \times Y \to [0, \infty]$ be lower semicontinuous. Then the two functions*

$$x \mapsto \int f(x, y)\, d\nu(y) \qquad \textit{and} \qquad y \mapsto \int f(x, y)\, d\mu(x) \tag{11}$$

*are again lower semicontinuous and*

$$\int_{X \times Y} f\, d(\mu \otimes \nu) = \int_X \int_Y f(x, y)\, d\nu(y)\, d\mu(x) = \int_Y \int_X f(x, y)\, d\mu(x)\, d\nu(y). \tag{12}$$

*If $\mu$ and $\nu$ are $\sigma$-finite Radon measures and $f: X \times Y \to [0, \infty]$ is Borel measurable, then the two functions in* (11) *are again Borel functions and* (12) *continues to hold.*

Proof. We know from the preceding corollary that the restriction of $\mu \otimes \nu$ to $\mathscr{B}(X) \otimes \mathscr{B}(Y)$ is a product measure in the usual sense. Let us first consider the simple case where $f = 1_{A \times B}$ for Borel sets $A \subseteq X$ and $B \subseteq Y$. Then $\int f(x, y)\, d\nu(y) = \nu(B) \cdot 1_A(x)$, $\int f(x, y)\, d\mu(x) = \mu(A) \cdot 1_B(y)$ are certainly measurable on $X$ (resp. $Y$) and (12) obviously holds. This result extends immediately to the case where $f$ is the indicator function of a set in the algebra spanned by the "measurable rectangles" $A \times B$, $A \in \mathscr{B}(X)$ and $B \in \mathscr{B}(Y)$, so that it holds, in particular, for $f = 1_U$ where $U = \bigcup_{i=1}^n (G_i \times H_i)$ and $G_i \subseteq X$, $H_i \subseteq Y$ are open sets. In this case, however, $f$ is also lower semicontinuous and we have asserted that the partial integrations in (11) yield again lower semicontinuous functions. To show this we have to make use of the sections of a subset of $X \times Y$, defined for an arbitrary $V \subseteq X \times Y$ by

$$V_x := \{y \in Y \mid (x, y) \in V\}$$

and

$$V^y := \{x \in X \mid (x, y) \in V\}.$$

All the properties of sections which we shall need follow from the simple fact that $V_x$ is the preimage of $V$ under the continuous mapping $y \mapsto (x, y)$ and the corresponding fact for $V^y$. If all the sections of $V$ are Borel sets in $X$ (resp. $Y$) (which certainly is true for $V \in \mathscr{B}(X \times Y)$), then the two functions in (11) are well defined for $f = 1_V$ and

$$\int f(x, y)\, d\nu(y) = \nu(V_x), \qquad \int f(x, y)\, d\mu(x) = \mu(V^y).$$

Now let us continue to assume $f = 1_U$ with $U = \bigcup_{i=1}^n (G_i \times H_i)$, $G_i$ and $H_i$ being open. For given $t \in \mathbb{R}_+$ let

$$D_t := \left\{\alpha \subseteq \{1, \ldots, n\} \mid \nu\left(\bigcup_{i \in \alpha} H_i\right) > t\right\},$$

then

$$\{x \in X \mid \nu(U_x) > t\} = \bigcup_{\alpha \in D_t} \bigcap_{i \in \alpha} G_i$$

is an open set, hence $\nu(U_x)$ is lower semicontinuous as a function of $x$ and, of course, $y \mapsto \mu(U^y)$ is also lower semicontinuous.

If $V \subseteq X \times Y$ is an arbitrary open set, then $V$ is the union of an upwards filtering family of open sets $U_\gamma$ of the above simple type, i.e. each $U_\gamma$ is a finite union of open rectangles. In this case

$$\nu(V_x) = \sup \nu((U_\gamma)_x) \qquad \text{and} \qquad \mu(V^y) = \sup \mu((U_\gamma)^y)$$

are again lower semicontinuous, and then Theorem 1.5 shows that (12) remains valid for $f = 1_V$. The extension to an arbitrary nonnegative lower semicontinuous function $f$ is now easily obtained using the approximating functions $f_n$ as defined in (9) and using once more Theorem 1.5.

Let us now assume that $\mu$ and $\nu$ both are finite measures and put $Z := X \times Y$. Then the set system

$$\mathscr{D} := \{V \in \mathscr{B}(Z) \mid \nu(V_x) \text{ and } \mu(V^y) \text{ are Borel measurable and (12) is valid for } f = 1_V\}$$

has the following three properties:

(i) $Z \in \mathscr{D}$

(ii) $A \in \mathscr{D} \Rightarrow A^c \in \mathscr{D}$

(iii) $A_1, A_2, \ldots \in \mathscr{D}$ pairwise disjoint $\Rightarrow \bigcup_{i=1}^\infty A_i \in \mathscr{D}$.

This means that $\mathscr{D}$ is a so-called *Dynkin class* and the main theorem about these classes is as follows (cf. Bauer 1978, Satz 2.4): If $\Omega$ is a nonempty set and $\mathscr{E}$ is a family of subsets of $\Omega$ closed under finite intersections, then the smallest Dynkin class containing $\mathscr{E}$ equals the $\sigma$-algebra generated by $\mathscr{E}$.

Applying this result in our special situation where $\Omega = Z$ and where $\mathscr{E}$ is the family of all open subsets of $Z$, we may conclude that $\mathscr{D} = \mathscr{B}(Z)$, so that (12) is indeed valid for all $f = 1_V$, $V \in \mathscr{B}(Z)$, and then, by the usual extension, for all Borel measurable $f: Z \to [0, \infty]$.

The extension to the case where $\mu$ and $\nu$ are $\sigma$-finite is completely straightforward and therefore omitted. □

**1.13.** It is, of course, a natural question to ask if equality in (12) holds for more general functions than just nonnegative lower semicontinuous ones. The following example shows that one cannot, in general, hope for too much.

Let $X$ be the unit interval with usual topology and with Lebesgue measure $\mu$, and let $Y$ be the unit interval with discrete topology (i.e. every subset is open in $Y$). On $Y$, we consider the counting measure $\nu$, i.e. $\nu(B) = \text{card}(B)$ for all $B \subseteq Y$. Both measures $\mu$ and $\nu$ are Radon measures, so that Theorem 1.12 may be applied. The diagonal $\Delta := \{(x, x) | 0 \leqq x \leqq 1\}$ is closed in $X \times Y$, hence $f := 1_\Delta$ is a bounded nonnegative upper semicontinuous function, in particular $f$ is Borel measurable. But

$$\iint 1_\Delta(x, y)\, d\mu(x)\, d\nu(y) = 0$$

and

$$\iint 1_\Delta(x, y)\, d\nu(y)\, d\mu(x) = 1.$$

**1.14.** Another important method of generating new Radon measures from given ones is the formation of *image measures*. Let $X$ and $Y$ be two Hausdorff spaces, let $\mu$ be a Radon measure on $X$ and suppose that the mapping $f: X \to Y$ is continuous. Then a set function $\mu^f$ may be defined on the Borel sets of $Y$ by

$$\mu^f(B) := \mu(f^{-1}(B)), \qquad B \in \mathscr{B}(Y)$$

and it is immediate that $\mu^f$ is $\sigma$-additive, i.e. $\mu^f$ is a Borel measure on $Y$, called the *image of $\mu$ under $f$*.

The simple example of Lebesgue measure on the real line and a constant function shows that the image of a Radon measure need not again be of this type. We have, however, the following positive result which will be sufficient in many cases of interest.

**1.15. Proposition.** *Let $X$ and $Y$ be Hausdorff spaces, let $\mu$ be a Radon measure on $X$ and suppose that $f: X \to Y$ is continuous.*

*If $\mu(f^{-1}(K)) < \infty$ for each compact set $K \subseteq Y$ then $\mu^f$ is a Radon measure.*

*This condition holds if either $\mu(X) < \infty$ or if $f$ is proper, i.e. $f^{-1}(K)$ is compact for each compact set $K \subseteq Y$.*

PROOF. We have only to verify condition (ii) of Definition 1.1 for $\mu^f$, since condition (i) is part of the assumptions. Let $B \in \mathscr{B}(Y)$ be given. For any $a < \mu^f(B) = \mu(f^{-1}(B))$ there exists a compact set $K \subseteq f^{-1}(B)$ such that $a < \mu(K)$. Now $C := f(K)$ is a compact subset of $B$ and

$$\mu^f(C) = \mu(f^{-1}(C)) \geqq \mu(K) > a$$

which shows condition (ii). □

**1.16.** Later on in this book we will work repeatedly with the so-called *convolution* of finite Radon measures on a Hausdorff topological semigroup or group. We are now going to give the precise definition of this notion. Let $S$ denote a Hausdorff topological abelian semigroup, i.e. $S$ is a Hausdorff space and there is a composition law $+: S \times S \to S$ which is assumed to be associative, commutative and continuous. For a detailed discussion of this subject see Chapter 4. Let $\mu$ and $\nu$ be two finite Radon measures on $S$. Then their *convolution* $\mu * \nu$ is defined by

$$\mu * \nu := (\mu \otimes \nu)^+,$$

i.e. as the image of the product measure $\mu \otimes \nu$ under the composition law. By the preceding proposition $\mu * \nu$ is again a finite Radon measure on $S$ and it is not difficult to see that

$$\mu * \nu = \nu * \mu$$

and

$$(\mu * \nu) * \kappa = \mu * (\nu * \kappa)$$

hold for all $\mu, \nu, \kappa \in M_+^b(S)$, the set of finite Radon measures on $S$; another way to express this is, of course, that $(M_+^b(S), *)$ is again an abelian semigroup. We shall see later (cf. 3.4) that $M_+^b(S)$ is even a topological semigroup in a naturally chosen topology.

A special case deserves mention. Suppose $S$ is an abstract abelian semigroup, i.e. no topology is given on $S$ in advance. If we then declare every subset of $S$ as open, i.e. if we equip $S$ with the so-called discrete topology, $S$ becomes a topological semigroup in which the compact subsets are just the finite ones. Every molecular measure

$$\mu = \sum_{i=1}^{n} \alpha_i \varepsilon_{s_i}$$

(where $\{s_1, \ldots, s_n\} \subseteq S$, $\{\alpha_1, \ldots, \alpha_n\} \subseteq \mathbb{R}_+$) is, of course, a finite Radon measure on $S$, so that the convolution of molecular measures is well defined.

In fact, if $\nu = \sum_{j=1}^{m} \beta_j \varepsilon_{t_j}$ is a second molecular measure on $S$ then

$$\mu * \nu = \sum_{i=1}^{n} \sum_{j=1}^{m} \alpha_i \beta_j \varepsilon_{s_i + t_j}.$$

We finish this section by proving the so-called "*localization principle*" for Radon measures which will turn out later to be very important for the proofs of several integral representation theorems.

The following lemma will be needed in the proof of the localization principle.

**1.17. Lemma.** *Let $X$ be a Hausdorff space and let $C \subseteq X$ be a compact subset covered by finitely many open sets $G_1, \ldots, G_n$, i.e. $C \subseteq G_1 \cup \cdots \cup G_n$. Then there are compact subsets $C_i \subseteq G_i$, $1 \leqq i \leqq n$, such that $C = C_1 \cup \cdots \cup C_n$.*

Proof. We use induction on $n$. For $n = 2$ we have $C \subseteq G_1 \cup G_2$, hence the disjoint compact sets $C \cap G_1^c$, $C \cap G_2^c$ can be separated by open sets $U_1$, $U_2$, i.e.

$$C \cap G_1^c \subseteq U_1, \qquad C \cap G_2^c \subseteq U_2 \qquad \text{and} \qquad U_1 \cap U_2 = \varnothing.$$

But then $C_1 := C \cap U_1^c \subseteq G_1$, $C_2 := C \cap U_2^c \subseteq G_2$ and, of course, $C = C_1 \cup C_2$.

Assuming the assertion for $n$, let now $C \subseteq G_1 \cup \cdots \cup G_{n+1} = (G_1 \cup \cdots \cup G_n) \cup G_{n+1}$. Then $C = K \cup C_{n+1}$ where $K \subseteq G_1 \cup \cdots \cup G_n$ and $C_{n+1} \subseteq G_{n+1}$ are compact. By assumption $K = C_1 \cup \cdots \cup C_n$ for compact sets $C_i \subseteq G_i$, $i \leqq n$, thus finishing the proof. □

**1.18. Theorem.** *Let $(G_\alpha)_{\alpha \in D}$ be an open covering of the Hausdorff space $X$ and on each $G_\alpha$ let a Radon measure $\mu_\alpha$ be given such that $\mu_\alpha(B) = \mu_\beta(B)$ for each pair of indices $\alpha$, $\beta \in D$ and for each Borel set $B \subseteq G_\alpha \cap G_\beta$. Then there is a uniquely determined Radon measure $\mu$ on $X$ such that $\mu(B) = \mu_\alpha(B)$ if $B$ is a Borel set contained in $G_\alpha$.*

Proof. Let $C \subseteq X$ be compact. We say that $C = \bigcup_{i=1}^{n} A_i$ is a *decomposition* of $C$ if $(A_i)$ is a finite family of pairwise disjoint Borel sets such that for each $i = 1, \ldots, n$ the (compact) closure $\bar{A}_i$ is contained in some $G_{\alpha_i}$, $\alpha_i \in D$. Decompositions always exist, because by compactness $C \subseteq \bigcup_{i=1}^{n} G_{\alpha_i}$ for suitable $\alpha_1, \ldots, \alpha_n \in D$, and by Lemma 1.17 there exist compact sets $C_i \subseteq G_{\alpha_i}$ with $C = \bigcup_{i=1}^{n} C_i$. Finally we put $A_1 := C_1$, $A_i := C_i \setminus (C_1 \cup \cdots \cup C_{i-1})$ for $i = 2, \ldots, n$.

If we have two decompositions of a compact set $C$, $C = \bigcup_{i=1}^{n} A_i = \bigcup_{j=1}^{m} B_j$ with $\bar{A}_i \subseteq G_{\alpha_i}$, $\bar{B}_j \subseteq G_{\beta_j}$, then

$$\sum_{i=1}^{n} \mu_{\alpha_i}(A_i) = \sum_{i=1}^{n} \mu_{\alpha_i}\left(\bigcup_{j=1}^{m} A_i \cap B_j\right) = \sum_{i=1}^{n} \sum_{j=1}^{m} \mu_{\alpha_i}(A_i \cap B_j)$$

$$= \sum_{j=1}^{m} \sum_{i=1}^{n} \mu_{\beta_j}(A_i \cap B_j) = \sum_{j=1}^{m} \mu_{\beta_j}\left(\bigcup_{i=1}^{n} A_i \cap B_j\right) = \sum_{j=1}^{m} \mu_{\beta_j}(B_j),$$

so that

$$\lambda(C) := \sum_{i=1}^{n} \mu_{\alpha_i}(A_i)$$

is a well-defined function $\lambda: \mathscr{K} \to [0, \infty[$. We show that $\lambda$ is a Radon content. Indeed, if $C_1, C_2$ and $L$ are compact subsets of $X$ with $C_1 \subseteq C_2, L \subseteq C_2 \backslash C_1$, and if $C_2 = A_1 \cup \cdots \cup A_n$ is a decomposition, then $C_1 = (A_1 \cap C_1) \cup \cdots \cup (A_n \cap C_1)$ and $L = (A_1 \cap L) \cup \cdots \cup (A_n \cap L)$ are decompositions, too, and

$$\begin{aligned}\lambda(C_2) - \lambda(C_1) &= \sum_{i=1}^{n} [\mu_{\alpha_i}(A_i) - \mu_{\alpha_i}(A_i \cap C_1)] \\ &= \sum_{i=1}^{n} \mu_{\alpha_i}(A_i \backslash C_1) \geqq \sum_{i=1}^{n} \mu_{\alpha_i}(A_i \cap L) = \lambda(L).\end{aligned}$$

On the other hand, given $\varepsilon > 0$, there exist compact subsets $K_i \subseteq A_i \backslash C_1$ such that $\mu_{\alpha_i}(A_i \backslash C_1) - \mu_{\alpha_i}(K_i) < \varepsilon/n$, and $K = \bigcup_{i=1}^{n} K_i$ is a decomposition of the compact set $K \subseteq C_2 \backslash C_1$; hence

$$\begin{aligned}\lambda(K) &= \sum_{i=1}^{n} \mu_{\alpha_i}(K_i) > \sum_{i=1}^{n} \mu_{\alpha_i}(A_i \backslash C_1) - \varepsilon \\ &= \lambda(C_2) - \lambda(C_1) - \varepsilon.\end{aligned}$$

Let $\mu$ be the unique extension of $\lambda$ to a Radon measure on $X$. Then if $B$ is a Borel subset of some $G_\alpha$, we have $\mu(C) = \mu_\alpha(C)$ for each compact subset $C \subseteq B$ and therefore $\mu(B) = \mu_\alpha(B)$.

If $v$ is another Radon measure on $X$ such that $v(B) = \mu_\alpha(B)$ for $B \in \mathscr{B}(X)$, $B \subseteq G_\alpha$, $\alpha \in D$, and if $C = \bigcup_{i=1}^{n} A_i$ is a decomposition of the compact set $C \subseteq X$, then

$$v(C) = \sum_{i=1}^{n} v(A_i) = \sum_{i=1}^{n} \mu_{\alpha_i}(A_i) = \lambda(C) = \mu(C)$$

implying, of course, equality of $\mu$ and $v$. □

**1.19. Exercise.** Let $X$ be a Hausdorff space with a countable base $\mathscr{D}$ (i.e. each open set in $X$ is the union of some subfamily of $\mathscr{D}$), then $\mathscr{B}(X)$ equals the $\sigma$-algebra generated by $\mathscr{D}$. If $Y$ is a further Hausdorff space with a countable base, then $\mathscr{B}(X \times Y) = \mathscr{B}(X) \otimes \mathscr{B}(Y)$.

**1.20. Exercise.** Let $\mathbb{R}_s$ be the real line equipped with the Sorgenfrey topology (i.e. a neighbourhood base of $x \in \mathbb{R}$ is given by $\{[x, a[\,|\, x < a < \infty\}$). Then $\mathbb{R}_s$ is a Hausdorff space, $\mathscr{B}(\mathbb{R}_s) = \mathscr{B}(\mathbb{R})$, but $\mathscr{B}(\mathbb{R}_s) \otimes \mathscr{B}(\mathbb{R}_s) \subsetneqq \mathscr{B}(\mathbb{R}_s \times \mathbb{R}_s)$. *Hint*: The topology induced by $\mathbb{R}_s^2$ on the second diagonal

$$\tilde{\Delta} := \{(x, -x) \,|\, x \in \mathbb{R}\}$$

is discrete.

**1.21. Exercise.** Show that any Borel measure $\mu$ on $\mathbb{R}^n$ which is finite on compact sets, is already a Radon measure.

**1.22. Exercise.** There are $\sigma$-finite measures on $\mathbb{R}$ which are not Radon measures.

**1.23. Exercise.** If $\mu$ is a Radon measure on $X$ and $A \in \mathscr{B}(X)$, then $\mu_A: \mathscr{B}(X) \to [0, \infty]$ defined by $\mu_A(B) := \mu(A \cap B)$ is again a Radon measure.

**1.24. Exercise.** If $\mu$ and $\nu$ are Radon measures on $X$ and $Y$, then $\text{supp}(\mu \otimes \nu) = \text{supp}(\mu) \times \text{supp}(\nu)$. If $\mu$ is finite and $f: X \to Y$ is continuous, then $\text{supp}(\mu^f) = \overline{f(\text{supp}(\mu))}$. If $X = Y$ is an abelian topological semigroup and $\mu$, $\nu$ are both finite, then $\text{supp}(\mu * \nu) = \overline{\text{supp}(\mu) + \text{supp}(\nu)}$.

**1.25. Exercise.** Let $X$ be a Hausdorff space and let the set function $\mu: \mathscr{B}(X) \to [0, \infty]$ be finitely additive, finite on compact sets and inner regular, i.e. $\mu(B) = \sup\{\mu(K) | B \supseteq K \in \mathscr{K}(X)\}$ for all $B \in \mathscr{B}(X)$. Then $\mu$ is already $\sigma$-additive and hence a Radon measure.

**1.26. Exercise.** If $\mu$ is a Radon measure on $X$, $\nu$ is a Radon measure on $Y$, and $f: X \times Y \to Y \times X$ is defined by $f(x, y) = (y, x)$ then $(\mu \otimes \nu)^f = \nu \otimes \mu$.

**1.27. Exercise.** Let $\mu$ and $\nu$ be two finite Radon measures on a completely regular Hausdorff space $X$. If $\int f\, d\mu \leqq \int f\, d\nu$ for each bounded nonnegative continuous function then $\mu \leqq \nu$, i.e. $\mu(B) \leqq \nu(B)$ for all Borel sets $B \subseteq X$.

**1.28. Exercise.** Let $X$ be a Hausdorff space, let $(\mu_n)$ be a sequence of Radon measures on $X$ and let the set function $\mu: \mathscr{B}(X) \to [0, \infty]$ be defined by

$$\mu(B) = \sum_{n=1}^{\infty} \mu_n(B) \qquad \text{for} \quad B \in \mathscr{B}(X).$$

Show that if $\mu(K) < \infty$ for all compact sets $K \subseteq X$ then $\mu$ is a Radon measure on $X$ and

$$\int f\, d\mu = \sum_{n=1}^{\infty} \int f\, d\mu_n$$

for all Borel measurable functions $f: X \to [0, \infty]$ and all $\mu$-integrable functions $f: X \to \mathbb{C}$.

**1.29. Exercise.** Let $X$ be a Hausdorff space, let $(\mu_\alpha)_{\alpha \in A}$ be an increasing net of Radon measures on $X$ (i.e. $\alpha, \beta \in A$, $\alpha \leqq \beta \Rightarrow \mu_\alpha(B) \leqq \mu_\beta(B)$ for all $B \in \mathscr{B}(X)$), and let the set function $\mu: \mathscr{B}(X) \to [0, \infty]$ be defined by

$$\mu(B) = \sup_{\alpha \in A} \mu_\alpha(B) = \lim_{\alpha \in A} \mu_\alpha(B) \qquad \text{for} \quad B \in \mathscr{B}(X).$$

Show that if $\mu(K) < \infty$ for all compact sets $K \subseteq X$ then $\mu$ is a Radon measure on $X$ and

$$\int f\, d\mu = \sup_{\alpha \in A} \int f\, d\mu_\alpha = \lim_{\alpha \in A} \int f\, d\mu_\alpha$$

for all Borel measurable functions $f: X \to [0, \infty]$. Furthermore,

$$\int f\, d\mu = \lim_{\alpha \in A} \int f\, d\mu_\alpha$$

for all $\mu$-integrable functions $f: X \to \mathbb{C}$. (Note that 1.28 is a special case of 1.29.)

**1.30. Exercise.** Let $X = \mathbb{N} \cup \{\infty\}$ be the Appert-Varadarajan space, i.e. all sets $\{n\} \subseteq \mathbb{N}$ are open and a subset $G \subseteq X$ containing $\infty$ is open if and only if its "density" $\lim_{n\to\infty}(1/n)|G \cap \{1, \ldots, n\}|$ equals one. Show that $X$ is a normal Hausdorff space in which only finite sets are compact. Therefore the counting measure on $X$ is a Radon measure which is not locally finite.

**1.31. Exercise.** Let $m$ denote Lebesgue measure on $[0, 1]$ and let $X_1, X_2$ be disjoint nonmeasurable subsets of $[0, 1]$ both with outer Lebesgue measure 1. (For the existence of such sets see Oxtoby (1971, pp. 23–24).) On $X_i$ we consider the Borel $\sigma$-algebra $\mathscr{B}(X_i) = \{X_i \cap B \mid B \in \mathscr{B}([0, 1])\}$, $i = 1, 2$. Show that

$$\Phi(A_1, A_2) := m(B_1 \cap B_2) \qquad \text{for} \quad A_i \in \mathscr{B}(X_i)$$

is independent of the choice of $B_i \in \mathscr{B}([0, 1])$ such that $A_i = X_i \cap B_i, i = 1, 2$, and that $\Phi$ is a bimeasure on $\mathscr{B}(X_1) \times \mathscr{B}(X_2)$. Show that there is no measure $\mu$ on $\mathscr{B}(X_1) \otimes \mathscr{B}(X_2)$ such that

$$\mu(A_1 \times A_2) = \Phi(A_1, A_2) \qquad \text{for} \quad A_i \in \mathscr{B}(X_i), \qquad i = 1, 2.$$

*Hint*: Consider the decreasing sequence of sets in $\mathscr{B}(X_1) \otimes \mathscr{B}(X_2)$

$$E_n = \bigcup_{k=0}^{2^n - 1} \left(X_1 \cap \left[\frac{k}{2^n}, \frac{k+1}{2^n}\right[\right) \times \left(X_2 \cap \left[\frac{k}{2^n}, \frac{k+1}{2^n}\right[\right).$$

# §2. The Riesz Representation Theorem

In the introduction to §1 we have mentioned the close connection between certain linear functionals and Radon measures on locally compact spaces, a connection made precise in the famous Riesz representation theorem. There is, however, a much more general integral representation theorem due to Pollard and Topsøe (1975) which implies not only numerous topological representation theorems but also, for example, the abstract Daniell extension theorem. We shall not prove this result in full generality, but instead confine our presentation to the topological setting.

Let $X$ be a Hausdorff space and let $\mathscr{C}$ be a convex cone of continuous $\mathbb{R}_+$-valued functions on $X$, separating the points in $X$ (i.e. if $x \neq y$ then $f(x) \neq f(y)$ for some $f \in \mathscr{C}$). We consider a mapping

$$T: \mathscr{C} \to [0, \infty],$$

and we shall assume that $\mathscr{C}$ and $T$ fulfil the following conditions:

(i) If $f, g \in \mathscr{C}$ then $f \wedge g \in \mathscr{C}$, $(f - g)^+ \in \mathscr{C}$ and $f \wedge 1 \in \mathscr{C}$.
(ii) $T(f + g) = T(f) + T(g)$ for all $f, g \in \mathscr{C}$.
(iii) For each $f \in \mathscr{C}$ we have

$$T(f) = \sup\{T(g) \mid f \geqq g \in \mathscr{C}, g \text{ bounded}, T(g) < \infty\}.$$

(iv) For each compact set $K \subseteq X$ there is some $f \in \mathscr{C}$ with $1_K \leqq f$ and $T(f) < \infty$.
(v) Given $f \in \mathscr{C}$ such that $T(f) < \infty$ and given $\varepsilon > 0$ there is a compact set $K \subseteq X$ such that $g \in \mathscr{C}$, $g \leqq f$ and $g \mid K = 0$ implies $T(g) < \varepsilon$.

Note that these conditions are obviously all fulfilled in the "classical" setting where $\mathscr{C} = C_+^c(X)$ is the set of all nonnegative continuous functions with compact support on a locally compact space $X$, and where $T$ is the restriction to $\mathscr{C}$ of some positive linear functional on $C^c(X)$.

An immediate consequence of (i) and (ii) is the monotonicity of $T$, i.e. we have $T(f) \leqq T(g)$ if $f \leqq g$, and this again, together with the additivity of $T$, implies

$$T(\alpha f) = \alpha T(f) \qquad \text{for all} \quad f \in \mathscr{C} \quad \text{and all} \quad \alpha \in \mathbb{R}_+.$$

One might consider condition (v) as the "crucial" one among (i) to (v); it follows from Proposition 1.7 that (v) necessarily holds if $T(f) = \int f \, d\mu$ for some Radon measure $\mu$ on $X$.

For the proof of the announced representation theorem we need the lemma below.

**2.1. Lemma.** *Assuming the above conditions on $\mathscr{C}$ the following properties hold*:

(a) *For all $x \neq y$ in $X$ there is some $f \in \mathscr{C}$ such that $f(x) = 1$ and $f(y) = 0$.*
(b) *For all $x \neq y$ in $X$ there are disjoint neighbourhoods $U$ of $x$ and $V$ of $y$ such that $f \mid U = 1$ and $f \mid V = 0$ for some $f \in \mathscr{C}$.*
(c) *If $K$ and $L$ are disjoint compact subsets of $X$ then $f \mid K = 1$ and $f \mid L = 0$ for some $f \in \mathscr{C}$.*
(d) *If $K$ and $L$ are disjoint compact subsets of $X$ then there are two functions $f, g \in \mathscr{C}$ such that $1_K \leqq f \leqq 1$, $1_L \leqq g \leqq 1$ and $f \wedge g = 0$.*

PROOF. (a) With the cone $\mathscr{C}$ being point separating we find some $h \in \mathscr{C}$ with $h(x) \neq h(y)$ and without restriction we may assume $\max\{h(x), h(y)\} = 1$.

Using property (iv) there is some $g \in \mathscr{C}$ such that $1_{\{x,y\}} \leqq g$. Now if $h(x) < h(y)$ then

$$f := \frac{1}{1 - h(x)} (g \wedge 1 - h)^+$$

belongs to $\mathscr{C}$ and $f(x) = 1, f(y) = 0$; if on the other hand $h(x) > h(y)$ then put

$$h' := \frac{1}{1 - h(y)} (g \wedge 1 - h)^+,$$

and in this case $f := (g \wedge 1 - h')^+$ has the desired properties.

(b) Using (a) we find two functions $g, h \in \mathscr{C}$ such that $g(x) = 1, g(y) = 0$, $h(x) = 0$ and $h(y) = 1$. The two sets

$$U := \{g > \tfrac{3}{4}, h < \tfrac{1}{4}\}, \qquad V := \{g < \tfrac{1}{4}, h > \tfrac{3}{4}\}$$

are open disjoint neighbourhoods of $x$ (resp. $y$), and for the functions $g' := \frac{4}{3}(g \wedge \frac{3}{4})$, $h' := \frac{4}{3}(h \wedge \frac{3}{4})$ we see that $g'|U = 1 = h'|V$ whereas $g'|V < \frac{1}{3}$ and also $h'|U < \frac{1}{3}$. Now it is clear that

$$f := [g' - (h' - g')^+]^+$$

is one on $U$ and zero on $V$.

(c) For each pair $x \in K$, $y \in L$ there are disjoint neighbourhoods $U_x$ of $x$, $V_y$ of $y$ and functions $f_{x,y} \in \mathscr{C}$ such that $f_{x,y}|U_x = 0, f_{x,y}|V_y = 1$.

Let us first fix a point $y \in L$, then there is a finite subset $\{x_1, \ldots, x_n\} \subseteq K$ such that $K \subseteq U_{x_1} \cup \cdots \cup U_{x_n}$. The function

$$f_y := \min\{f_{x_1,y}, \ldots, f_{x_n,y}\}$$

belongs to $\mathscr{C}$, equals one in some neighbourhood $W_y$ of $y$ and is zero on all of $K$. Using again (iv) we choose some $g \in \mathscr{C}$ with $1_{K \cup L} \leqq g$; then

$$h_y := (g \wedge 1 - f_y)^+$$

is zero on $W_y$ and one on $K$. Now by compactness of $L$ we find $\{y_1, \ldots, y_m\} \subseteq L$ with $L \subseteq W_{y_1} \cup \cdots \cup W_{y_m}$ and finally $f := \min\{h_{y_1}, \ldots, h_{y_m}\}$ has the desired properties.

(d) By (c) and (i) there is some $h \in \mathscr{C}$ such that $h \leqq 1$, $h|K = 1$ and $h|L = 0$; again we choose $\varphi \in \mathscr{C}$ with $1_{K \cup L} \leqq \varphi \leqq 1$. Putting

$$f := [h - (\varphi - h)^+]^+$$

and

$$g := [(\varphi - h)^+ - h]^+,$$

we have indeed $f|K = 1, g|L = 1$ and $f \wedge g = 0$. □

**2.2. Theorem.** *Let $X$ be a Hausdorff space, let $\mathscr{C}$ denote a point separating convex cone of continuous functions $f: X \to [0, \infty[$, and let $T: \mathscr{C} \to [0, \infty]$ together with $\mathscr{C}$ fulfil the above conditions* (i) – (v). *Then there is a uniquely* *determined Radon measure $\mu$ on $X$ such that*

$$T(f) = \int f \, d\mu \qquad \text{for all} \quad f \in \mathscr{C}.$$

*The measure $\mu$ is furthermore locally finite.*

PROOF. We define $\lambda\colon \mathscr{K}(X) \to [0, \infty[$ by

$$\lambda(K) := \inf\{T(f) \mid 1_K \leqq f \in \mathscr{C}\}$$

and we shall show that $\lambda$ is a Radon content. It is very easy to see that $\lambda$ is subadditive, i.e.

$$\lambda(K \cup L) \leqq \lambda(K) + \lambda(L)$$

for all $K, L \in \mathscr{K}(X)$. If, furthermore, $K \cap L = \varnothing$, then Lemma 2.1 ensures existence of $f, g \in \mathscr{C}$ such that $1_K \leqq f$, $1_L \leqq g$ and $f \wedge g = 0$. For any function $h \in \mathscr{C}$ with $1_{K \cup L} \leqq h$ we then have

$$\begin{aligned} \lambda(K) + \lambda(L) &\leqq T(h \wedge f) + T(h \wedge g) \\ &= T(h \wedge f + h \wedge g) \\ &\leqq T(h) \end{aligned}$$

and therefore

$$\lambda(K) + \lambda(L) = \lambda(K \cup L),$$

i.e. $\lambda$ is additive on disjoint compact sets, and this implies for compact sets $C_1 \subseteq C_2$ that

$$\sup\{\lambda(K) \mid K \subseteq C_2 \backslash C_1, K \in \mathscr{K}(X)\} \leqq \lambda(C_2) - \lambda(C_1).$$

To show the other direction we choose, given $\varepsilon > 0$, some $f \in \mathscr{C}$ such that $1_{C_1} \leqq f$ and $T(f) \leqq \lambda(C_1) + \varepsilon$. We also fix some number $\alpha \in ]0, 1[$ and define $K_\alpha := C_2 \cap \{f \leqq \alpha\}$. Certainly $K_\alpha$ is a compact subset of $C_2 \backslash C_1$, and if $1_{K_\alpha} \leqq g \in \mathscr{C}$ then

$$1_{C_2} \leqq g + \frac{1}{\alpha} f$$

implying

$$\begin{aligned} \lambda(C_2) &\leqq \lambda(K_\alpha) + \frac{1}{\alpha} T(f) \leqq \lambda(K_\alpha) + \frac{1}{\alpha}[\lambda(C_1) + \varepsilon] \\ &\leqq \sup\{\lambda(K) \mid K \subseteq C_2 \backslash C_1, K \in \mathscr{K}(X)\} + \frac{1}{\alpha}[\lambda(C_1) + \varepsilon]. \end{aligned}$$

Taking now on the right-hand side the limit for $\alpha \to 1$ and $\varepsilon \to 0$ we see that $\lambda$ indeed is a Radon content.

Let $\mu$ denote the Radon extension of $\lambda$. We have to show that $\mu$ represents $T$, i.e. that $T(f) = \int f \, d\mu$ for all $f \in \mathscr{C}$. To see that $\int f \, d\mu \leqq T(f)$ it is certainly enough to prove that

$$h = \sum_{i=1}^{n} a_i 1_{K_i} \leqq f$$

implies

$$\sum_{i=1}^{n} a_i \mu(K_i) \leqq T(f),$$

whenever $K_1, \ldots, K_n$ are pairwise disjoint compact sets and $a_i > 0$ for all $i = 1, \ldots, n$.

Again we use Lemma 2.1 providing us for each $1 \leqq i < j \leqq n$ with functions $f_{ij}, g_{ij} \in \mathscr{C}$ such that

$$1_{K_i} \leqq f_{ij} \leqq 1, \qquad 1_{K_j} \leqq g_{ij} \leqq 1 \quad \text{and} \quad f_{ij} \wedge g_{ij} = 0.$$

For convenience put $f_{ii} = g_{ii} = 1$ and let

$$f_i := \left(\min_{k \leqq i} g_{ki}\right) \wedge \left(\min_{j \geqq i} f_{ij}\right) \wedge \left(\frac{1}{a_i} f\right), \qquad i = 1, \ldots, n.$$

Then $f_i | K_i = 1$ and $f_i \wedge f_j = 0$ for $i \neq j$. Furthermore,

$$h \leqq \sum_{i=1}^{n} a_i f_i \leqq f,$$

whence

$$\sum_{i=1}^{n} a_i \mu(K_i) \leqq \sum_{i=1}^{n} a_i T(f_i) = T\left(\sum_{i=1}^{n} a_i f_i\right) \leqq T(f),$$

and therefore, as already remarked, $\int f \, d\mu \leqq T(f)$.

The reverse inequality remains to be shown, i.e. $T(f) \leqq \int f \, d\mu$ for each $f \in \mathscr{C}$. By (iii) we may assume $0 \leqq f \leqq 1$ and $T(f) < \infty$. Given $\varepsilon > 0$ we choose a compact set $K$ as indicated in (v). There also exists a function $h \in \mathscr{C}$ with $1_K \leqq h \leqq 1$ and $T(h) < \infty$, and then for $n$ suitably chosen we have $(1/n)T(h) < \varepsilon$.

Consider now the compact sets

$$K_j := K \cap \left\{f \geqq \frac{j}{n}\right\}, \qquad j = 1, \ldots, n$$

leading to the inequality

$$\left(f - \frac{1}{n}\right) 1_K \leqq \sum_{j=1}^{n} \frac{1}{n} 1_{K_j} \leqq f$$

and thus to

$$\int f \, d\mu \geqq \frac{1}{n} \sum_{j=1}^{n} \mu(K_j) \geqq \inf\left\{T(g) | g \in \mathscr{C}, g \geqq \sum_{j=1}^{n} \frac{1}{n} 1_{K_j}\right\},$$

where the last inequality is an immediate consequence of the definition of $\mu$ on $\mathscr{K}(X)$, i.e. the definition of the Radon content $\lambda$. (In view of what has already been shown, this last inequality is in fact an equality.)

Let us now choose a $g \in \mathscr{C}$, $g \geqq (1/n) \sum_{j=1}^{n} 1_{K_j}$, such that

$$T(g) \leqq \frac{1}{n} \sum_{j=1}^{n} \mu(K_j) + \varepsilon$$

and without restriction $g \leqq f$. Then

$$f' := \left(f - g - \frac{h}{n}\right)^{+} \leqq f, \qquad f' | K = 0$$

and therefore $T(f') < \varepsilon$ by choice of $K$. But

$$f - g \leqq f' + \frac{h}{n},$$

and so

$$T(f - g) = T(f) - T(g) \leqq 2\varepsilon,$$

and finally

$$T(f) \leqq T(g) + 2\varepsilon \leqq \frac{1}{n} \sum_{j=1}^{n} \mu(K_j) + 3\varepsilon \leqq \int f \, d\mu + 3\varepsilon.$$

We now show uniqueness of the representing measure. Let $\nu$ be some Radon measure on $X$ which also represents $T$, i.e. $\int f \, d\nu = T(f) = \int f \, d\mu$ for all $f \in \mathscr{C}$. Given $K \in \mathscr{K}(X)$ and $1_K \leqq f \in \mathscr{C}$ we have

$$\nu(K) \leqq \int f \, d\nu = T(f)$$

and therefore $\nu(K) \leqq \mu(K)$ by construction of $\mu$. For the other direction we choose some $h \in \mathscr{C}$ such that $1_K \leqq h \leqq 1$ and $T(h) < \infty$. By 1.7 the set function $B \mapsto \int_B h \, d\nu$ is a finite Radon measure on $X$, hence given $\varepsilon > 0$ there is a compact set $L$ disjoint with $K$ such that

$$\int_{(K \cup L)^c} h \, d\nu < \varepsilon.$$

Once more by Lemma 2.1 there is some $f \in \mathscr{C}$ with $f | K = 1, f | L = 0$ and without restriction $f \leqq h$, implying

$$\mu(K) \leqq T(f) = \int f \, d\nu = \int_K f \, d\nu + \int_L f \, d\nu + \int_{(K \cup L)^c} f \, d\nu$$

$$\leqq \nu(K) + \int_{(K \cup L)^c} h \, d\nu < \nu(K) + \varepsilon.$$

Hence $\mu$ and $\nu$ agree on compact sets and are therefore equal.

The local finiteness of $\mu$ results as an immediate consequence of property (iv). □

If $X$ is a Hausdorff space we let $C(X)$ denote the vector space of all real-valued continuous functions on $X$. Three subspaces of $C(X)$ deserve particular interest: the space $C^0(X)$ of continuous functions "tending to zero at infinity" in the sense that $\{|f| \geqq \varepsilon\}$ is compact for each $\varepsilon > 0$, the space $C^c(X)$ of continuous functions with compact support, where the support of $f \in C(X)$ is defined by

$$\operatorname{supp}(f) := \overline{\{f \neq 0\}},$$

and the space $C^b(X)$ of bounded continuous functions, being a Banach space with respect to the supremum norm. Clearly,

$$C^c(X) \subseteq C^0(X) \subseteq C^b(X) \subseteq C(X)$$

and the four spaces coincide in case $X$ is compact. If $X$ is locally compact, then it follows from Urysohn's lemma that $C^0(X)$ is the uniform closure of $C^c(X)$.

As a corollary we now get the classical Riesz representation theorem.

**2.3. Corollary.** *Let $X$ be locally compact. Then there is a bijection between all positive linear functionals $T$ on $C^c(X)$ and all Radon measures $\mu$ on $X$ given by*

$$T(f) = \int f \, d\mu, \qquad f \in C^c(X).$$

In the following we also have to consider $\sigma$-additive set-functions which may assume negative values. If $\mu_1$ and $\mu_2$ are two finite measures on a measurable space $(X, \mathscr{B})$ then $\mu := \mu_1 - \mu_2$ is of this type and will be called a *finite signed measure*. Conversely, the Jordan–Hahn decomposition theorem tells that any $\sigma$-additive set function $\mu: \mathscr{B} \to \mathbb{R}$ is representable as the difference of two finite measures $\mu_1$ and $\mu_2$, even in such a way that $\mu_1$ and $\mu_2$ are concentrated on disjoint measurable sets. It follows in particular that a $\sigma$-additive real-valued function defined on a $\sigma$-algebra is bounded. Any difference of two Radon measures will be called a *signed Radon measure.* This is perfectly well-defined when dealing with finite Radon measures. If infinite Radon measures are involved, the difference is as a set function only well-defined on the family of all relatively compact Borel sets.

In the first chapter we have defined for any topological vector space $E$ the topological dual $E'$ consisting of all scalar-valued continuous linear functions on $E$. We shall now identify $E'$ in some cases where $E$ is a certain space of continuous functions. At first we shall treat the case where $X$ is compact and $C(X)$ denotes the space of all real-valued continuous functions on $X$. The space $C(X)$ will be given the sup-norm; it is well known that $C(X)$ in this norm is a Banach space, and that $(C(X))'$ is a Banach space, too, if we define the norm of $T \in (C(X))'$ by

$$\|T\| := \sup_{\|f\| \leqq 1} |T(f)|.$$

The set $M(X)$ of all signed Radon measures on $X$ will be given the *total variation* norm; if $\mu \in M(X)$ then

$$\|\mu\| := \sup\{|\mu(A)| + |\mu(A^c)| \mid A \in \mathscr{B}(X)\}.$$

**2.4. Theorem.** *If $X$ is compact then $(C(X))' = M(X)$ in the sense that there is a bijective linear isometry $\mu \mapsto T_\mu$ from $M(X)$ onto $(C(X))'$, given by the natural mapping*

$$T_\mu(f) = \int f\, d\mu,$$

*where $\int f\, d\mu := \int f\, d\mu_1 - \int f\, d\mu_2$ is independent of the choice of $\mu_1, \mu_2 \in M_+(X)$ such that $\mu = \mu_1 - \mu_2$.*

PROOF. At first we remark that $T_\mu \in (C(X))'$ for $\mu \in M(X)$. For let $\mu = \mu_1 - \mu_2$ with (positive) Radon measures $\mu_1, \mu_2$ being concentrated on disjoint Borel sets; then

$$\left|\int f\, d\mu\right| \leqq \int |f|\, d\mu_1 + \int |f|\, d\mu_2 \leqq \|f\| \cdot \|\mu\|$$

showing also $\|T_\mu\| \leqq \|\mu\|$.

Let now $T \in (C(X))'$ be given. We then define $T^+ : C_+(X) \to \mathbb{R}$ by

$$T^+(f) := \sup\{T(h) \mid f \geqq h \in C_+(X)\}, \qquad f \in C_+(X).$$

It follows that $0 \leqq T^+(f) \leqq \|T\|\ \|f\| < \infty$, and for $f, g \in C_+(X)$ we have

$$T^+(f + g) \geqq T^+(f) + T^+(g).$$

Let $f, g, h \in C_+(X)$ such that $h \leqq f + g$. We put

$$h'(x) := \begin{cases} \dfrac{h(x)f(x)}{f(x) + g(x)} & \text{if } f(x) + g(x) > 0, \\ 0 & \text{if } f(x) = g(x) = 0, \end{cases}$$

$$h''(x) := \begin{cases} \dfrac{h(x)g(x)}{f(x) + g(x)} & \text{if } f(x) + g(x) > 0, \\ 0 & \text{if } f(x) = g(x) = 0, \end{cases}$$

then $h', h'' \in C_+(X)$, $h' \leqq f$, $h'' \leqq g$ and $h' + h'' = h$, implying

$$T(h) \leqq T^+(f) + T^+(g)$$

so that finally $T^+$ is additive on $C_+(X)$.

We put $T^- := T^+ - T$ which also is additive, nonnegative and positively homogeneous on $C_+(X)$. By Corollary 2.3 there are two Radon measures $\mu_1, \mu_2$ on $X$ such that

$$T^+(f) = \int f\, d\mu_1, \qquad T^-(f) = \int f\, d\mu_2, \qquad f \in C_+(X),$$

implying for $f \in C(X)$

$$\begin{aligned} T(f) &= T(f^+) - T(f^-) = T^+(f^+) - T^-(f^+) - [T^+(f^-) - T^-(f^-)] \\ &= \int f^+ \, d(\mu_1 - \mu_2) - \int f^- \, d(\mu_1 - \mu_2) \\ &= \int f \, d\mu, \qquad \mu := \mu_1 - \mu_2 \end{aligned}$$

so that $T = T_\mu$. By Corollary 2.3 we also get immediately that there is only one signed Radon measure $\mu$ with this property.

Let $\mu = \mu^+ - \mu^-$ denote the Jordan–Hahn decomposition of $\mu$, i.e. $\mu^+(B) = \mu(B \cap D)$ and $\mu^-(B) = \mu(B \cap D^c)$ where $D \in \mathscr{B}(X)$ is chosen in such a way that both $\mu^+$ and $\mu^-$ are nonnegative (see, for example, Billingsley (1979, p. 373)). Then $\mu^+$ and $\mu^-$ are Radon measures and furthermore

$$\begin{aligned} \int f \, d\mu_1 &= \sup\left\{\int h \, d\mu^+ - \int h \, d\mu^- \,|\, 0 \leqq h \leqq f, h \in C(X)\right\} \\ &\leqq \int f \, d\mu^+ \end{aligned}$$

for all $f \in C_+(X)$, hence $\mu_1 \leqq \mu^+$ by Exercise 1.27 and similarly $\mu_2 \leqq \mu^-$. Consequently we get $\mu_1(D^c) = 0 = \mu_2(D)$ and therefore $\mu_1 = \mu^+$, $\mu_2 = \mu^-$ as well as $\|\mu\| = \mu_1(D) + \mu_2(D^c)$.

We still have to show the reverse inequality $\|T_\mu\| \geqq \|\mu\|$. Given $\varepsilon > 0$ we choose compact sets $K^+ \subseteq D$, $K^- \subseteq D^c$ such that

$$\mu_1(D \backslash K^+) < \varepsilon, \qquad \mu_2(D^c \backslash K^-) < \varepsilon.$$

By Urysohn's lemma there is a continuous function $f: X \to [-1, 1]$ such that $f \,|\, K^+ \equiv 1$ and $f \,|\, K^- \equiv -1$, implying

$$\left|\int f \, d\mu\right| \geqq \mu_1(K^+) + \mu_2(K^-) - 2\varepsilon \geqq \|\mu\| - 4\varepsilon.$$

This shows $\|T_\mu\| \geqq \|\mu\|$ and finishes the proof of the theorem. □

We shall now consider the case where $X$ is locally compact. For a given compact subset $K \subseteq X$ we denote by $C_K(X)$ the vector space of all continuous functions on $X$ whose support is contained in $K$; of course, $C_K(X)$ is a Banach space with respect to the supremum norm and

$$C^c(X) = \bigcup_{K \in \mathscr{K}(X)} C_K(X).$$

A linear functional $T$ on $C^c(X)$ is called continuous if all the restrictions $T \,|\, C_K(X)$ are continuous in the usual sense.

**2.5. Theorem.** *For a locally compact Hausdorff space $X$ the continuous linear functionals $T$ on $C^c(X)$ are in a bijective linear relation with the signed*

*Radon measures $\mu$ on $X$ via the natural formula*

$$T(f) = \int f \, d\mu.$$

PROOF. If $\mu$ is a signed Radon measure, i.e. $\mu = \mu_1 - \mu_2$ for Radon measures $\mu_1$, $\mu_2$, then $T(f) = \int f \, d\mu := \int f \, d\mu_1 - \int f \, d\mu_2$ is well defined and continuous in the above sense; the unicity of the representing measure $\mu$ is obvious by Corollary 2.3.

Now let a continuous linear functional $T\colon C^c(X) \to \mathbb{R}$ be given. Imitating the proof of Theorem 2.4 we define for $f \in C^c_+(X)$

$$T^+(f) := \sup\{T(h) \mid f \geqq h \in C^c_+(X)\}.$$

Any function $h$ occurring here belongs to $C_{\text{supp}(f)}(X)$, implying that for some constant $\alpha > 0$

$$|T(h)| \leqq \alpha \|h\| \leqq \alpha \|f\|,$$

hence

$$0 \leqq T^+(f) < \infty.$$

The same arguments used already in the preceding proof show that $T^+$ and also $T^- := T^+ - T$ are both additive. Again an application of Corollary 2.3 finishes the proof. □

Let $X$ be a locally compact space and $V \subseteq C(X)$ a linear subspace of continuous functions. Motivated by the classical moment problem Choquet introduced a simple sufficient condition on $V$ in order that every positive linear functional $L\colon V \to \mathbb{R}$ can be represented by a Radon measure on $X$ as in the Riesz representation theorem where $V = C^c(X)$, cf. Choquet (1962, 1969).

For functions $f, g\colon X \to \mathbb{R}$ we write $f \in o(g)$ if formally $f/g$ vanishes at infinity on $X$, but due to possible zeros of $g$ the precise meaning of $f \in o(g)$ is the following:

*For every $\varepsilon > 0$ there exists a compact subset $K \subseteq X$ such that $|f(x)| \leqq \varepsilon |g(x)|$ for $x \in X \setminus K$.*

**2.6. Definition.** A convex cone $C \subseteq C_+(X)$ of nonnegative continuous functions is called an *adapted cone* if:

(i) For every $x \in X$ there exists $f \in C$ such that $f(x) > 0$.
(ii) For every $f \in C$ there exists $g \in C$ such that $f \in o(g)$.

A linear subspace $V \subseteq C(X)$ is called an *adapted space* if:

(iii) $V = V_+ - V_+$, where $V_+ = V \cap C_+(X)$.
(iv) $V_+$ is an adapted cone.

If $f \in C^c_+(X)$ then $f \in o(f)$ and for $f \in C^0_+(X)$ we have $f \in o(\sqrt{f})$, which shows that $C^c(X)$ and $C^0(X)$ are adapted spaces. If $p$ and $q$ are two real polynomials in one variable then $p \in o(q)$ if and only if $\deg(p) < \deg(q)$. It follows easily that the space of polynomials in one variable is an adapted space of continuous functions on $X = \mathbb{R}$. Similarly the space of polynomials in $k$ variables is adapted on $X = \mathbb{R}^k$.

The main property of adapted spaces is given in the following:

**2.7. Theorem.** *Let $V$ be an adapted space of continuous functions on a locally compact space $X$. For every positive linear functional $L: V \to \mathbb{R}$ there exists a Radon measure $\mu$ on $X$ such that $V \subseteq \mathscr{L}^1(\mu)$ and*

$$L(f) = \int f \, d\mu \qquad \textit{for all} \quad f \in V.$$

PROOF. We define

$$\tilde{V} = \{f \in C(X) \,|\, |f| \leqq g \text{ for some } g \in V_+\}.$$

Then $\tilde{V}$ is a subspace of $C(X)$ containing $V$, and a simple compactness argument combined with (i) shows that $C^c(X) \subseteq \tilde{V}$. We claim that $\tilde{V} = \tilde{V}_+ + V$. In fact, if $f \in \tilde{V}$ and $g \in V_+$ is such that $|f| \leqq g$, then $f = (g + f) + (-g)$ shows the assertion. By Corollary 1.2.7 it follows that $L$ can be extended to a positive linear functional $\tilde{L}: \tilde{V} \to \mathbb{R}$, and by the Riesz representation theorem there exists a Radon measure $\mu$ on $X$ representing $\tilde{L} | C^c(X)$.

For $g \in V_+$ and $\varphi \in C^c(X)$ satisfying $0 \leqq \varphi \leqq g$ we have

$$L(g) = \tilde{L}(g) \geqq \tilde{L}(\varphi) = \int \varphi \, d\mu.$$

By Urysohn's lemma the family $\mathscr{F} := \{\varphi \in C^c(X) | 0 \leqq \varphi \leqq g\}$ filters upwards to $g$, so by Theorem 1.5

$$\int g \, d\mu = \sup\left\{\int \varphi \, d\mu \,|\, \varphi \in \mathscr{F}\right\} \leqq L(g) < \infty,$$

hence $g \in \mathscr{L}^1(\mu)$. To see that $\int g \, d\mu = L(g)$ we choose $h \in V_+$ such that $g \in o(h)$. Let $\varepsilon > 0$ be given. There exists a compact set $K \subseteq X$ such that

$$g(x) \leqq \varepsilon h(x) \qquad \text{for} \quad x \in X \backslash K.$$

We choose $\varphi \in C^c(X)$ such that $1_K \leqq \varphi \leqq 1$ and find

$$0 \leqq g - g\varphi \leqq \varepsilon h,$$

hence

$$0 \leqq L(g) - \tilde{L}(g\varphi) \leqq \varepsilon L(h),$$

or

$$L(g) \leqq \int g\varphi \, d\mu + \varepsilon L(h) \leqq \int g \, d\mu + \varepsilon L(h),$$

which suffices since $\varepsilon > 0$ is arbitrary and independent of $h$. The equality

$$L(g) = \int g \, d\mu$$

extends clearly from $V_+$ to $V$. □

**2.8. Exercise.** Let $X$ be locally compact. Show that the dual $(C^0(X))'$ of the Banach space $C^0(X)$ equals $M^b(X)$, the space of all finite signed Radon measures on $X$ (which is again a Banach space with respect to the total variation norm) in the sense, that there exists a linear bijective isometry from $M^b(X)$ onto $(C^0(X))'$. *Hint*: Use the one-point compactification of $X$.

**2.9. Exercise.** Show that the following conditions are equivalent for a locally compact space $X$:

(i) $X$ is $\sigma$-compact, i.e. $X$ is a countable union of compact sets.
(ii) There is a strictly positive function $f \in C^0(X)$.
(iii) There is a function $f \in C(X)$ with $f(x) \to \infty$ for $x \to \infty$, i.e. such that $\{f \leqq a\}$ is compact for all $a \in \mathbb{R}$.
(iv) $C(X)$ is an adapted space.

**2.10. Exercise.** Try to find a finite signed Radon measure $\mu$ on $\mathbb{R}$ such that $\operatorname{supp}(\mu^+) = \operatorname{supp}(\mu^-) = \mathbb{R}$, where $\mu = \mu^+ - \mu^-$ is the Jordan–Hahn decomposition of $\mu$.

**2.11. Exercise.** Let $X$ be locally compact and consider on $C^c(X)$ the following four families $(p_i)_{i \in I_j}, j = 1, \ldots, 4$, of seminorms:

$$I_1 = \{A \subseteq X \mid \varnothing \neq A \text{ finite}\}, \qquad p_A(f) := \max_{x \in A} |f(x)|,$$

$$I_2 = \mathscr{K}(X), \qquad p_K(f) := \max_{x \in K} |f(x)|,$$

$$I_3 = \{\infty\}, \qquad p_\infty(f) := \sup_{x \in X} |f(x)|,$$

$$\{p_i \mid i \in I_4\} = \{p \mid p \text{ is a seminorm on } C^c(X) \text{ whose restriction to } C_K(X) \text{ is continuous for all } K \in \mathscr{K}(X)\},$$

where $C_K(X) := \{f \in C^c(X) \mid \operatorname{supp}(f) \subseteq K\}$ is considered as a Banach space with respect to the supremum norm. Let $\mathcal{O}_1, \ldots, \mathcal{O}_4$ denote the corresponding locally convex topologies on $C^c(X)$ (which are all Hausdorff).

Show that the topological duals of $C^c(X)$ with respect to these four topologies are given (by natural identification) in the following way:

$$(C^c(X), \mathcal{O}_1)' = \{\mu \in M(X) \mid \operatorname{supp}(\mu) \text{ is finite}\},$$

$$(C^c(X), \mathcal{O}_2)' = \{\mu \in M(X) \mid \operatorname{supp}(\mu) \text{ is compact}\},$$

$$(C^c(X), \mathcal{O}_3)' = \{\mu \in M(X) \mid \|\mu\| < \infty\},$$

$$(C^c(X), \mathcal{O}_4)' = M(X),$$

where $\operatorname{supp}(\mu) := \operatorname{supp}(\mu^+) \cup \operatorname{supp}(\mu^-)$ if $\mu = \mu^+ - \mu^-$ is the Jordan–Hahn decomposition of $\mu$.

**2.12. Exercise.** Let $V$ be an adapted space of continuous functions on a locally compact space $X$, let $F \subseteq X$ be a closed subset and put

$$V_+^F = \{f \in V \mid f(x) \geqq 0 \text{ for all } x \in F\}.$$

Show that any linear functional $L: V \to \mathbb{R}$ which is nonnegative on $V_+^F$ can be represented as

$$L(f) = \int f\, d\mu \qquad \text{for} \quad f \in V,$$

where $\mu \in M_+(X)$ is supported by $F$.

**2.13. Exercise.** Let $X$, $Y$ be locally compact spaces and $f: X \to Y$ a continuous surjective mapping such that $f^{-1}(K)$ is compact in $X$ for each compact subset $K \subseteq Y$. Show that each $\nu \in M_+(Y)$ is of the form $\mu^f$ for some $\mu \in M_+(X)$. *Hint*: Use Corollary 1.2.7 and the Riesz representation theorem.

**2.14. Exercise.** Let $X$ and $Y$ be locally compact spaces and $T: C^c(X) \times C^c(Y) \to \mathbb{R}$ a bilinear mapping which is positive in the sense that $T(f, g) \geqq 0$ if $f \in C_+^c(X)$ and $g \in C_+^c(Y)$. Show that

$$T(f, g) = \int f \otimes g\, d\mu$$

for some uniquely determined $\mu \in M_+(X \times Y)$.

## §3. Weak Convergence of Finite Radon Measures

The theory of weak convergence of finite Radon measures is a well-developed theory which is of great importance in probability theory, in particular when dealing with stochastic processes. We will later need only a very few basic facts which we are going to develop in this section.

Let $X$ be a Hausdorff space and denote by $M_+^b(X)$ the set of all finite Radon measures on $X$, i.e. all Radon measures $\mu$ with $\mu(X) < \infty$. The *weak topology* on $M_+^b(X)$ is the coarsest topology such that the functions $\mu \mapsto \int f\, d\mu$ become lower semicontinuous for every bounded lower semicontinuous $f: X \to \mathbb{R}$. The family of sets

$$G_{f,t} := \left\{\mu \in M_+^b(X) \,\middle|\, \int f\, d\mu > t\right\}$$

is a subbase for the weak topology when $f$ ranges over the bounded lower semicontinuous functions on $X$ and $t \in \mathbb{R}$.

The following result is part of the so-called *portmanteau* theorem (cf. Topsøe 1970, Theorem 8.1).

**3.1. Theorem.** *For $\mu \in M_+^b(X)$ and a net $(\mu_\alpha)_{\alpha \in A}$ in $M_+^b(X)$ the following properties are equivalent:*

(i) $\mu_\alpha \to \mu$ *weakly, i.e. in the weak topology;*
(ii) $\liminf \mu_\alpha(G) \geqq \mu(G)$ *for all open* $G \subseteq X$ *and* $\lim \mu_\alpha(X) = \mu(X)$;
(iii) $\limsup \mu_\alpha(F) \leqq \mu(F)$ *for all closed* $F \subseteq X$ *and* $\lim \mu_\alpha(X) = \mu(X)$;
(iv) $\liminf \int f\, d\mu_\alpha \geqq \int f\, d\mu$ *for all bounded lower semicontinuous* $f: X \to \mathbb{R}$;
(v) $\limsup \int f\, d\mu_\alpha \leqq \int f\, d\mu$ *for all bounded upper semicontinuous* $f: X \to \mathbb{R}$.

*If* (i)–(v) *are fulfilled, then* $\lim \int f\, d\mu_\alpha = \int f\, d\mu$ *for each bounded continuous* $f: X \to \mathbb{R}$, *and this property implies* (i)–(v) *if* $X$ *is in addition a completely regular space.*

Proof. By definition (i) is equivalent with (iv), and the equivalence of (ii) and (iii) (resp. (iv) and (v)) is immediate.

"(ii) ⇒ (iv)" Let $f: X \to \mathbb{R}$ be lower semicontinuous and assume without restriction $0 \leqq f \leqq 1$. From the obvious inequality

$$\left(f - \frac{1}{n}\right)^+ \leqq \frac{1}{n} \sum_{i=1}^{n-1} 1_{\{f > i/n\}} \leqq f$$

we get

$$\liminf \int f\, d\mu_\alpha \geqq \frac{1}{n} \sum_i \liminf \mu_\alpha\left(\left\{f > \frac{i}{n}\right\}\right) \geqq \frac{1}{n} \sum_i \mu\left(\left\{f > \frac{i}{n}\right\}\right),$$

where the last expression converges to $\int f\, d\mu$ as $n$ tends to $\infty$. The implication "(iv) ⇒ (ii)" is obvious.

If $f: X \to \mathbb{R}$ is bounded and continuous and $\mu_\alpha \to \mu$ weakly, then by (iv) and (v) $\int f\, d\mu_\alpha \to \int f\, d\mu$. Now suppose that $X$ is completely regular and $\lim \int f\, d\mu_\alpha = \int f\, d\mu$ for all continuous bounded $f: X \to \mathbb{R}$. To show (ii) let $G \subseteq X$ be open and let $K \subseteq G$ be compact. As an immediate consequence of the very definition of complete regularity we find a continuous function $f: X \to [0, 1]$ such that $f | K \equiv 1$ and $f | G^c \equiv 0$. Then

$$\mu_\alpha(G) \geqq \int f\, d\mu_\alpha \to \int f\, d\mu \geqq \mu(K),$$

hence $\liminf \mu_\alpha(G) \geqq \mu(K)$ and finally $\liminf \mu_\alpha(G) \geqq \mu(G)$, which had to be proved. □

**3.2. Proposition.** *The space $M_+^b(X)$ of finite Radon measures is a Hausdorff space in the weak topology.*

Proof. Let $(\mu_\alpha)$ be a net in $M_+^b(X)$ converging to $\mu_1$ as well as to $\mu_2$ weakly. Then $\mu_1(X) = \mu_2(X)$ and, of course, to show $\mu_1 = \mu_2$ it is enough to prove that $\mu_1(B) \leqq \mu_2(B)$ for all Borel subsets $B \subseteq X$.

If $A \subseteq X$, then the above theorem implies

$$\mu_1(\mathring{A}) \leqq \liminf \mu_\alpha(\mathring{A}) \leqq \limsup \mu_\alpha(\bar{A}) \leqq \mu_2(\bar{A}).$$

Now let $B \in \mathscr{B}(X)$ and $\varepsilon > 0$ be given. There exist compact sets $K_1 \subseteq B$, $K_2 \subseteq B^c$ such that $\mu_1(B \backslash K_1) < \varepsilon$ and $\mu_2(B^c \backslash K_2) < \varepsilon$. A simple compactness argument shows the existence of two open sets $G_1, G_2 \subseteq X$ separating $K_1$ and $K_2$, i.e. $K_1 \subseteq G_1$, $K_2 \subseteq G_2$ and $G_1 \cap G_2 = \varnothing$. Then

$$K_1 \subseteq G_1 \subseteq \bar{G}_1 \subseteq G_2^c \subseteq K_2^c$$

and hence

$$\mu_1(B) - \varepsilon \leqq \mu_1(K_1) \leqq \mu_2(K_2^c) \leqq \mu_2(B) + \varepsilon.$$

This holds for all $\varepsilon > 0$ so that $\mu_1(B) \leqq \mu_2(B)$. □

In §1 it was shown that given two Radon measures $\mu$, $\nu$ on spaces $X$ and $Y$, there is a unique product Radon measure $\mu \otimes \nu$ on $X \times Y$ characterized by

$$\mu \otimes \nu(C \times D) = \mu(C)\nu(D) \qquad \text{for all compact} \quad C \subseteq X, \quad D \subseteq Y,$$

and giving for all measurable rectangles the "right" value. In particular the product of two finite Radon measures is finite again (as it should be), and it is only natural to guess that $\mu \otimes \nu$ depends continuously (in the weak topology) on both of its arguments.

**3.3. Theorem.** *Let $X$ and $Y$ be two Hausdorff spaces. Then the mapping $(\mu, \nu) \mapsto \mu \otimes \nu$ from $M_+^b(X) \times M_+^b(Y)$ to $M_+^b(X \times Y)$ is weakly continuous.*

PROOF. In a first step we show that the mapping

$$X \times M_+^b(Y) \to M_+^b(X \times Y)$$

$$(x, \nu) \mapsto \varepsilon_x \otimes \nu$$

is continuous. Assume that $x_\alpha \to x$ and $\nu_\alpha \to \nu$. Let $G_1, \ldots, G_n \subseteq X$ and $H_1, \ldots, H_n \subseteq Y$ be open and put $U := \bigcup_{i=1}^n (G_i \times H_i)$. We show first that

$$\liminf \varepsilon_{x_\alpha} \otimes \nu_\alpha(U) \geqq \varepsilon_x \otimes \nu(U).$$

This holds trivially if $x \notin \bigcup_{i=1}^n G_i$. Suppose now that

$$I := \{i \leqq n \mid x \in G_i\} \neq \varnothing.$$

Then there exists some $\alpha_0$ such that $x_\alpha \in \bigcap_{i \in I} G_i$ for all $\alpha \geqq \alpha_0$ and for those $\alpha$ we get

$$\varepsilon_{x_\alpha} \otimes \nu_\alpha(U) \geqq \varepsilon_{x_\alpha} \otimes \nu_\alpha\left(\bigcup_{i \in I} (G_i \times H_i)\right) = \nu_\alpha\left(\bigcup_{i \in I} H_i\right),$$

hence

$$\liminf \varepsilon_{x_\alpha} \otimes \nu_\alpha(U) \geqq \liminf \nu_\alpha\left(\bigcup_{i \in I} H_i\right)$$

$$\geqq \nu\left(\bigcup_{i \in I} H_i\right) = \varepsilon_x \otimes \nu(U).$$

Every open set $U \subseteq X \times Y$ has the form $U = \bigcup_{\lambda \in \Lambda} (G_\lambda \times H_\lambda)$ for suitable open sets $G_\lambda \subseteq X$ and $H_\lambda \subseteq Y$. By Theorem 1.5 we can find, given $\varepsilon > 0$, finitely many $\lambda_1, \ldots, \lambda_n \in \Lambda$ such that

$$\varepsilon_x \otimes \nu \left( \bigcup_{i=1}^{n} (G_{\lambda_i} \times H_{\lambda_i}) \right) > \varepsilon_x \otimes \nu(U) - \varepsilon.$$

This implies

$$\liminf \varepsilon_{x_\alpha} \otimes \nu_\alpha(U) \geq \liminf \varepsilon_{x_\alpha} \otimes \nu_\alpha \left( \bigcup_{i=1}^{n} (G_{\lambda_i} \times H_{\lambda_i}) \right)$$

$$\geq \varepsilon_x \otimes \nu \left( \bigcup_{i=1}^{n} (G_{\lambda_i} \times H_{\lambda_i}) \right) > \varepsilon_x \otimes \nu(U) - \varepsilon.$$

Hence $\liminf \varepsilon_{x_\alpha} \otimes \nu_\alpha(U) \geq \varepsilon_x \otimes \nu(U)$ and 3.1 implies $\varepsilon_{x_\alpha} \otimes \nu_\alpha \to \varepsilon_x \otimes \nu$.

The second step will now be an easy consequence of the first one. Let $f : X \times Y \to [0, \infty]$ be lower semicontinuous. From the continuity of $(x, \nu) \mapsto \varepsilon_x \otimes \nu$ we get that

$$X \times M_+^b(Y) \to [0, \infty]$$

$$(x, \nu) \mapsto \int_Y f(x, y) \, d\nu(y) = \sup_n \int_Y (n \wedge f(x, y)) \, d\nu(y)$$

is also lower semicontinuous, and using the fact that $M_+^b(Y)$ is again a Hausdorff space as well as the Fubini theorem for lower semicontinuous functions (1.12), we may repeat this argument and conclude that

$$M_+^b(X) \times M_+^b(Y) \to [0, \infty],$$

$$(\mu, \nu) \mapsto \int_X \int_Y f(x, y) \, d\nu(y) \, d\mu(x) = \int_{X \times Y} f \, d(\mu \otimes \nu)$$

is lower semicontinuous, too. If now $f : X \times Y \to \mathbb{R}$ is lower semicontinuous and bounded, then $f + c \geq 0$ for some $c \in \mathbb{R}$, and

$$(\mu, \nu) \mapsto \int f \, d(\mu \otimes \nu) = \int (f + c) \, d(\mu \otimes \nu) - c\mu(X)\nu(Y)$$

is again lower semicontinuous, the second term on the right being a continuous function of $\mu$ and $\nu$. This finishes our proof. $\qquad \square$

**3.4. Corollary.** *If $S$ is a Hausdorff topological semigroup, then so is $M_+^b(S)$ with respect to convolution.*

PROOF. The mapping $(\mu, \nu) \mapsto \mu * \nu$ is continuous as composition of $(\mu, \nu) \mapsto \mu \otimes \nu$ and the mapping $\Phi : M_+^b(S \times S) \to M_+^b(S)$ defined by $\Phi(\kappa) := \kappa^+$, where $+$ denotes the semigroup operation and $\kappa^+$ is the image of $\kappa$ under $+$; cf. 1.15, 1.16 and Exercise 3.7 below. $\qquad \square$

It is a general principle in mathematics to approximate complicated functions (or other objects) by simpler ones. Among the Radon measures the so-called molecular measures (i.e. finite positive linear combinations of one-point (or "atomic") measures) are considered to be "simple" objects. They are dense in the set of all finite Radon measures even in a stronger sense than with respect to the weak topology.

**3.5. Proposition.** *For every Hausdorff space $X$ the set of molecular measures is a dense subset of $M_+^b(X)$ with respect to the pointwise convergence on $\mathscr{B}(X)$.*

PROOF. As the directed set we choose the family $A$ of all finite Borel partitions of $X$, i.e. the set of all $\alpha = \{B_1, \ldots, B_n\} \subseteq \mathscr{B}(X)$ such that $B_i \neq \emptyset$, $B_i \cap B_j = \emptyset$ for $i \neq j$ and $\bigcup_{i=1}^n B_i = X$, ordered by refinement. For such $\alpha$ we put, given any $\mu \in M_+^b(X)$, $\mu_\alpha := \sum_{i=1}^n \mu(B_i)\varepsilon_{x_i}$, where $x_i \in B_i$ is chosen arbitrarily. Now let $B \in \mathscr{B}(X)\setminus\{\emptyset, X\}$ be given; then for all $\alpha \in A$ finer than $\alpha_0 = \{B, B^c\}$ we have $\mu_\alpha(B) = \mu(B)$; and $\mu_\alpha(B) = \mu(B)$ for all $\alpha \in A$ when $B \in \{\emptyset, X\}$. □

**3.6. Exercise.** Let $(\mu_\alpha)$ and $\mu$ be finite Radon measures on the Hausdorff space $X$ and let them all be concentrated on the Borel subset $Y \subseteq X$. Then $\mu_\alpha \to \mu$ weakly in $M_+^b(X)$ if and only if $\mu_\alpha | Y \to \mu | Y$ weakly in $M_+^b(Y)$.

**3.7. Exercise.** Let $X$ and $Y$ be two Hausdorff spaces and let $f: X \to Y$ be a continuous mapping. Then for any $\mu \in M_+^b(X)$ the image measure $\mu^f$ belongs to $M_+^b(Y)$, cf. Proposition 1.15. Show that the transformation $\mu \mapsto \mu^f$ from $M_+^b(X)$ to $M_+^b(Y)$ is continuous.

**3.8. Exercise.** Let $X$ be a Hausdorff space and $M_+^1(X)$ the set of Radon probability measures on $X$, i.e. $M_+^1(X) = \{\mu \in M_+^b(X) | \mu(X) = 1\}$. Let $E := \{\varepsilon_x | x \in X\}$ be the set of all one-point measures. Show that every $\{0, 1\}$-valued measure $\mu \in M_+^1(X)$ already belongs to $E$. Show further that $E$ is a weakly closed subset of $M_+^1(X)$ homeomorphic to $X$ and that $E = \text{ex}(M_+^1(X))$. (For the notion of an extreme point see 5.1.)

**3.9. Exercise.** Let $\mu$ be a Radon measure on the Hausdorff space $X$. For $K \in \mathscr{K}(X)$ define $\mu^K(B) := \mu(B \cap K)$ and $\mu_K(B) := \mu(B \cap K)/\mu(K)$ (if $\mu(K) > 0$). Show that the net $(\mu^K)$ converges to $\mu$ pointwise on $\mathscr{B}(X)$ and that

$$\lim_K \int g \, d\mu^K = \int g \, d\mu$$

for every $\mu$-integrable function $g: X \to \mathbb{C}$. Show that if $\mu(X) < \infty$ then $(\mu^K)$ (resp. $(\mu_K)$) converges weakly to $\mu$ (resp. $\mu/\mu(X)$).

**3.10. Exercise.** Let $(\rho_\alpha)$ denote a net of probability Radon measures on the product $X \times Y$ of two Hausdorff spaces, and denote by $\mu_\alpha$ (resp. $\nu_\alpha$) the marginal distribution of $\rho_\alpha$ on $X$ (resp. $Y$) (i.e. the image measures under the two canonical projections). If $(\mu_\alpha)$ converges to $\mu$ and $(\nu_\alpha)$ converges to some one-point measure $\varepsilon_y$, then $(\rho_\alpha)$ tends to $\mu \otimes \varepsilon_y$.

**3.11. Exercise.** Let $(S, +, 0)$ denote a Hausdorff topological abelian semigroup with neutral element 0 (cf. 1.16) and let $\mu \in M_+^b(S)$. Then

$$\sum_{n=0}^{\infty} \frac{\mu^{*n}}{n!} = \lim_{n\to\infty} (\varepsilon_0 + \mu/n)^{*n}$$

holds in the sense that both limits exist and agree for all Borel subsets of $S$. Their common value is often called the *exponential* of $\mu$ and abbreviated $\exp(\mu)$. In particular both limits exist and agree with respect to the weak topology.

**3.12. Exercise.** Let $S$ and $T$ be two Hausdorff topological semigroups and let $h: S \to T$ be a continuous homomorphism. Then for $\mu, \nu \in M_+^b(S)$ we have

$$(\mu * \nu)^h = \mu^h * \nu^h.$$

## §4. Vague Convergence of Radon Measures on Locally Compact Spaces

In this section $X$ denotes a locally compact Hausdorff space. The vector space $C^c(X)$ of continuous functions $f: X \to \mathbb{R}$ with compact support and the vector space $M(X)$ of signed Radon measures on $X$ form a dual pair under the bilinear form

$$\langle \mu, f \rangle = \int f \, d\mu, \qquad \mu \in M(X), \qquad f \in C^c(X).$$

**4.1. Definition.** The *vague topology* on $M(X)$ is the weak topology $\sigma(M(X), C^c(X))$, i.e. the coarsest topology in which the mappings $\mu \mapsto \langle \mu, f \rangle$ are continuous, when $f$ ranges over $C^c(X)$, cf. 1.3.10. In particular the vague topology is a Hausdorff topology.

We first remark that for any lower semicontinuous function $f: X \to [0, \infty]$ the function

$$\mu \mapsto \int f \, d\mu$$

is lower semicontinuous on $M_+(X)$ with the vague topology.

In fact, from Urysohn's lemma it follows that any such $f$ is the supremum of the upward filtering family of functions $\varphi \in C^c(X)$ satisfying $0 \leqq \varphi \leqq f$, so by Theorem 1.5

$$\int f\,d\mu = \sup\left\{\int \varphi\,d\mu \,|\, \varphi \in C^c(X), 0 \leqq \varphi \leqq f\right\}.$$

A combination of this remark and Theorem 3.1 immediately gives the following relationship between the weak topology on $M_+^b(X)$ and the restriction of the vague topology to $M_+^b(X)$:

**4.2. Proposition.** *Let $(\mu_\alpha)_{\alpha \in A}$ be a net on $M_+^b(X)$ and let $\mu \in M_+^b(X)$. Then $(\mu_\alpha)$ converges weakly to $\mu$ if and only if $(\mu_\alpha)$ converges vaguely to $\mu$ and* $\lim \mu_\alpha(X) = \mu(X)$.

**4.3. Corollary.** *The vague topology and the weak topology coincide on the set $M_+^1(X)$ of Radon probability measures.*

Under some extra assumptions on a net $(\mu_\alpha)_{\alpha \in A}$ in $M_+(X)$ the vague convergence implies the convergence of $(\int f\,d\mu_\alpha)_{\alpha \in A}$ for certain functions $f \in C(X)\backslash C^c(X)$. As an important example we have

**4.4. Proposition.** *Let $(\mu_\alpha)_{\alpha \in A}$ be a net on $M_+^b(X)$ converging vaguely to $\mu \in M_+^b(X)$. If*

$$c := \sup_\alpha \mu_\alpha(X) < \infty,$$

*then $\mu(X) \leqq c$, and for all $f \in C^0(X)$ we have*

$$\lim_\alpha \int f\,d\mu_\alpha = \int f\,d\mu. \tag{1}$$

Proof. For any $\varphi \in C^c(X)$ satisfying $0 \leqq \varphi \leqq 1$ we have

$$\langle \mu, \varphi \rangle = \lim_\alpha \langle \mu_\alpha, \varphi \rangle \leqq c,$$

hence

$$\mu(X) = \sup\{\langle \mu, \varphi \rangle \,|\, 0 \leqq \varphi \leqq 1, \varphi \in C^c(X)\} \leqq c.$$

Let $f \in C^0(X)$. For any $\varepsilon > 0$ there exists $\varphi \in C^c(X)$ such that $\|f - \varphi\|_\infty \leqq \varepsilon$, and therefore we find

$$\left|\int f\,d\mu - \int f\,d\mu_\alpha\right| \leqq 2c\varepsilon + \left|\int \varphi\,d\mu - \int \varphi\,d\mu_\alpha\right|,$$

and now it is easy to see that (1) holds. □

The following result characterizes the *relatively compact* subsets $M \subseteq M_+(X)$ in the vague topology, i.e. the subsets $M$ for which the vague closure $\overline{M}$ is vaguely compact.

**4.5. Theorem.** *A subset $M \subseteq M_+(X)$ is relatively vaguely compact if and only if*

$$\sup\{\langle \mu, \varphi \rangle \mid \mu \in M\} < \infty$$

*for each $\varphi \in C_+^c(X)$.*

PROOF. The space of functions of $C^c(X)$ into $\mathbb{R}$ can be considered as the product space $\Pi := \mathbb{R}^{C^c(X)}$ which will be equipped with the product topology. The subset $\Lambda$ of positive linear functionals of $C^c(X)$ into $\mathbb{R}$ is closed in $\Pi$, and because of the Riesz representation theorem (2.3) there is a bijection of $M_+(X)$ onto $\Lambda$ which is a homeomorphism when $M_+(X)$ carries the vague topology and $\Lambda$ carries the topology inherited from $\Pi$. Therefore, $M \subseteq M_+(X)$ is relatively compact in the vague topology if and only if the corresponding set of positive linear functionals is relatively compact in $\Pi$. By the Tychonoff theorem this is the case if and only if $\{\langle \mu, \varphi \rangle \mid \mu \in M\}$ is relatively compact in $\mathbb{R}$ for each $\varphi \in C^c(X)$, which gives the conditions of the theorem since a subset of $\mathbb{R}$ is relatively compact precisely if it is bounded. □

**4.6. Proposition.** *Let $c > 0$. The set*

$$\{\mu \in M_+^b(X) \mid \mu(X) \leqq c\}$$

*is vaguely compact.*

PROOF. The set in question is closed by Proposition 4.4 and relatively compact by Theorem 4.5. □

**4.7. Corollary.** *Suppose $X$ is a compact space. Then $M_+^1(X)$ is compact in the weak (or vague) topology.*

PROOF. By Proposition 4.6 we have that $M_+^1(X)$ is relatively compact in the vague topology, but also closed since $1 \in C^c(X)$. The proof is finished by the observation in Corollary 4.3. □

The following result was established by Choquet (1962) in his treatment of the moment problem.

**4.8. Proposition.** *Let $V$ be an adapted space of continuous functions on a locally compact space $X$ and let $L: V \to \mathbb{R}$ be a positive linear functional. The set of representing measures, i.e. the set*

$$C = \left\{\mu \in M_+(X) \,\middle|\, \int f\,d\mu = L(f) \quad \text{for all} \quad f \in V\right\}$$

*is convex and compact in the vague topology.*

PROOF. It is clear that $C$ is convex. For $\varphi \in C_+^c(X)$ there exists $f \in V_+$ such that $\varphi \leqq f$ (cf. the proof of Theorem 2.7), hence

$$\langle \mu, \varphi \rangle \leqq \int f\,d\mu = L(f) \qquad \text{for} \quad \mu \in C,$$

so $C$ is relatively compact by Theorem 4.5. Let $(\mu_\alpha)_{\alpha\in A}$ be a net from $C$ converging vaguely to $\mu \in M_+(X)$ and let $f \in V_+$. By the remark following 4.1 we get

$$\int f\,d\mu \leqq \liminf \int f\,d\mu_\alpha = L(f).$$

We choose $g \in V_+$ such that $f \in o(g)$, and let $\varepsilon > 0$ be given. There exists a compact set $K \subseteq X$ such that $f(x) \leqq \varepsilon g(x)$ for $x \in X \backslash K$, and if $\varphi \in C_+^c(X)$ is such that $1_K \leqq \varphi \leqq 1$ we have $f(1-\varphi) \leqq \varepsilon g$. From

$$\int f\,d(\mu - \mu_\alpha) = \int f(1-\varphi)\,d(\mu - \mu_\alpha) + \int f\varphi\,d(\mu - \mu_\alpha)$$

we therefore get

$$\left|\int f\,d\mu - L(f)\right| \leqq \varepsilon \int g\,d\mu + \varepsilon L(g) + \left|\int f\varphi\,d(\mu - \mu_\alpha)\right|,$$

and since $f\varphi \in C^c(X)$ the last term tends to zero, so we have

$$\left|\int f\,d\mu - L(f)\right| \leqq 2\varepsilon L(g).$$

Since $\varepsilon > 0$ is arbitrary we get $\int f\,d\mu = L(f)$ for all $f \in V_+$, and then for all $f \in V$, thus proving that $\mu \in C$. □

Let $\mu$ be a signed Radon measure on $X$ and let $G \subseteq X$ be an open subset. Then $G$ is itself locally compact and if $f \in C^c(G)$ is extended to $X$ by

$$\tilde{f}(x) = \begin{cases} f(x), & x \in G, \\ 0, & x \in X \backslash G, \end{cases}$$

then $\tilde{f} \in C^c(X)$. The mapping $f \mapsto \int \tilde{f}\,d\mu$ is a continuous linear functional on $C^c(G)$, hence by Theorem 2.5 represented by a signed Radon measure on $G$, denoted $\mu|G$, and called the *restriction* of $\mu$ to $G$. In case of a (nonnegative) Radon measure $\mu$ on $X$ the restriction $\mu|G$ is of course the usual set-theoretical restriction, introduced already after Remark 1.6.

Vague convergence of Radon measures is a local concept as the following result shows:

**4.9. Theorem.** *Let $(G_\alpha)_{\alpha\in D}$ be an open covering of a locally compact space $X$. A net $(\mu_i)_{i\in I}$ of signed Radon measures on $X$ converges vaguely to $\mu \in M(X)$ if and only if $(\mu_i|G_\alpha)_{i\in I}$ converges vaguely to $(\mu|G_\alpha)$ for each $\alpha \in D$.*

PROOF. The "only if" part is obvious, so suppose that $(\mu_i|G_\alpha)_{i\in I}$ converges to $(\mu|G_\alpha)$ for each $\alpha \in D$. Let $f \in C^c(X)$ have the compact support $C$, and choose $\alpha_1, \ldots, \alpha_n \in D$ such that $C \subseteq G_{\alpha_1} \cup G_{\alpha_2} \cup \cdots \cup G_{\alpha_n}$. By Lemma 1.17 there exist compact sets $C_k \subseteq G_{\alpha_k}$ $k = 1, \ldots, n$ such that $C = C_1 \cup \cdots \cup C_n$. By Urysohn's lemma there exist functions $\varphi_k \in C^c(X)$ such that $1_{C_k} \leqq \varphi_k \leqq 1_{G_{\alpha_k}}$

and $\mathrm{supp}(\varphi_k) \subseteq G_{\alpha_k}$, $k = 1, \ldots, n$. The functions

$$f_k(x) := \begin{cases} f(x) \dfrac{\varphi_k(x)}{\sum_{j=1}^{n} \varphi_j(x)}, & x \in C, \\ 0, & x \in X \setminus C, \end{cases}$$

belong to $C^c(X)$, $\mathrm{supp}(f_k) \subseteq G_{\alpha_k}$ and $f = \sum_{k=1}^n f_k$, so we have

$$\int f \, d\mu_i = \sum_{k=1}^{n} \int f_k \, d(\mu_i | G_{\alpha_k}) \qquad \text{for} \quad i \in I,$$

and the assertion follows. □

The following result will be used occasionally. For the proof, see e.g. Bauer (1978, p. 233).

**4.10. Proposition.** *Suppose that the locally compact space $X$ has a countable base for the topology. Then the vague topology on $M_+(X)$ is metrizable.*

**4.11. Exercise.** Let $X$ be locally compact. For a net $(\mu_\alpha)$ on $M_+(X)$ and $\mu \in M_+(X)$ the following conditions are equivalent:

(i) $\mu_\alpha \to \mu$ vaguely;
(ii) $\limsup \mu_\alpha(K) \leqq \mu(K)$ for each compact $K \subseteq X$ and $\liminf \mu_\alpha(G) \geqq \mu(G)$ for all relatively compact open sets $G \subseteq X$;
(iii) $\lim \mu_\alpha(B) = \mu(B)$ for all relatively compact Borel sets $B \subseteq X$ such that $\mu(\partial B) = 0$.

**4.12. Exercise.** Show that the set of all Radon measures on a locally compact space taking only values in $\overline{\mathbb{N}}_0 = \{0, 1, 2, \ldots, \infty\}$ is vaguely closed.

**4.13. Exercise.** Let the Radon measures $\mu_\alpha$ tend vaguely to $\mu$ and assume that all the $\mu_\alpha$ are concentrated on the closed subset $Y \subseteq X$. Then $\mu$ is concentrated on $Y$, too, and $\mu_\alpha$ tends vaguely to $\mu$ also on the locally compact space $Y$.

**4.14. Exercise.** Show that $(\mu, \nu) \mapsto \mu \otimes \nu$ is vaguely continuous as a mapping from $M_+(X) \times M_+(Y)$ to $M_+(X \times Y)$ for two locally compact spaces $X$ and $Y$. *Hint*: Use the Stone–Weierstrass theorem.

**4.15. Exercise.** Property (ii) in Exercise 4.11 above makes sense on any Hausdorff space $X$ and hence can be used to define vague convergence of Radon measures in this generality. Show however that if $M_+(X)$ with respect to vague convergence is a Hausdorff space, then $X$ is necessarily locally compact. *Hint*: If $x_0 \in X$ has no relatively compact neighbourhood, then for a certain net $(x_\alpha)$ in $X$ we have $x_\alpha \to x_0$, $\varepsilon_{x_\alpha} \to \varepsilon_{x_0}$ and $\varepsilon_{x_\alpha} \to 0$.

# §5. Introduction to the Theory of Integral Representations

Throughout this section $E$ denotes a locally convex Hausdorff topological vector space over $\mathbb{R}$. The theory below can be applied to complex vector spaces by restricting the multiplication with scalars to real scalars.

The basic observation, which should be kept in mind when reading the following general theorems, is that a convex polyhedron is the convex hull of its corners, or equivalently, that any point in the convex polyhedron is the centre of gravity of a molecular probability measure on the corners. The basic notion is that of an extreme point of a set, which is a generalization of a corner of a convex polyhedron and defined below. We shall limit ourselves here to those parts of the theory of integral representations which will be needed in the sequel. A detailed exposition can be found in Alfsen (1971) or Phelps (1966).

**5.1. Definition.** Let $A \subseteq B \subseteq E$ be subsets of $E$. Then $A$ is called an *extreme subset* of $B$ if for all $x, y \in B$ and $\lambda \in ]0, 1[$:

$$\lambda x + (1 - \lambda)y \in A \Rightarrow x, y \in A.$$

A point $a \in A$ is called an *extreme point* of $A$ if $\{a\}$ is an extreme subset of $A$.

It is easy to see that if $A$ is convex then $a \in A$ is an extreme point if and only if for all $x, y \in A$:

$$a = \tfrac{1}{2}(x + y) \Rightarrow x = y = a.$$

The set of extreme points of $A$ is denoted ex($A$) and in some literature called the *extreme boundary* of $A$.

Notice that "extreme subset of" is a transitive relation in the set of subsets of $E$. The above definitions of course make sense for an arbitrary real vector space without topology.

Let $x_1, \ldots, x_n$ be points in $E$ and let $\lambda_1, \ldots, \lambda_n$ be numbers $\in [0, 1]$ with with $\sum_{i=1}^n \lambda_i = 1$. The corresponding *convex combination* of $x_1, \ldots, x_n$ is the point

$$b = \sum_{i=1}^{n} \lambda_i x_i.$$

If $\mu$ denotes the molecular measure on $E$ with mass $\lambda_i$ at $x_i$, i.e.

$$\mu = \sum_{i=1}^{n} \lambda_i \varepsilon_{x_i},$$

then the point $b$ is called the *barycentre* of $\mu$. For any (continuous) linear functional $f$ on $E$ we have

$$f(b) = \sum_{i=1}^{n} \lambda_i f(x_i) = \int f \, d\mu,$$

and this is the motivation for the following:

**5.2. Definition.** Let $X$ be a compact subset of $E$ and let $\mu \in M_+^1(X)$ be a Radon probability measure on $X$. A point $b \in E$ is called the *barycentre* of $\mu$ if and only if

$$f(b) = \int_X f \, d\mu \qquad \text{for all} \quad f \in E'.$$

**Remark.** There exists at most one barycentre of $\mu \in M_+^1(X)$. In fact, if $b_1$, $b_2$ are barycentres of $\mu$ we have $f(b_1) = f(b_2)$ for all $f \in E'$, and since $E'$ separates the points of $E$ we find $b_1 = b_2$. As well as barycentre one also encounters the words *centre of gravity* and *resultant*. The barycentre is "the value of" the vector integral

$$\int_X x \, d\mu(x),$$

but such a vector integral need not always converge in $E$, so a barycentre need not exist. We have, however, the following result:

**5.3. Proposition.** *Let $X$ be a compact subset of $E$ such that $K = \overline{\operatorname{conv}}(X)$ is compact. Then for every $\mu \in M_+^1(X)$ the barycentre exists and belongs to $K$. Conversely, every point $x \in K$ is barycentre of some $\mu \in M_+^1(X)$.*

PROOF. For $f \in E'$ and $\mu \in M_+^1(X)$ we put

$$H_f = \left\{x \in E \mid f(x) = \int f \, d\mu\right\}.$$

The intersection of the $H_f$'s for $f \in E'$ is the set of barycentres of $\mu$, hence either empty or a singleton. We shall show

$$\bigcap_{f \in E'} H_f \cap K \neq \varnothing.$$

The set $H_f \cap K$ is a closed subset of $K$, so by a result from general topology it suffices to prove that

$$\bigcap_{i=1}^{n} H_{f_i} \cap K \neq \varnothing \tag{1}$$

for an arbitrary finite subset $\{f_1, \ldots, f_n\} \subseteq E'$. For such a subset we define a continuous linear mapping $T: E \to \mathbb{R}^n$ by

$$T(x) = (f_1(x), \ldots, f_n(x)),$$

and claim that

$$p := \left( \int f_1 \, d\mu, \ldots, \int f_n \, d\mu \right) \in T(K),$$

which shows (1). If the contrary is true, there exists by the separation theorem 1.2.3 a linear form $\varphi: \mathbb{R}^n \to \mathbb{R}$ such that

$$\sup \varphi(T(K)) < \varphi(p). \tag{2}$$

The linear form $\varphi$ is given as $\varphi(x) = \langle a, x \rangle$ for some $a = (a_1, \ldots, a_n) \in \mathbb{R}^n$, and defining $g = \sum_{i=1}^n a_i f_i \in E'$, (2) can be expressed

$$\sup g(K) < \int_K g \, d\mu,$$

which is impossible, $\mu$ being a probability.

We next have to prove that any $x \in \overline{\text{conv}}(X)$ is the barycentre of some $\mu \in M_+^1(X)$. This is clear if $x \in \text{conv}(X)$, in fact such a point is the barycentre of a molecular measure as remarked earlier. For $x \in \overline{\text{conv}}(X)$ there exist nets $(x_\alpha)_{\alpha \in A}$ of points from $\text{conv}(X)$ converging to $x$ and $(\mu_\alpha)_{\alpha \in A}$ of molecular measures from $M_+^1(X)$ such that $x_\alpha$ is the barycentre of $\mu_\alpha$ for each $\alpha \in A$. By Corollary 4.7 there exist $\mu \in M_+^1(X)$ and a subnet $(\mu_{\alpha_\beta})$ converging weakly to $\mu$. For $f \in E'$ we then have

$$\int f \, d\mu = \lim_\beta \int f \, d\mu_{\alpha_\beta} = \lim_\beta f(x_{\alpha_\beta}) = f(x),$$

which shows that $x$ is the barycentre of $\mu$. □

**5.4. Remark.** If $E$ is *complete* then $\overline{\text{conv}}(X)$ is compact for every compact $X \subseteq E$, so every $\mu \in M_+^1(X)$ has a barycentre in this case. This applies in particular to Fréchet spaces and Banach spaces. For details, see, e.g. Robertson and Robertson (1964).

Already Minkowski proved that a compact convex set $K$ in $\mathbb{R}^n$ is the convex hull of $\text{ex}(K)$. In 1940 Krein and Milman found a far-reaching generalization of Minkowski's result:

**5.5. Theorem.** *Every compact convex set $K$ in $E$ is the closed convex hull of its extreme points, i.e.*

$$K = \overline{\text{conv}}(\text{ex}(K)).$$

PROOF. We first show that any nonempty compact set $C$ has extreme points.

We form the family $\mathscr{F}$ of nonempty, closed extreme subsets of $C$. Notice that $C \in \mathscr{F}$. A Zorn's lemma argument shows that $\mathscr{F}$ contains a minimal element $M$ with respect to inclusion. To see that $M$ has only one point, which is then an extreme point of $C$, we assume the existence of $x, y \in M$,

$x \neq y$, and choose $f \in E'$ such that $f(x) > f(y)$. Then

$$M_0 = \{z \in M \mid f(z) = \sup f(M)\}$$

is easily seen to be an extreme subset of $M$, hence $M_0 \in \mathscr{F}$. Since $M_0$ is a proper subset of $M$ we are led to a contradiction.

Clearly $\overline{\text{conv}}(\text{ex}(K))$ is a compact convex subset of $K$. If they are not equal, there exists by the separation theorem 1.2.3 an $f \in E'$ such that

$$\sup f(K) > \sup f(\overline{\text{conv}}(\text{ex}(K))). \tag{3}$$

The set

$$M = \{x \in K \mid f(x) = \sup f(K)\}$$

is a nonempty compact subset of $K$, and by the first part of the proof $\text{ex}(M) \neq \varnothing$. Since $M$ is an extreme subset of $K$ we have $\text{ex}(M) \subseteq \text{ex}(K)$, but this is impossible due to (3). □

For every subset $A \subseteq E$ we have $\overline{\text{conv}}(A) = \overline{\text{conv}}(\bar{A})$. An equivalent formulation of the Krein-Milman theorem is therefore that $K = \overline{\text{conv}}(\overline{\text{ex}(K)})$. Using Proposition 5.3 we can reformulate the theorem in the following way:

**5.6. Theorem.** *Let $K$ be a compact convex set in $E$. Every $x \in K$ is the barycentre of a measure $\mu \in M_+^1(\overline{\text{ex}(K)})$.*

A natural and important question in connection with the above theorem is whether $\mu$ can be chosen such that $\mu$ is *concentrated* on the set of extreme points, i.e. such that $\mu(K \setminus \text{ex}(K)) = 0$.

The answer is *yes* if $K$ is *metrizable*, and this is the content of Choquet's theorem. The answer is *no* in general if $K$ is *nonmetrizable*, simply because $\text{ex}(K)$ can be a nonmeasurable subset of $K$ in this case. There is however a satisfactory solution to the question in the nonmetrizable case also, related to the notion of a boundary measure.

For a compact convex subset $K$ of $E$ we define a partial ordering $\lhd$ on $M_+^1(K)$ by

$$\mu \lhd \nu \Leftrightarrow \int f \, d\mu \leqq \int f \, d\nu$$

for all continuous convex functions $f: K \to \mathbb{R}$.

**5.7. Definition.** A measure $\mu \in M_+^1(K)$ is called a *boundary measure* if it is maximal with respect to the ordering $\lhd$.

A boundary measure $\mu$ is *pseudo-concentrated* on $\text{ex}(K)$ in the sense that $\mu(G) = 0$ for all $G_\delta$-sets $G \subseteq K$ such that $G \cap \text{ex}(K) = \varnothing$. (We recall that a subset of a topological space is called a $G_\delta$-set if it is the intersection of countably many open sets.)

In the metrizable case the set $\mathrm{ex}(K)$ of extreme points is a Borel set, in fact even a $G_\delta$-set. Furthermore, any compact subset of $K$ is a $G_\delta$-set. It follows that a boundary measure $\mu$ on $K$ is concentrated on $\mathrm{ex}(K)$ in the ordinary sense.

We can now formulate the generalization of Theorem 5.6 going back to Choquet in the metrizable case and to Bishop and de Leeuw in the general case.

**5.8. Theorem.** *Let $K$ be a compact convex set in $E$. Every $x \in K$ is the barycentre of a boundary measure $\mu \in M_+^1(K)$.*

**Remark.** In our applications of the present theory the compact convex sets $K$ which we consider always have the property that $\mathrm{ex}(K)$ is closed, so we may apply Theorem 5.6 instead of Theorem 5.8.

We will mention briefly the very important modern notion of a *simplex.*

**5.9. Definition.** A compact convex set $K \subseteq E$ is called a *simplex* if every $x \in K$ is the barycentre of precisely one boundary measure, and it is called a *Bauer simplex* if furthermore $\mathrm{ex}(K)$ is closed.

**5.10. Example.** Let $X$ be a compact Hausdorff space and let $E = M(X)$ be the vector space of signed Radon measures with the vague topology. The set $K = M_+^1(X)$ is a compact convex set, and by Exercise 3.8 $\mathrm{ex}(K) = \{\varepsilon_x \mid x \in X\}$ is a compact set. It is easily seen that $K$ is a Bauer simplex.

In applications we often want to give an integral representation of the elements in a convex cone $C$. This is possible if $C$ has a compact *base* $B$, i.e. a compact convex subset $B \subseteq C \setminus \{0\}$ such that for any $x \in C \setminus \{0\}$ there exists a unique number $\lambda > 0$ with $\lambda x \in B$.

The following general result was proved by Neumann (1983) and will be applied later.

Let $X$ be a nonempty set, let $E = \mathbb{C}^X$ be the vector space of functions $f: X \to \mathbb{C}$ with the topology of pointwise convergence and let $K \subseteq E$ be a compact convex set. We assume that $\gamma: X \to X$ is a mapping and define the following subsets of $K$:

$$K_\gamma = \{f \in K \mid |f(x)|^2 \leqq f(\gamma(x)) \text{ for all } x \in X\},$$

$$\Gamma = \{f \in K \mid |f(x)|^2 = f(\gamma(x)) \text{ for all } x \in X\}.$$

**5.11. Proposition.** *With the above notation $K_\gamma$ is a compact convex set and $\Gamma \subseteq \mathrm{ex}(K_\gamma)$.*

Proof. The function $z \mapsto |z|^2$ is convex from $\mathbb{C}$ to $\mathbb{R}$ and therefore $K_\gamma$ is a convex set which is clearly closed, hence compact. Let $f \in \Gamma$ and suppose

$f = \frac{1}{2}(g + h)$ with $g, h \in K_\gamma$. For $x \in X$ we then have

$$\begin{aligned}\tfrac{1}{2}(|g(x)|^2 + |h(x)|^2) &\leqq \tfrac{1}{2}(g(\gamma(x)) + h(\gamma(x))) = f(\gamma(x)) \\ &= |f(x)|^2 = \tfrac{1}{4}|g(x) + h(x)|^2\end{aligned}$$

which implies $|g(x) - h(x)|^2 \leqq 0$, hence $g = h$. This shows that $f$ is an extreme point of $K_\gamma$. □

**5.12. Corollary.** *If* $\operatorname{ex}(K) \subseteq \Gamma$ *then* $\operatorname{ex}(K) = \Gamma$.

PROOF. If $\operatorname{ex}(K) \subseteq \Gamma$, we get by the Krein-Milman theorem that $K = K_\gamma$, hence $\Gamma \subseteq \operatorname{ex}(K)$ by 5.11. □

**5.13. Exercise.** Let $K$ be a metrizable compact convex set in $E$ and let $d: K \times K \to [0, \infty[$ be a metric defining the topology of $K$. For each $n \in \mathbb{N}$ define

$$F_n = \left\{x \in K \,|\, x = \tfrac{1}{2}(y + z),\ y, z \in K,\ d(y, z) \geqq \frac{1}{n}\right\}.$$

Show that

$$\operatorname{ex}(K) = \bigcap_{n=1}^{\infty} K \backslash F_n$$

and deduce that $\operatorname{ex}(K)$ is a $G_\delta$-set.

**5.14. Exercise.** Let $K$ be a compact convex set in $E$ and let $b: M_+^1(K) \to E$ be the mapping which to $\mu \in M_+^1(K)$ associates the barycentre $b(\mu)$ of $\mu$. Show that $b$ is continuous when $M_+^1(K)$ carries the weak topology and that $b$ is affine.

**5.15. Exercise.** Let $K$ be the set of *stochastic* $n \times n$ matrices, i.e. $A = (a_{ij}) \in K$ if and only if $a_{ij} \geqq 0$ and $\sum_{j=1}^{n} a_{ij} = 1$ for $i = 1, \ldots, n$. Show that $K$ is a compact convex set in the vector space of all real $n \times n$ matrices with the canonical topology. Show that $A$ is an extreme point of $K$ if and only if each row has $n - 1$ zero entries.

Show that the set $\Omega_n$ of *doubly stochastic* $n \times n$ matrices is a compact convex subset of $K$, where a stochastic matrix $A$ is called doubly stochastic if also the sum of each column is one. Show that the extreme points of $\Omega_n$ are the permutation matrices arising from the unit matrix by permutations of the columns. (This result is due to G. Birkhoff.) *Hint*: Consider the matrix as a chessboard. If a doubly stochastic matrix $A$ is not a permutation matrix there exists a closed circuit for a rook with each move starting on a position $ij$ with $0 < a_{ij} < 1$. Adding successively $\varepsilon, -\varepsilon, \varepsilon, \ldots$ and $-\varepsilon, \varepsilon, -\varepsilon, \ldots$ at each position of the circuit with $\varepsilon > 0$ sufficiently small, we get two matrices $A_+$ and $A_-$ in $\Omega_n$ such that $A = \frac{1}{2}(A_+ + A_-)$.

**5.16. Exercise.** Let $K$ be the unit ball $\{x \in H \mid \|x\| \leqq 1\}$ of a Hilbert space $H$. Show that $\text{ex}(K)$ is the topological boundary of $K$, i.e.

$$\text{ex}(K) = \{x \in H \mid \|x\| = 1\}.$$

**5.17. Exercise.** Let $(X, \mathscr{A})$ be a measurable space and let $E$ be the Banach space over $\mathbb{K}$ of bounded $\mathscr{A}$-measurable functions $f: X \to \mathbb{K}$ with the uniform norm $\|f\| = \sup\{|f(x)| \mid x \in X\}$. Show that the set of extreme points of the unit ball $\{f \in E \mid \|f\| \leqq 1\}$ of $E$ is the set of functions $f \in E$ with $|f(x)| = 1$ for all $x \in X$.

Formulate and prove a similar result when $X$ is a compact Hausdorff space and $E = C(X)$.

**5.18. Exercise.** Let $K$ denote the set of convex functions $f: ]0, \infty[ \to [0, 1]$ considered as a subset of $\mathbb{R}^{]0, \infty[}$ with the topology of pointwise convergence. Show that $K$ is a metrizable compact convex set. Show that every $f \in K$ is decreasing, continuous and differentiable from the left and the right. For $f \in K$ and $t > 0$ we define

$$f_t(x) = \begin{cases} f(t) + (x - t)f'_-(t) & \text{for } 0 < x \leqq t, \\ f(x) & \text{for } x > t. \end{cases}$$

Show that $f_t, f - f_t \in K$ and deduce that the extreme points of $K$ are the following functions $\varphi_0 \equiv 0$, $\varphi_\infty \equiv 1$, $\varphi_t(x) = (1 - x/t)^+$, $0 < t < \infty$. Show finally that $K$ is a Bauer simplex. (This way of finding the extreme points of $K$ is taken from Johansen (1967).)

**5.19. Exercise.** Show that the Riesz representation theorem on a compact Hausdorff space is a special case of the Krein–Milman theorem.

**5.20. Exercise.** (Douglas 1964). Let $V$ be an adapted space of continuous functions on a locally compact space $X$ and let $C$ be the convex set of representing measures for a positive linear functional $L: V \to \mathbb{R}$. Show that $\mu \in C$ is an extreme point of $C$ if and only if $V$ is dense in $\mathscr{L}^1(X, \mu)$.

**5.21. Exercise.** Let $K$ and $L$ be compact convex subsets of locally convex spaces, and let $f: K \to L$ be continuous, affine and onto. Show that $\text{ex}(L) \subseteq f(\text{ex}(K))$, and use this result to determine the set of extreme points of

$$L := \left\{ \left( \int t\, d\mu(t), \int t^2\, d\mu(t), \cdots, \int t^n\, d\mu(t) \right) \,\middle|\, \mu \in M_+^1([0, 1]) \right\}.$$

## Notes and Remarks

We have not assumed a Radon measure to be locally finite. The reason for this is simply that we do not need this condition to derive any of the main results and, on the other hand, on many spaces a Radon measure is automatically locally finite. Certainly this is the case for locally compact spaces but it holds, for example, also for metric spaces: Let $\mu$ be a Radon measure on the metric space $X$ and suppose there is some $x \in X$ such that $\mu(G) = \infty$ for each open set $G$ containing $x$, then in particular $\mu(B_n) = \infty$ if $B_n$ is the open ball of radius $1/n$ around $x$, and hence $\mu(K_n) \geqq n$ for a suitable compact set $K_n \subseteq B_n$, where without restriction $x \in K_n$. Now the point is that $K := \bigcup_{n=1}^{\infty} K_n$ is again compact, because if $K \subseteq \bigcup_{\lambda \in \Lambda} G_\lambda$ for a family of open sets $(G_\lambda)$, then for some $\lambda_0$ we have $x \in G_{\lambda_0}$, and then for some $n_0$ $x \in B_{n_0} \subseteq G_{\lambda_0}$ implying $\bigcup_{n=n_0}^{\infty} K_n \subseteq G_{\lambda_0}$. The fact that $\mu(K) = \infty$ shows that our assumption was wrong.

Exercise 1.30 shows that non locally finite Radon measures may occur, and this depends on the fact that each compact subset is finite. As another example where this is the case we mention the fine topology of potential theory, for instance on $\mathbb{R}^3$, cf. Helms (1969). The fine topology is by definition the coarsest topology on $\mathbb{R}^3$ in which all superharmonic functions are continuous. The fine topology is completely regular (Brelot 1971, p. 5), and the fact that every finely compact set is finite is proved in Helms (1969, p. 208).

Next we want to relate our approach to Radon measures with the "classical" one as developed, for example, in Bourbaki (1965–1969). There, as already mentioned in the introduction, the "functional point of view" is prevalent, a (Radon) measure $\mu$ being by definition a positive linear form on $C^c(X)$, if $X$ is locally compact. Two set functions are then considered, *la mesure extérieure* $\mu^*$ defined by

$$\mu^*(G) = \sup\{\langle \mu, f\rangle \mid f \in C^c(X), 0 \leqq f \leqq 1_G\} \qquad \text{for open} \quad G \subseteq X$$

and

$$\mu^*(A) = \inf\{\mu^*(G) \mid A \subseteq G, G \text{ open}\} \qquad \text{for} \quad A \subseteq X,$$

and $\mu^{\bullet}$ derived from *l'intégrale superieure essentielle* and given by

$$\mu^{\bullet}(A) = \sup\{\mu^*(A \cap K) \mid K \in \mathscr{K}(X)\} \qquad \text{for} \quad A \subseteq X.$$

Both $\mu^*$ and $\mu^{\bullet}$ are Borel measures, i.e. $\sigma$-additive when restricted to $\mathscr{B}(X)$, and $\mu^{\bullet}$ is a Radon measure in our sense whereas $\mu^*$ is not so in general. One has $\mu^{\bullet} \leqq \mu^*$ and they agree on open sets and on Borel sets $B$ with $\mu^*(B) < \infty$, and in particular on compact sets. It follows that for locally compact and $\sigma$-compact spaces one has $\mu^{\bullet}(B) = \mu^*(B)$ for all $B \in \mathscr{B}(X)$, so for these spaces Bourbaki's notion of a Radon measure is equivalent to ours. This holds in particular for compact spaces. Bourbaki (1965–1969, Ch. IV, §1, Ex. 5) gives an example where $\mu^{\bullet}(F) = 0$ and $\mu^*(F) = \infty$ for a certain closed subset $F$ in some locally compact space, and this shows that $\mu^*$ is not a Radon measure in our sense.

Bourbaki defines a (Radon) premeasure on a Hausdorff space $X$ as a mapping $W$ which to every compact subset $K \subseteq X$ associates a Radon measure $W_K$ on $K$ such that $W_K | L = W_L$ if $L$ is a compact subset of $K$. Then the following set function $W^{\bullet}$ is considered

$$W^{\bullet}(A) = \sup\{(W_K)^{\bullet}(A \cap K) \mid K \in \mathscr{K}(X)\} \qquad \text{for} \quad A \subseteq X.$$

The restriction of $W^{\cdot}$ to $\mathscr{B}(X)$ is a Radon measure in our sense. Bourbaki calls $W$ a (Radon) measure if $W^{\cdot}$ is in addition locally finite. It follows that Bourbaki's notion of (Radon) measures is equivalent with locally finite Radon measures in our sense.

Let $X$ be a Hausdorff space and consider Borel measures $\nu\colon \mathscr{B}(X) \to [0, \infty]$ satisfying

(a) $\nu(K) < \infty \quad$ for $\quad K \in \mathscr{K}(X)$;
(b) $\nu(G) = \sup\{\nu(K) \mid K \subseteq G, K \in \mathscr{K}(X)\} \quad$ for open $\quad G \subseteq X$;
(c) $\nu(B) = \inf\{\nu(G) \mid B \subseteq G, G \text{ open}\} \quad$ for $\quad B \in \mathscr{B}(X)$.

There is a one-to-one correspondence between locally finite Radon measures $\mu$ on $X$ and Borel measures $\nu$ satisfying (a), (b) and (c).

In fact, if $\mu$ is a locally finite Radon measure then

$$\mu^*(B) = \inf\{\mu(G) \mid B \subseteq G, G \text{ open}\} \qquad \text{for} \quad B \in \mathscr{B}(X)$$

is a Borel measure satisfying (a), (b) and (c), and if $\nu$ has these properties then

$$\nu^{\cdot}(B) = \sup\{\nu(K) \mid K \subseteq B, K \in \mathscr{K}(X)\} \qquad \text{for} \quad B \in \mathscr{B}(X)$$

is a locally finite Radon measure. Furthermore $(\mu^*)^{\cdot} = \mu$ and $(\nu^{\cdot})^* = \nu$. For the proof of these assertions see Schwartz (1973), where a third equivalent definition of a locally finite Radon measure is given, namely as a pair $(m, M)$ of Borel measures satisfying certain conditions realized by $(\mu, \mu^*)$ and $(\nu^{\cdot}, \nu)$ in the above notation. Notice that a Borel measure $\nu$ satisfying (b) and (c) is locally finite if and only if (a) holds. In the locally compact case the exterior measure $\mu^*$ satisfies (a), (b) and (c) and $(\mu^*)^{\cdot} = \mu^{\cdot}$, $(\mu^{\cdot})^* = \mu^*$.

The generalized monotone convergence theorem (1.5) involves only the values of the underlying Radon measure $\mu$ on open sets, so it holds for any Borel measure $\nu$ which agrees with $\mu$ on open sets. In particular, we have

$$\int f \, d\mu = \int f \, d\nu$$

for each lower semicontinuous function $f \geqq 0$ on $X$. For a continuous real-valued function $f$ integrability with respect to $\mu$ and $\nu$ are equivalent and $\int f \, d\mu = \int f \, d\nu$ in case of integrability.

Bourbaki (Ch. IX, §3) also considers the possibility of "extending" a set function $\lambda\colon \mathscr{K}(X) \to [0, \infty[$ to a Radon measure, however the crucial property (1) of our §1, the defining property of a Radon content, which goes back to Kisyński (1968), is not discussed there. Théorème 1 of §3 in Bourbaki should be compared with our Lemma 1.3. Theorem 1.4 is due to Kisyński.

Replacing the Hausdorff space $X$ by an abstract set and the family $\mathscr{K}(X)$ of compact subsets of $X$ by a suitable set system called "compact paving", Topsøe (1978) proved an abstract measure extension theorem which not only contains Theorem 1.4 but also, for example, Carathéodory's classical result.

It should be mentioned that on many "nice" spaces a finite Borel measure is automatically a Radon measure. This holds in particular on Polish spaces, i.e. separable and completely metrizable spaces, cf. for example, Bauer (1978, Satz 41.3), but it can even be shown for so-called *analytic spaces*, i.e. Hausdorff spaces which are the continuous image of some Polish space; see Dellacherie and Meyer (1978, Chap. III) for a proof.

The last-mentioned book also contains a proof of the bimeasure theorem for finite Radon bimeasures on separable metric spaces. A slightly more general version may be found in Morando (1969), but both results are in fact a special case of a theorem of Marczewski and Ryll-Nardzewski (1953) about nondirect products of measures.

The Riesz representation theorem is certainly a cornerstone of functional analysis. Our proof is based on Pollard and Topsøe (1975) and we refer to the references given there, in particular to Batt (1973) for further information on this important topic. The theory of adapted spaces has had important applications in potential theory, see Sibony (1967–1968).

The theory of weak convergence of finite (or probability) Radon measures is mainly motivated by its applications in probability theory and mathematical statistics. For a thorough treatment on metric spaces we refer to Billingsley (1968) and Parthasarathy (1967). Later on Topsøe (1970) discovered that a satisfactory theory of weak convergence can be developed on arbitrary Hausdorff spaces. For a completely regular space $X$ the weak topology on $M_+^b(X)$ is induced by the weak topology $\sigma(M^b(X), C^b(X))$. However, for Hausdorff spaces in general it is not possible to extend the weak topology from $M_+^b(X)$ to $M^b(X)$ in such a way that $M^b(X)$ is a Hausdorff topological vector space. In fact, if such an extension was possible then $M^b(X)$ would be completely regular and so would $\{\varepsilon_x | x \in X\}$, which is homeomorphic to $X$ by Exercise 3.8. A particularly important topic for probabilistic applications is the characterization of relatively compact subsets of $M_+^1(X)$ in the weak topology, and we should mention the striking result due to Prohorov: For Polish spaces $X$ a subset $M \subseteq M_+^1(X)$ is weakly relatively compact if and only if for each $\varepsilon > 0$ there is a compact set $K \subseteq X$ such that $\sup_{\mu \in M} \mu(X \setminus K) < \varepsilon$, a condition on $M$ called uniform tightness, see Billingsley (1968, Theorems 6.1 and 6.2). Theorem 3.3 may be found in Ressel (1977); Exercise 3.10 is a generalization of Slutsky's theorem, cf. Ressel (1982b).

For a locally compact space $X$ the space $C^c(X)$ is often equipped with the inductive limit topology of the Banach spaces $C_K(X)$, $K \in \mathscr{K}(X)$, appearing before Theorem 2.5. With this topology, which is equal to the topology given in Exercise 2.11 by the family $I_4$, $C^c(X)$ is a barrelled space, and the topological dual space is $M(X)$ with the vague topology. This approach is the starting point in Bourbaki (1965–1969), who also seems to be the first who has systematically studied the vague topology. Theorem 4.5 is a special case of the Alaoglu–Bourbaki theorem, cf. Exercise 1.3.11. In the special case of $X = \mathbb{R}$ Theorem 4.5 is sometimes called Helly's selection theorem. In fact it follows from 4.5 and 4.10 that any sequence of measures in $M_+^b(\mathbb{R})$ with bounded total mass has a vaguely convergent subsequence.

The result in Exercise 4.15 is due to Topsøe.

The importance of the theory of integral representations lies undoubtedly in the fact that it gives a unified approach to a great number of classical formulas and theorems, cf. Phelps (1966). Let us just mention here Herglotz' formula for nonnegative harmonic functions in a ball, the far more general Martin representation, and the theorems of Bernstein and Bochner. In Chapter 4 we shall use the theory to prove integral representation theorems for positive definite functions on abelian semigroups.

The idea of considering a point in a metrizable compact convex set $K$ as the barycentre of a probability measure concentrated on $\text{ex}(K)$ is due to Choquet, and the whole theory is often called Choquet theory.

In our applications of the theory we use only the special case where $\text{ex}(K)$ is closed, in which case the representation theorem is equivalent with the Krein–Milman theorem. Therefore we have given a complete proof of the latter and only indicated the general results, which can be found in many books, for example, Alfsen (1971) and Phelps (1966).

CHAPTER 3

# General Results on Positive and Negative Definite Matrices and Kernels

## §1. Definitions and Some Simple Properties of Positive and Negative Definite Kernels

When dealing with positive and negative definite kernels a certain amount of confusion often arises concerning terminology. A positive definite kernel defined on a finite set is usually called a positive semidefinite matrix. Sometimes it is only called "positive", which may be misleading. When working on groups, the name positive definite function is used traditionally. In our previous papers on abelian semigroups we also followed this tradition. Instead of calling a kernel $\psi$ negative definite, some authors call the kernel $-\psi$ "conditionally positive definite" or "almost positive." In this book we use mainly the larger class of "semidefinite" kernels of all kinds and therefore prefer to avoid the prefix "semi" which otherwise would appear several hundred times.

Adapting the above point of view, an $n \times n$ matrix $A = (a_{jk})$ of complex numbers is called *positive definite* if and only if

$$\sum_{j,k=1}^{n} c_j \overline{c_k} a_{jk} \geqq 0$$

for all $\{c_1, \ldots, c_n\} \subseteq \mathbb{C}$.

It is well known that this is the case if and only if $A$ is *hermitian* (i.e. $a_{jk} = \overline{a_{kj}}$ for $j, k = 1, \ldots, n$) and the eigenvalues of $A$ are all $\geqq 0$.

Similarly $A$ is called *negative definite* if and only if $A$ is hermitian and

$$\sum_{j,k=1}^{n} c_j \overline{c_k} a_{jk} \leqq 0$$

for all $\{c_1, \ldots, c_n\} \subseteq \mathbb{C}$ with the extra condition $\sum_{j=1}^n c_j = 0$. (This definition requires $n \geqq 2$. Any $1 \times 1$ matrix $A = (a_{11})$ with real $a_{11}$ is called negative definite.)

**1.1. Definition.** Let $X$ be a nonempty set. A function $\varphi: X \times X \to \mathbb{C}$ is called a *positive definite kernel* if and only if

$$\sum_{j,k=1}^{n} c_j \bar{c}_k \varphi(x_j, x_k) \geqq 0$$

for all $n \in \mathbb{N}$, $\{x_1, \ldots, x_n\} \subseteq X$ and $\{c_1, \ldots, c_n\} \subseteq \mathbb{C}$. We call the function $\varphi$ a *negative definite kernel* if and only if it is *hermitian* (i.e. $\varphi(y, x) = \overline{\varphi(x, y)}$ for all $x, y \in X$) and

$$\sum_{j,k=1}^{n} c_j \bar{c}_k \varphi(x_j, x_k) \leqq 0$$

for all $n \geqq 2$, $\{x_1, \ldots, x_n\} \subseteq X$ and $\{c_1, \ldots, c_n\} \subseteq \mathbb{C}$ with $\sum_{j=1}^n c_j = 0$.

If the above inequalities are strict whenever $x_1, \ldots, x_n$ are different and at least one of the $c_1, \ldots, c_n$ does not vanish, then the kernel $\varphi$ is called *strictly positive* (resp. *strictly negative*) *definite*.

**1.2. Remark.** In the above definitions it is enough to consider mutually different elements $x_1, \ldots, x_n \in X$. In fact, if $x_1, \ldots, x_n \in X$ are arbitrary and $x_{\alpha_1}, \ldots, x_{\alpha_p}$ are the mutually different elements among the $x_i$'s, then

$$\sum_{j,k=1}^{n} c_j \bar{c}_k \varphi(x_j, x_k) = \sum_{j,k=1}^{p} d_j \bar{d}_k \varphi(x_{\alpha_j}, x_{\alpha_k}),$$

where

$$d_k := \sum_{\{i \mid x_i = x_{\alpha_k}\}} c_i, \qquad k = 1, \ldots, p.$$

Furthermore, if $\sigma: X \to X$ is a bijection, then $\varphi$ is a positive (resp. negative) definite kernel if and only if $\varphi \circ (\sigma \times \sigma)$ is a positive (resp. negative) definite kernel.

If $X$ is a finite set, say $X = \{x_1, \ldots, x_n\}$, then plainly $\varphi$ is positive (resp. negative) definite if and only if the $n \times n$ matrix

$$(\varphi(x_j, x_k))_{1 \leqq j,k \leqq n}$$

is positive (resp. negative) definite.

We now list some simple properties and examples of positive and negative definite kernels.

**1.3.** A kernel $\varphi$ on $X \times X$ is positive (resp. negative) definite if and only if for every finite subset $X_0 \subseteq X$ the restriction of $\varphi$ to $X_0 \times X_0$ is positive (resp. negative) definite.

**1.4.** If $\varphi$ is positive definite, then $\varphi(x, x) \geqq 0$ for all $x \in X$, i.e. $\varphi$ is nonnegative on the *diagonal* $\Delta := \{(x, x) \mid x \in X\}$.

**1.5.** Let $\begin{pmatrix} a & b \\ c & d \end{pmatrix}$ be a positive definite $2 \times 2$ matrix. Then

$$0 \leqq (1, 1)\begin{pmatrix} a & b \\ c & d \end{pmatrix}\begin{pmatrix} 1 \\ 1 \end{pmatrix} = a + b + c + d$$

implying $\operatorname{Im} b = -\operatorname{Im} c$. Further,

$$0 \leqq (1, i)\begin{pmatrix} a & b \\ c & d \end{pmatrix}\begin{pmatrix} 1 \\ -i \end{pmatrix} = a - ib + ic + d$$

implying $\operatorname{Re} b = \operatorname{Re} c$, i.e. $b = \bar{c}$. It follows immediately that any positive definite kernel is hermitian.

**1.6.** A real-valued kernel $\varphi$ on $X \times X$ is positive (resp. negative) definite if and only if $\varphi$ is *symmetric* (i.e. $\varphi(x, y) = \varphi(y, x)$ for all $x, y \in X$) and

$$\sum_{j,k=1}^{n} c_j c_k \varphi(x_j, x_k) \geqq 0 \qquad (\text{resp. } \leqq 0)$$

for all $n \in \mathbb{N}$, $\{x_1, \ldots, x_n\} \subseteq X$ and $\{c_1, \ldots, c_n\} \subseteq \mathbb{R}$ (resp. $\sum_{j=1}^{n} c_j = 0$ in addition). For, if $c_j = a_j + ib_j$, $a_j$ and $b_j$ being real, then

$$\sum_{j,k=1}^{n} c_j \bar{c}_k \varphi(x_j, x_k) = \sum_{j,k=1}^{n} (a_j a_k + b_j b_k)\varphi(x_j, x_k)$$

$$+ i \sum_{j,k=1}^{n} (b_j a_k - a_j b_k)\varphi(x_j, x_k),$$

and the last sum is zero if $\varphi$ is symmetric.

**1.7.** A $2 \times 2$ matrix $\begin{pmatrix} a & b \\ c & d \end{pmatrix}$ is negative definite if and only if $a, d \in \mathbb{R}$, $b = \bar{c}$ and

$$0 \geqq (1, -1)\begin{pmatrix} a & b \\ c & d \end{pmatrix}\begin{pmatrix} 1 \\ -1 \end{pmatrix} = a - b - c + d,$$

and this inequality is equivalent with

$$a + d \leqq 2 \operatorname{Re} b.$$

Therefore, we have for any negative definite kernel $\psi$ the inequality

$$\psi(x, x) + \psi(y, y) \leqq 2 \operatorname{Re} \psi(x, y).$$

**1.8.** Let $\begin{pmatrix} a & \bar{b} \\ b & d \end{pmatrix}$ be a hermitian $2 \times 2$ matrix. Then for $z, w \in \mathbb{C}$ we have

$$(w, z)\begin{pmatrix} a & \bar{b} \\ b & d \end{pmatrix}\begin{pmatrix} \bar{w} \\ \bar{z} \end{pmatrix} = a|w|^2 + 2\,\mathrm{Re}(bz\bar{w}) + d|z|^2$$

$$= a\left|w + \frac{b}{a}z\right|^2 + \frac{|z|^2}{a}(ad - |b|^2) \qquad (\text{for } a \neq 0).$$

The matrix is therefore positive definite if and only if $a \geqq 0$, $d \geqq 0$ and

$$\det\begin{pmatrix} a & \bar{b} \\ b & d \end{pmatrix} = ad - |b|^2 \geqq 0.$$

Hence for any positive definite kernel $\varphi$ we have

$$|\varphi(x, y)|^2 \leqq \varphi(x, x) \cdot \varphi(y, y).$$

**1.9.** If $f: X \to \mathbb{C}$ is an arbitrary function, then $\varphi(x, y) := f(x)\overline{f(y)}$ is positive definite, because

$$\sum_{j,k=1}^{n} c_j \bar{c}_k \varphi(x_j, x_k) = \left|\sum_{j=1}^{n} c_j f(x_j)\right|^2 \geqq 0.$$

The kernel $\psi(x, y) := f(x) + \overline{f(y)}$ is negative definite, for if $\sum_{j=1}^{n} c_j = 0$, then even $\sum_{j,k=1}^{n} c_j \bar{c}_k \psi(x_j, x_k) = 0$. In particular, a constant kernel $(x, y) \mapsto c$ is positive definite if and only if $c \geqq 0$ and negative definite if and only if $c \in \mathbb{R}$.

**1.10.** The kernel $\psi(x, y) = (x - y)^2$ on $\mathbb{R} \times \mathbb{R}$ is negative definite, $c_1 + \cdots + c_n = 0$ implying (for real numbers $c_j$, see 1.6)

$$\sum_{j,k=1}^{n} c_j c_k (x_j - x_k)^2 = -2\left(\sum_{j=1}^{n} c_j x_j\right)^2 \leqq 0.$$

**1.11.** If $X$ is a nonempty set, then the family of all positive (resp. negative) definite kernels on $X \times X$ is a convex cone, closed in the topology of pointwise convergence.

A very important property of positive definite kernels is their closure under pointwise multiplication which was proved by Schur (1911) (in the case of matrices):

**1.12. Theorem.** *Let $\varphi_1, \varphi_2: X \times X \to \mathbb{C}$ be positive definite kernels. Then $\varphi_1 \cdot \varphi_2: X \times X \to \mathbb{C}$ is positive definite, too.*

Proof. It suffices to prove that if $A = (a_{jk})$ and $B = (b_{jk})$ are positive definite $n \times n$ matrices, then $C := (a_{jk} b_{jk})$ is positive definite.

Now it is well known from linear algebra (and also follows from 3.1 below) that there are $n$ functions $f_1, \ldots, f_n\colon \{1, \ldots, n\} \to \mathbb{C}$ such that

$$a_{jk} = \sum_{p=1}^{n} f_p(j)\overline{f_p(k)}, \qquad \text{for} \quad j, k = 1, \ldots, n.$$

Let $c_1, \ldots, c_n \in \mathbb{C}$ be arbitrary, then

$$\sum_{j,k=1}^{n} c_j \bar{c}_k a_{jk} b_{jk} = \sum_{p=1}^{n} \sum_{j,k=1}^{n} c_j f_p(j) \overline{c_k f_p(k)} b_{jk} \geqq 0. \qquad \square$$

**1.13. Corollary.** *Let $\varphi_1\colon X \times X \to \mathbb{C}$ and $\varphi_2\colon Y \times Y \to \mathbb{C}$ be positive definite kernels. Then their tensor product $\varphi_1 \otimes \varphi_2\colon (X \times Y) \times (X \times Y) \to \mathbb{C}$ defined by $\varphi_1 \otimes \varphi_2(x_1, y_1, x_2, y_2) = \varphi_1(x_1, x_2) \cdot \varphi_2(y_1, y_2)$ is also positive definite.*

Proof. If $\tilde{\varphi}_1\colon (X \times Y) \times (X \times Y) \to \mathbb{C}$ is defined by $\tilde{\varphi}_1(x_1, y_1, x_2, y_2) = \varphi_1(x_1, x_2)$ and analogously $\tilde{\varphi}_2(x_1, y_1, x_2, y_2) = \varphi_2(y_1, y_2)$, then $\varphi_1 \otimes \varphi_2 = \tilde{\varphi}_1 \cdot \tilde{\varphi}_2$ and is therefore positive definite. $\square$

**1.14. Corollary.** *Let $\varphi\colon X \times X \to \mathbb{C}$ be positive definite such that $|\varphi(x, y)| < \rho$ for all $(x, y) \in X \times X$. Then if $f(z) = \sum_{n=0}^{\infty} a_n z^n$ is holomorphic in $\{z \in \mathbb{C} \mid |z| < \rho\}$ and $a_n \geqq 0$ for all $n \geqq 0$, the composed kernel $f \circ \varphi$ is again positive definite. In particular if $\varphi$ is positive definite, then so is* $\exp(\varphi)$.

Proof. By Theorem 1.12 for each $n \in \mathbb{N}$ the kernel $\varphi^n$ is positive definite, therefore $\sum_{n=0}^{N} a_n \varphi^n$ is positive definite for all $N \in \mathbb{N}$ and so is its pointwise limit $f \circ \varphi$. $\square$

**1.15. Remark.** In contrast to Theorem 1.12 above the ordinary matrix product of two positive definite matrices is positive definite if and only if the two matrices commute. This follows from the simultaneous diagonalization of these matrices. In particular the matrix exponential of any positive definite matrix again has this property. By using the Jordan decomposition one can show that the matrix exponential of every symmetric real matrix is positive definite (even strictly).

The following remarkable criterion for strict positive definiteness is often useful.

**1.16. Theorem.** *Let $A = (a_{jk})$ be some hermitian $n \times n$ matrix. Then $A$ is strictly positive definite if and only if*

$$\det((a_{jk})_{j,k \leqq p}) > 0$$

*for $p = 1, \ldots, n$.*

Proof. Suppose first $A$ to be strictly positive definite. As in the proof of 1.12 we choose $n$ vectors $z_1, \ldots, z_n \in \mathbb{C}^n$ such that

$$a_{jk} = \langle z_j, z_k \rangle, \qquad j, k = 1, \ldots, n,$$

which of course can also be written as $A = BB^*$, where $B$ is the $n \times n$ matrix with rows $z_1, \ldots, z_n$. This implies $\det A = |\det B|^2 \geqq 0$ and certainly $A$ cannot be singular so that in fact $\det A > 0$. Obviously the same reasoning can be applied to the submatrices $(a_{jk})_{j,k \leqq p}$ for $p = 1, \ldots, n$.

For the other direction, we proceed by induction on $n$. The case $n = 1$ being trivially true, let us suppose that the theorem holds for $n - 1$. By assumption $a_{11} > 0$ and we subtract $a_{1k}/a_{11}$ times the first column from the $k$th column, $k = 2, \ldots, n$. The new matrix $(a'_{jk})_{j,k \leqq n}$ (where the first column remained unchanged whereas for $k \geqq 2$ we have $a'_{jk} = a_{jk} - (a_{1k}/a_{11})a_{j1}$) has the same principal minors as $(a_{jk})$, i.e.

$$\det((a_{jk})_{j,k \leqq p}) = \det((a'_{jk})_{j,k \leqq p}), \qquad p = 1, \ldots, n$$

and if we now change the matrix $(a'_{jk})$ to $B$, where

$$B = \begin{pmatrix} a_{11} & 0 & \ldots & 0 \\ 0 & a'_{22} & \ldots & a'_{2n} \\ \vdots & \vdots & & \vdots \\ 0 & a'_{n2} & \ldots & a'_{nn} \end{pmatrix}$$

then still $\det((b_{jk})_{j,k \leqq p}) = \det((a_{jk})_{j,k \leqq p})$ for all $p$, and $B$ is furthermore hermitian. Now

$$\det((b_{jk})_{j,k \leqq p}) = a_{11} \cdot \det\begin{pmatrix} a'_{22} & \ldots & a'_{2p} \\ \vdots & & \vdots \\ a'_{p2} & \ldots & a'_{pp} \end{pmatrix} > 0$$

for $p = 2, 3, \ldots, n$ implying by assumption that the $(n - 1) \times (n - 1)$ matrix

$$\begin{pmatrix} a'_{22} & \ldots & a'_{2n} \\ \vdots & & \vdots \\ a'_{n2} & \ldots & a'_{nn} \end{pmatrix}$$

is strictly positive definite. For $c_1, \ldots, c_n \in \mathbb{C}$ we have

$$\begin{aligned} \sum_{j,k=1}^{n} c_j \bar{c}_k a_{jk} &= \sum_{j,k=2}^{n} c_j \bar{c}_k \left( a'_{jk} + \frac{a_{1k} a_{j1}}{a_{11}} \right) + \sum_{j=2}^{n} c_j \bar{c}_1 a_{j1} \\ &\quad + \sum_{k=2}^{n} c_1 \bar{c}_k a_{1k} + |c_1|^2 a_{11} \\ &= \sum_{j,k=2}^{n} c_j \bar{c}_k a'_{jk} + \frac{1}{a_{11}} \\ &\quad \times \left[ \left| \sum_{j=2}^{n} c_j a_{j1} \right|^2 + 2a_{11} \operatorname{Re}\left( c_1 \sum_{k=2}^{n} \bar{c}_k a_{1k} \right) + (|c_1| a_{11})^2 \right] \\ &= \sum_{j,k=2}^{n} c_j \bar{c}_k a'_{jk} + \frac{1}{a_{11}} \left| \sum_{j=1}^{n} c_j a_{j1} \right|^2 . \end{aligned}$$

If $(c_2, \ldots, c_n) \neq 0$ then the first sum is $>0$, and for $(c_2, \ldots, c_n) = 0$ but $c_1 \neq 0$ the second term is strictly positive. Hence $A$ is a strictly positive definite matrix. □

One might expect that a corresponding result for positive definite matrices holds if the determinants in the above theorem are only supposed to be nonnegative. The simple counterexample given by the $2 \times 2$ matrix $\begin{pmatrix} 0 & 0 \\ 0 & -1 \end{pmatrix}$ destroys this hope. However, the following result seems to be rather satisfactory, at least theoretically.

**1.17. Theorem.** *Let $\varphi: X \times X \to \mathbb{C}$ be a hermitian kernel. Then $\varphi$ is positive definite if and only if*

$$\det((\varphi(x_j, x_k))_{j,k \leqq n}) \geqq 0$$

*for all $n \in \mathbb{N}$ and all $\{x_1, \ldots, x_n\} \subseteq X$.*

PROOF. If $\varphi$ is positive definite, then as in the beginning of the proof of the above theorem we see that all determinants in question are nonnegative.

Let us on the other hand assume this condition. We define a slightly perturbed kernel $\varphi_\varepsilon := \varphi + \varepsilon \cdot 1_\Delta$, where $\varepsilon > 0$ and $\Delta$ is the diagonal in $X \times X$. For mutually different elements $x_1, \ldots, x_n \in X$ it is easily seen that

$$\det((\varphi_\varepsilon(x_j, x_k))_{j,k \leqq n}) = \sum_{p=0}^{n} d_p \, \varepsilon^p,$$

where $d_n = 1$ and

$$d_p = \sum_{\substack{A \subseteq \{1,\ldots,n\} \\ |A| = n-p}} \det((\varphi(x_j, x_k))_{j,k \in A}) \geqq 0$$

for $p = 0, 1, \ldots, n-1$. Therefore $\det((\varphi_\varepsilon(x_j, x_k))_{j,k \leqq n}) \geqq \varepsilon^n > 0$ implying that $\varphi_\varepsilon$ is a strictly positive definite kernel. Hence the pointwise limit $\varphi = \lim_{\varepsilon \to 0} \varphi_\varepsilon$ is positive definite. □

The special case $n = 2$ has already been derived in 1.8.

**1.18. Exercise.** If $\varphi$ is a positive definite kernel, then also $\operatorname{Re} \varphi$, $\bar{\varphi}$ and $|\varphi|^2$ are positive definite, but not necessarily $|\varphi|$. If $\psi$ is negative definite, then so are $\operatorname{Re} \psi$ and $\bar{\psi}$.

**1.19. Exercise.** For $z \in \mathbb{C}$ define $M_z := \begin{pmatrix} 1 & z & z \\ \bar{z} & 1 & z \\ \bar{z} & \bar{z} & 1 \end{pmatrix}$. Then $M_z$ is positive definite if and only if $|z| \leqq 1$ and $[3 - 2\operatorname{Re}(z)]|z|^2 \leqq 1$. For $-1 \leqq z < -\frac{1}{2}$ the matrix $M_z$ is not positive definite.

**1.20. Exercise.** Let $H$ be a complex (pre-) Hilbert space. Then its scalar product $\langle \cdot, \cdot \rangle$ is a positive definite kernel. The squared distance $\psi(x, y) := \|x - y\|^2$ is negative definite.

**1.21. Exercise.** Let $\varphi: X \times X \to \mathbb{R}$ be a symmetric kernel. Then $\varphi$ is positive definite if (and only if)

$$\sum c_j c_k \varphi(x_j, x_k) \geqq 0$$

for all $n \in \mathbb{N}$, $\{x_1, \ldots, x_n\} \subseteq X$ and $\{c_1, \ldots, c_n\} \subseteq \mathbb{Z}$.

**1.22. Exercise.** Show that for each $a \in \mathbb{R}$ the kernel $\psi_a(x, y) := (a + x - y)^2$ on $\mathbb{R} \times \mathbb{R}$ fulfils the inequalities

$$\sum_{j,k=1}^{n} c_j c_k \psi_a(x_j, x_k) \leqq 0$$

for all $n \geqq 1$, all $x_1, \ldots, x_n \in \mathbb{R}$ and all $c_1, \ldots, c_n \in \mathbb{R}$ with $\sum c_j = 0$. Nevertheless $\psi_a$ is negative definite only for $a = 0$.

**1.23. Exercise.** Let $X$ be a nonempty set and let $T \subseteq X \times X$ contain the diagonal. Then the kernel $1_T$ is positive definite if and only if $T$ is an equivalence relation.

**1.24. Exercise.** Let $X = [a, b]$ be a compact interval and let $\varphi: [a, b] \times [a, b] \to \mathbb{C}$ be continuous. Then $\varphi$ is positive definite if and only if

$$\int_a^b \int_a^b c(x)\overline{c(y)}\varphi(x, y)\, dx\, dy \geqq 0$$

for each continuous function $c: X \to \mathbb{C}$.

**1.25. Exercise.** An invertible square matrix is positive definite if and only if its inverse has this property (and in this case both matrices are strictly positive definite).

**1.26. Exercise.** Let $A = (a_{jk})$ be a real negative definite $n \times n$ matrix. Then

$$\operatorname{tr} A \leqq \frac{2}{n-1} \sum_{j<k} a_{jk}.$$

## §2. Relations Between Positive and Negative Definite Kernels

There are many interesting and important relations between positive and negative definite kernels some of which were first known in special cases only, say for positive (resp. negative) definite functions on groups. Later on they turned out to hold for more general kernels, too. Of course, it is trivially true that $-\varphi$ is negative definite whenever $\varphi$ is positive definite. The remarkable part in the following lemma is therefore, that it is an "if and only if" statement.

**2.1. Lemma.** *Let $X$ be a nonempty set, $x_0 \in X$, and let $\psi: X \times X \to \mathbb{C}$ be a hermitian kernel. Put $\varphi(x, y) := \psi(x, x_0) + \overline{\psi(y, x_0)} - \psi(x, y) - \psi(x_0, x_0)$. Then $\varphi$ is positive definite if and only if $\psi$ is negative definite. If $\psi(x_0, x_0) \geqq 0$ and $\varphi_0(x, y) := \psi(x, x_0) + \overline{\psi(y, x_0)} - \psi(x, y)$, then $\varphi_0$ is positive definite if and only if $\psi$ is negative definite.*

PROOF. For $c_1, \ldots, c_n \in \mathbb{C}$, $\sum_1^n c_j = 0$, and $x_1, \ldots, x_n \in X$ we have

$$\sum_{j,k=1}^{n} c_j \bar{c}_k \varphi(x_j, x_k) = \sum_{j,k=1}^{n} c_j \bar{c}_k \varphi_0(x_j, x_k)$$

$$= - \sum_{j,k=1}^{n} c_j \bar{c}_k \psi(x_j, x_k).$$

Hence positive definiteness of $\varphi$ or of $\varphi_0$ implies negative definiteness of $\psi$.

Suppose on the other hand that $\psi$ is negative definite. Let $x_1, \ldots, x_n \in X$ and $c_1, \ldots, c_n \in \mathbb{C}$ be given and put $c_0 := -\sum_{j=1}^n c_j$. Then

$$0 \geqq \sum_{j,k=0}^{n} c_j \bar{c}_k \psi(x_j, x_k) = \sum_{j,k=1}^{n} c_j \bar{c}_k \psi(x_j, x_k)$$

$$+ \sum_{j=1}^{n} c_j \bar{c}_0 \psi(x_j, x_0) + \sum_{k=1}^{n} c_0 \bar{c}_k \psi(x_0, x_k) + |c_0|^2 \psi(x_0, x_0)$$

$$= \sum_{j,k=1}^{n} c_j \bar{c}_k [\psi(x_j, x_k) - \psi(x_j, x_0) - \psi(x_0, x_k) + \psi(x_0, x_0)]$$

$$= - \sum_{j,k=1}^{n} c_j \bar{c}_k \varphi(x_j, x_k),$$

thus showing that $\varphi$ is positive definite.

Now if $\psi(x_0, x_0) \geqq 0$ then $\varphi_0 = \varphi + \psi(x_0, x_0)$ is positive definite. □

The following result, due mainly to Schoenberg, is of central importance.

**2.2. Theorem.** *Let $X$ be a nonempty set and let $\psi: X \times X \to \mathbb{C}$ be a kernel. Then $\psi$ is negative definite if and only if $\exp(-t\psi)$ is positive definite for all $t > 0$.*

PROOF. If $\exp(-t\psi)$ is positive definite, then $1 - \exp(-t\psi)$ is, of course, negative definite and so is therefore the pointwise limit

$$\psi = \lim_{0 < t \to 0} \frac{1}{t}(1 - \exp(-t\psi)).$$

Now suppose that $\psi$ is negative definite. For obvious reasons we need only show that $\exp(-t\psi)$ is positive definite for $t = 1$. We choose $x_0 \in X$ and with $\varphi$ as in the above lemma we have

$$-\psi(x, y) = \varphi(x, y) - \psi(x, x_0) - \overline{\psi(y, x_0)} + \psi(x_0, x_0),$$

where $\varphi$ is positive definite. Hence

$$\exp(-\psi(x, y)) = \exp(\varphi(x, y)) \cdot \exp(-\psi(x, x_0)) \cdot \overline{\exp(-\psi(y, x_0))} \cdot \exp(\psi(x_0, x_0))$$

and from 1.14, 1.9 and 1.12 we conclude that $\exp(-\psi)$ is positive definite. □

In the following let $\mathbb{C}_+ = \{z \in \mathbb{C} \mid \operatorname{Re} z \geqq 0\}$. It is known that a kernel $\psi: X \times X \to \mathbb{C}_+$ is negative definite if and only if $(t + \psi)^{-1}$ is positive definite for all $t > 0$. Instead of proving this we show the following more general result.

**2.3. Theorem.** *Let $\mu$ be a probability measure on the half-line $\mathbb{R}_+$ such that $0 < \int_0^\infty s\, d\mu(s) < \infty$, and let $\mathscr{L}\mu$ denote its Laplace transform, i.e. $\mathscr{L}\mu(z) = \int_0^\infty e^{-sz}\, d\mu(s)$, $z \in \mathbb{C}_+$. Then $\psi: X \times X \to \mathbb{C}_+$ is negative definite if and only if $\mathscr{L}\mu(t\psi)$ is positive definite for all $t > 0$.*

PROOF. If $\psi$ is negative definite then for $t > 0$ we have

$$\mathscr{L}\mu(t\psi) = \int_0^\infty \exp(-ts\psi)\, d\mu(s)$$

pointwise on $X \times X$, which is positive definite, being a mixture of the positive definite kernels $\exp(-ts\psi)$.

If on the other hand $\mathscr{L}\mu(t\psi)$ is positive definite for all $t > 0$, then for each $(x, y) \in X \times X$ we get

$$\frac{1}{t}[1 - \mathscr{L}\mu(t\psi(x, y))] = \int_0^\infty \frac{1 - \exp[-ts\psi(x, y)]}{t}\, d\mu(s)$$

$$\to \psi(x, y) \int_0^\infty s\, d\mu(s) \qquad \text{for} \quad t \to 0,$$

where we could apply Lebesgue's theorem because of

$$\frac{|1 - \exp[-ts\psi(x, y)]|}{t} \leqq |\psi(x, y)| s.$$

Being a pointwise limit of negative definite kernels, $\psi$ itself is negative definite, too. □

Choosing $\mu = \varepsilon_1$ in the above theorem, we get back Theorem 2.2 for $\mathbb{C}_+$-valued $\psi$, and the choice of $\mu = e^{-t}\, dt$ shows, as already mentioned, that $\psi: X \times X \to \mathbb{C}_+$ is negative definite if and only if $(t + \psi)^{-1}$ is positive definite for all $t > 0$.

**2.4. Remark.** If a probability measure $\mu$ on $\mathbb{R}_+$ has infinite first moment and $\psi: X \times X \to \mathbb{C}_+$ is negative definite, then $\mathscr{L}\mu(t\psi)$ is still positive definite for all $t > 0$, but, in general, the converse does not hold. It follows from later results (cf. 2.10 and 4.4.5) that $\exp(-\sqrt{z})$ is the Laplace transform of some probability measure $\mu$ on $\mathbb{R}_+$. Now a matrix $(a_{jk})$ of the form $a_{jk} = (s_j + s_k)^2$, where $s_1, \ldots, s_n \geqq 0$, is not negative definite in general, $c_1 + \cdots + c_n = 0$ implying $\sum c_j c_k a_{jk} = 2(\sum c_j s_j)^2$, but nevertheless for all $t > 0$

$$\mathscr{L}\mu(ta_{jk}) = \exp(-\sqrt{ta_{jk}}) = \exp(-\sqrt{t}s_j)\exp(-\sqrt{t}s_k)$$

is positive definite by 1.9.

**2.5. Remark.** If $\varphi$ is positive definite and $\varphi|\Delta \equiv c$ for some $c \in \mathbb{R}_+$ then obviously $c - \varphi$ is negative definite, bounded and vanishes on the diagonal $\Delta$. A similar statement in the other direction (which may be found in the literature) is not generally true; see, however, 4.3.15. For $x_1 = -1, x_2 = 0, x_3 = +1$, the $3 \times 3$ matrix

$$(a_{jk}) = ((x_j - x_k)^2) = \begin{pmatrix} 0 & 1 & 4 \\ 1 & 0 & 1 \\ 4 & 1 & 0 \end{pmatrix}$$

is negative definite, vanishes on the diagonal, and is bounded by 4, but for no real number $t$ the matrix $(t - a_{jk})$ is positive definite, because

$$\det(t - a_{jk}) = \begin{vmatrix} t & t-1 & t-4 \\ t-1 & t & t-1 \\ t-4 & t-1 & t \end{vmatrix} \equiv -8 \qquad \text{for all} \quad t \in \mathbb{R}.$$

Negative definite kernels are intimately related to so-called "infinitely divisible" positive definite kernels.

**2.6. Definition.** A positive definite kernel $\varphi$ is called *infinitely divisible* if for each $n \in \mathbb{N}$ there exists a positive definite kernel $\varphi_n$ such that $\varphi = (\varphi_n)^n$.

If $\psi$ is negative definite then $\varphi = e^{-\psi}$ is infinitely divisible since $\varphi_n = \exp(-(1/n)\psi)$ is positive definite and $(\varphi_n)^n = \varphi$. Furthermore, $\varphi$ has no zeros. Proposition 2.7 below shows, in particular, that every strictly positive infinitely divisible kernel has this form.

Let $\varphi$ be infinitely divisible. Then

$$|\varphi| = |\varphi_n|^n = (|\varphi_{2n}|^2)^n = [(|\varphi_{2kn}|^2)^k]^n, \qquad k, n \geqq 1,$$

so that the nonnegative kernel $|\varphi|$ is infinitely divisible inside the family of all nonnegative positive definite kernels, each (under this restriction uniquely determined) $n$th root $|\varphi_n|$ again being an infinitely divisible positive definite kernel. Let $\mathscr{N}_\infty(X)$ denote the closure of all real-valued negative definite kernels on $X \times X$ in the space $]-\infty, \infty]^{X \times X}$.

**2.7. Proposition.** *For a positive definite kernel $\varphi \geqq 0$ on $X \times X$ the following conditions are equivalent:*

(i) *$\varphi$ is infinitely divisible;*
(ii) $-\log \varphi \in \mathcal{N}_\infty(X)$;
(iii) *$\varphi^t$ is positive definite for all $t > 0$.*

PROOF. "(i) ⇒ (ii)" Let $\psi := -\log \varphi$, then

$$\psi = \lim n[1 - \exp(-\psi/n)] \in \mathcal{N}_\infty(X).$$

"(ii) ⇒ (iii)" Let $(\psi_\alpha)$ be a net of (finite) negative definite kernels converging pointwise to $\psi = -\log \varphi$. Then for any $t > 0$ we have $\exp(-t\psi_\alpha) \to \exp(-t\psi) = \varphi^t$, so that $\varphi^t$ is positive definite by Theorem 2.2.

"(iii) ⇒ (i)" Take $t = \frac{1}{2}, \frac{1}{3}, \frac{1}{4}, \ldots$. □

**2.8. Remarks.** (1) The above proof shows that $\mathcal{N}_\infty(X)$ is in fact the monotone sequential closure of the subset of all negative definite kernels, bounded above.

(2) For $z \in \mathbb{C}$, $|z| \leqq 1$ the $2 \times 2$ matrix $\begin{pmatrix} 1 & z \\ \bar{z} & 1 \end{pmatrix}$ is an infinitely divisible positive definite kernel with nonuniquely determined positive definite "roots".

We conclude this section by indicating a large class of functions which operate on negative definite kernels.

For $\mu \in M_+(]0, \infty[)$ we define $g: D(\mu) \to \mathbb{C}$ by

$$g(z) = \int_0^\infty (1 - e^{-\lambda z})\, d\mu(\lambda),$$

where $D(\mu)$ is the set of $z \in \mathbb{C}$ for which $\lambda \mapsto 1 - e^{-\lambda z}$ is $\mu$-integrable.

**2.9. Proposition.** *Let $\psi: X \times X \to \mathbb{C}$ be a negative definite kernel and let $\mu \in M_+(]0, \infty[)$. If $\psi(X \times X) \subseteq D(\mu)$ then $g \circ \psi$ is negative definite. Furthermore, for $x_0 \in X$ the kernel $(x, y) \mapsto g[\psi(x, x_0) + \overline{\psi(y, x_0)}] - g[\psi(x, y) + \psi(x_0, x_0)]$ is positive definite provided $(\psi(X \times X) + \psi(x_0, x_0)) \cup (\psi(X, x_0) + \psi(x_0, X)) \subseteq D(\mu)$.*

*If $\int_0^\infty \lambda(1 + \lambda)^{-1}\, d\mu(\lambda) < \infty$ and $\psi|\Delta \geqq 0$ then $g \circ \psi$ is negative definite and $g[\psi(x, x_0) + \overline{\psi(y, x_0)}] - g[\psi(x, y)]$ is positive definite for all $x_0 \in X$.*

PROOF. It suffices to prove the result for $g(z) = 1 - e^{-\lambda z}$ where $\lambda \in ]0, \infty[$, i.e. $\mu = \varepsilon_\lambda$, $D(\mu) = \mathbb{C}$. If $x_1, \ldots, x_n \in X$ and $c_1, \ldots, c_n \in \mathbb{C}$ such that $\sum c_j = 0$ we get

$$\sum_{j,k} c_j \bar{c}_k (1 - e^{-\lambda\psi(x_j, x_k)}) = -\sum_{j,k} c_j \bar{c}_k e^{-\lambda\psi(x_j, x_k)} \leqq 0$$

as an immediate consequence of Theorem 2.2. For any $x_0 \in X$ the kernel $\psi(x, x_0) + \overline{\psi(y, x_0)} - \psi(x, y) - \psi(x_0, x_0)$ is positive definite by Lemma 2.1, as is therefore the kernel

$$\exp(\lambda[\psi(x, x_0) + \overline{\psi(y, x_0)} - \psi(x, y) - \psi(x_0, x_0)]) - 1$$

by 1.14, and multiplying with the positive definite kernel

$$\exp(-\lambda\psi(x, x_0)) \exp(-\lambda\overline{\psi(y, x_0)})$$

gives

$$\begin{aligned}\exp(-\lambda[\psi(x, y) + \psi(x_0, x_0)]) &- \exp(-\lambda[\psi(x, x_0) + \overline{\psi(y, x_0)}]) \\ &= g[\psi(x, x_0) + \overline{\psi(y, x_0)}] - g[\psi(x, y) + \psi(x_0, x_0)],\end{aligned}$$

which is positive definite by Theorem 1.12.

If $\int_0^\infty \lambda(1 + \lambda)^{-1}\, d\mu(\lambda) < \infty$ then $\mathbb{C}_+ \subseteq D(\mu)$. Notice that $\psi|\Delta \geqq 0$ if and only if $\psi(X \times X) \subseteq \mathbb{C}_+$ because of 1.7. It follows that $\psi(X \times X) \subseteq D(\mu)$ and $\psi(X \times X) + \psi(X \times X) \subseteq D(\mu)$ so $g \circ \psi$ is negative definite and $g[\psi(x, x_0) + \overline{\psi(y, x_0)}] - g[\psi(x, y) + \psi(x_0, x_0)]$ is positive definite. Using the fact that the kernel $\psi(x, x_0) + \overline{\psi(y, x_0)} - \psi(x, y)$ is positive definite by Lemma 2.1, it is seen as above that $g[\psi(x, x_0) + \overline{\psi(y, x_0)}] - g[\psi(x, y)]$ is also positive definite. □

**2.10. Corollary.** *If $\psi: X \times X \to \mathbb{C}$ is negative definite and satisfies $\psi|\Delta \geqq 0$ then so are $\psi^\alpha$ for $0 < \alpha < 1$ and $\log(1 + \psi)$.*

Proof. The assertions follow by Proposition 2.9 and the formulas

$$z^\alpha = \frac{\alpha}{\Gamma(1 - \alpha)} \int_0^\infty (1 - e^{-\lambda z}) \frac{d\lambda}{\lambda^{\alpha+1}},$$

$$\log(1 + z) = \int_0^\infty (1 - e^{-\lambda z}) \frac{e^{-\lambda}}{\lambda}\, d\lambda$$

which are valid for $\operatorname{Re} z \geqq 0$. Each formula can be established by showing that both sides of the equation have equal derivatives. □

**2.11. Corollary.** *If $f: X \to \mathbb{C}$ satisfies $\operatorname{Re} f \geqq 0$ then for each $\alpha \in [1, 2]$ the kernel*

$$\psi_\alpha(x, y) = -(f(x) + \overline{f(y)})^\alpha$$

*is negative definite.*

Proof. An equivalent formulation is that the kernel $-(x + \bar{y})^\alpha$ is negative definite on $\mathbb{C}_+ \times \mathbb{C}_+$. This is clear when $\alpha = 1$ and $\alpha = 2$. Integrating with respect to $z$ in the formula for $z^\alpha$ $(0 < \alpha < 1)$ in the previous proof we get

$$-z^{\alpha+1} = \frac{\alpha(1 + \alpha)}{\Gamma(1 - \alpha)} \int_0^\infty (1 - e^{-\lambda z} - \lambda z) \frac{d\lambda}{\lambda^{\alpha+2}}, \qquad z \in \mathbb{C}_+$$

and the assertion follows since

$$\sum c_j \overline{c_k}(1 - e^{-\lambda(z_j + \overline{z_k})} - \lambda(z_j + \overline{z_k})) = -|\sum c_j e^{-\lambda z_j}|^2 \leqq 0,$$

whenever $z_1, \ldots, z_n \in \mathbb{C}_+$ and $c_1, \ldots, c_n \in \mathbb{C}$ with $\sum c_j = 0$. □

**2.12. Exercise.** Show by using only the results of this and the preceding section that the following kernels are positive definite.

(a) $\varphi(x, y) = \cos(x - y)$ on $\mathbb{R} \times \mathbb{R}$.
(b) $\varphi(x, y) = \cos(x^2 - y^2)$ on $\mathbb{R} \times \mathbb{R}$.
(c) $\varphi(x, y) = (1 + |x - y|)^{-1}$ on $\mathbb{R} \times \mathbb{R}$.
(d) $\varphi(x, y) = \exp[-(1 - e^{-\sqrt{|x-y|}})]$ on $\mathbb{R} \times \mathbb{R}$.
(e) $\varphi(x, y) = (1 + \sqrt{x + y})^{-1}$ on $\mathbb{R}_+ \times \mathbb{R}_+$.
(f) $\varphi(x, y) = t^{(x-y)^2}$ on $\mathbb{Z} \times \mathbb{Z}$, where $t \in [-1, 1]$.
(g) $\varphi(x, y) = (x + y)^{-1}$ on $]0, \infty[ \times ]0, \infty[$.
(h) $\varphi(x, y) = (|x|^\beta + |y|^\beta)^\alpha - |x - y|^{\alpha\beta}$ on $\mathbb{R} \times \mathbb{R}$, where $0 < \beta \leqq 2$, $0 < \alpha \leqq 1$.
(i) $\varphi(x, y) = (x^\beta + y^\beta)^\alpha - (x + y)^{\alpha\beta}$ on $\mathbb{R}_+ \times \mathbb{R}_+$, where $\alpha, \beta \in ]0, 1]$.
(j) $\varphi(A, B) = P(A \cap B) - P(A)P(B)$ on $\mathscr{A} \times \mathscr{A}$, where $(\Omega, \mathscr{A}, P)$ denotes a probability space.
(k) $\varphi(x, y) = x \wedge y - xy$ on $[0, 1] \times [0, 1]$.

**2.13. Exercise.** Show that the following kernels are negative definite.

(a) $\psi(x, y) = [\sin(x - y)]^2$ on $\mathbb{R} \times \mathbb{R}$.
(b) $\psi(x, y) = \|x - y\|^p$ on $H \times H$, $H$ being a Hilbert space, $0 < p \leqq 2$.
(c) $\psi(x, y) = 1_{]0, \infty[}(x + y)$ on $\mathbb{R}_+ \times \mathbb{R}_+$.
(d) $\psi(x, y) = \log(x + y)$ on $]0, \infty[ \times ]0, \infty[$.
(e) $\psi(x, y) = 1 - \langle x, y \rangle$ on $H \times H$, $H$ being a Hilbert space.
(f) $\psi(x, y) = \sqrt[3]{1 + \sqrt{|x - y|}}$ on $\mathbb{R} \times \mathbb{R}$.
(g) $\psi(x, y) = 1_{\{0\}}(xy)$ on $\mathbb{R} \times \mathbb{R}$.

**2.14. Exercise.** Let $\varphi$ be a positive definite kernel bounded by 1 and let $0 < \alpha < 1$. Then $(1 - \alpha\varphi)^{-1}$ is an infinitely divisible positive definite kernel.

**2.15. Exercise.** Let $\varphi$ be an infinitely divisible positive definite kernel on $X \times X$ and denote $T := \{\varphi \neq 0\}$. Then $1_T$ is positive definite, too. This does not hold in general without the assumption of infinite divisibility.

**2.16. Exercise.** For $X \neq \varnothing$ the set of infinitely divisible positive definite kernels on $X \times X$ is closed with respect to pointwise convergence. *Hint*: Use universal subnets.

**2.17. Exercise.** Any negative definite kernel with nonnegative real part is the pointwise limit of a sequence of bounded negative definite kernels with nonnegative real part.

**2.18. Exercise.** Let the function $g: \mathbb{C}_+ \to \mathbb{C}$ be given by

$$g(z) = \alpha + \beta z + \gamma \bar{z} + \int_{\mathbb{R}_+^2} [1 - \exp(-sz - t\bar{z})] \, d\mu(s, t)$$

where $\alpha, \beta, \gamma \in \mathbb{R}_+$ and $\mu \in M_+(\mathbb{R}_+^2)$ satisfies

$$\int_{\mathbb{R}_+^2} \frac{s+t}{1+s+t} \, d\mu(s, t) < \infty.$$

If $\psi$ is a negative definite kernel with $\operatorname{Re} \psi \geqq 0$ then so is the kernel $g \circ \psi$.

**2.19. Exercise.** Let $g: ]0, \infty[ \to [0, \infty[$ be a Borel measurable function such that $g(\lambda) = \lambda + O(\lambda^2)$ for $\lambda \to 0$. Show that if $\mu \in M_+(]0, \infty[)$ satisfies

$$\int_0^1 \lambda^2 \, d\mu(\lambda) < \infty \qquad \text{and} \qquad \int_1^\infty \max(1, g(\lambda)) \, d\mu(\lambda) < \infty,$$

then

$$\psi(z) = \int_0^\infty (1 - e^{-\lambda z} - g(\lambda) z) \, d\mu(\lambda)$$

is well defined for $z \in \mathbb{C}_+$, and $\psi(x + \bar{y})$ is a negative definite kernel on $\mathbb{C}_+ \times \mathbb{C}_+$.

**2.20. Exercise.** Let $X$ be nonempty, $x_0 \in X$, and let the linear transformation $T$ be defined on the real vector space of all hermitian kernels $\psi: X \times X \to \mathbb{C}$ by

$$(T\psi)(x, y) := \psi(x, x_0) + \overline{\psi(y, x_0)} - \psi(x, y) - \psi(x_0, x_0).$$

Denoting by $\mathscr{P}$ and $\mathscr{N}$ the cones of all positive (resp. negative) definite kernels on $X \times X$, the result of Lemma 2.1 is that $T^{-1}(\mathscr{P}) = \mathscr{N}$. Show that $\ker(T) = \{\psi \mid \exists f: X \to \mathbb{C} \text{ such that } \psi(x, y) = f(x) + \overline{f(y)}\}$ and show also that $T(\mathscr{N}) = \{\varphi \in \mathscr{P} \mid \varphi(x, x_0) = 0 \text{ for all } x \in X\}$.

**2.21. Exercise.** Let $\psi$ be a negative definite kernel with strictly positive real part. Then $1/\psi$ is positive definite. For the case $\operatorname{Re} \psi \geqq 0$, the result still holds if $\psi$ does not assume the value zero.

**2.22. Exercise.** Given the negative definite $n \times n$ matrix $(b_{jk})$ put

$$a_{jk} := \frac{1}{n}(b_j + \overline{b_k}) - b_{jk} - \frac{1}{n^2} b,$$

where $b_j = \sum_{k=1}^n b_{jk}$, $j = 1, \ldots, n$ and $b = \sum_{j,k=1}^n b_{jk}$. Show that $(a_{jk})$ is positive definite, and derive from this another proof that $(e^{-tb_{jk}})$ is positive definite for all $t > 0$.

**2.23. Exercise.** Let $A$ be a hermitian $n \times n$ matrix, let $v \in \mathbb{C}^n$ satisfy $\sum_{j=1}^n v_j = 1$ and define an $n \times n$ matrix $P_v$ with $ij$'s element $\delta_{ij} - v_j$. Show that $-A$ is negative definite if and only if $P_v A P_v^*$ is positive definite.

# §3. Hilbert Space Representation of Positive and Negative Definite Kernels

Around 1940 Schoenberg published three fundamental papers (1938a, b, 1942) all of which were very closely connected with positive and negative definite kernels. The main motivation for deriving these results at that time was to decide which metric (or more general semimetric) spaces $(X, d)$ can be imbedded into a Hilbert space $H$, i.e. when does there exist a mapping $\Phi: X \to H$ such that

$$\|\Phi(x) - \Phi(y)\| = d(x, y) \qquad \text{for all} \quad x, y \in X.$$

It turned out that this property of $(X, d)$ is equivalent to $d^2$ being negative definite. We are now going to give an easy derivation of this result using the so-called *reproducing kernel Hilbert space* (RKHS) associated with a positive definite kernel.

**3.1.** Let $X$ be a nonempty set and $\varphi: X \times X \to \mathbb{C}$ be positive definite. Let $H_0$ be the linear subspace of $\mathbb{C}^X$ generated by the functions $\{\varphi_x | x \in X\}$ where $\varphi_x(y) = \varphi(x, y)$. If $f = \sum c_j \varphi_{x_j}$ and $g = \sum d_k \varphi_{y_k}$ belong to $H_0$, then

$$\sum_k \bar{d}_k f(y_k) = \sum_{j,k} c_j \bar{d}_k \varphi(x_j, y_k) = \sum_j c_j \overline{g(x_j)} \tag{1}$$

evidently does not depend on the chosen representations of $f$, $g$ (which may be not unique) and is denoted $\langle f, g \rangle$. Then $\langle f, f \rangle = \sum c_j \bar{c}_k \varphi(x_j, x_k) \geqq 0$ by assumption and the form $\langle \cdot, \cdot \rangle$ is linear in its first and antilinear in its second argument, implying in particular

$$\sum_{j,k=1}^n z_j \bar{z}_k \langle f_j, f_k \rangle = \left\langle \sum_{j=1}^n z_j f_j, \sum_{j=1}^n z_j f_j \right\rangle \geqq 0$$

for $f_1, \ldots, f_n \in H_0$ and $z_1, \ldots, z_n \in \mathbb{C}$, i.e. $\langle \cdot, \cdot \rangle$ is a positive definite kernel on $H_0 \times H_0$. An immediate consequence of (1) is the *reproducing property*

$$\langle f, \varphi_x \rangle = f(x) \qquad \text{for all} \quad f \in H_0 \quad \text{and} \quad x \in X$$

which implies in particular $\langle \varphi_y, \varphi_x \rangle = \varphi(y, x)$ and by 1.8

$$|f(x)|^2 \leqq \langle f, f \rangle \cdot \varphi(x, x)$$

so that $\langle f, f \rangle = 0$ if and only if $f$ is identically zero. Therefore, $H_0$ is a pre-Hilbert space and its completion $H$ is a Hilbert space in which $H_0$ is a dense subspace. The transformation

$$H \to \mathbb{C}^X$$
$$f \mapsto (x \mapsto \langle f, \varphi_x \rangle)$$

being linear and injective, the Hilbert space $H$ can even be thought of as a linear subspace of $\mathbb{C}^X$, i.e. as a space of functions and not, as mostly, as a space of equivalence classes of functions. This Hilbert function space is usually called the RKHS associated with $\varphi$. If $\varphi$ is real-valued then, of course, $H$ can be chosen as a real function space.

*To sum up there is a Hilbert space $H \subseteq \mathbb{C}^X$ and a mapping $x \mapsto \varphi_x$ from $X$ to $H$ such that*

$$\varphi(x, y) = \langle \varphi_x, \varphi_y \rangle \quad \text{for } x, y \in X.$$

The Hilbert space representation of general negative definite kernels looks a little bit more complicated.

**3.2. Proposition.** *Let $X$ be a nonempty set and $\psi: X \times X \to \mathbb{C}$ be negative definite. Then there is a Hilbert space $H \subseteq \mathbb{C}^X$ and a mapping $x \mapsto \varphi_x$ from $X$ to $H$ such that*

$$\psi(x, y) = \|\varphi_x\|^2 + \|\varphi_y\|^2 - 2\langle \varphi_x, \varphi_y \rangle + f(x) + \overline{f(y)}, \tag{2}$$

*where $f: X \to \mathbb{C}$ is a certain complex function on $X$. If there is some $x_0 \in X$ such that $\psi(x, x_0) \in \mathbb{R}$ for all $x \in X$ and if $\psi$ vanishes on the diagonal $\Delta = \{(x, x) | x \in X\}$, then $f$ may be chosen to be zero. If $\psi$ is real-valued, then $H$ may be chosen as a real Hilbert space and equation* (2) *becomes*

$$\psi(x, y) = \|\varphi_x - \varphi_y\|^2 + f(x) + f(y)$$

*where $f: X \to \mathbb{R}$. The function $f$ is nonnegative whenever $\psi$ is.*

*If $\psi$ is real-valued and vanishes on $\Delta$ then $f = 0$ and $\sqrt{\psi}$ is a semimetric such that $x \mapsto \varphi_x$ is an isometry. If, furthermore, $\{\psi = 0\} = \Delta$ then $\sqrt{\psi}$ is a (hilbertian) metric on $X$.*

Proof. We fix some $x_0 \in X$ and define

$$\varphi(x, y) := \tfrac{1}{2}[\psi(x, x_0) + \overline{\psi(y, x_0)} - \psi(x, y) - \psi(x_0, x_0)],$$

which is a positive definite kernel by Lemma 2.1. Let $H$ be the associated RKHS for $\varphi$ and again put $\varphi_x(y) = \varphi(x, y)$. Then

$$\begin{aligned} \|\varphi_x - \varphi_y\|^2 &= \varphi(x, x) + \varphi(y, y) - 2 \operatorname{Re} \varphi(x, y) \\ &= \operatorname{Re} \psi(x, y) - \tfrac{1}{2}[\psi(x, x) + \psi(y, y)], \end{aligned}$$

$$\begin{aligned} \|\varphi_x\|^2 + \|\varphi_y\|^2 - 2\langle \varphi_x, \varphi_y \rangle &= \|\varphi_x - \varphi_y\|^2 - 2i \operatorname{Im}\langle \varphi_x, \varphi_y \rangle \\ &= \psi(x, y) - \tfrac{1}{2}[\psi(x, x) + \psi(y, y)] \\ &\quad - i \operatorname{Im}[\psi(x, x_0) + \overline{\psi(y, x_0)}]. \end{aligned}$$

Setting $f(x) := \tfrac{1}{2}\psi(x, x) + i \operatorname{Im} \psi(x, x_0)$, we therefore obtain

$$\psi(x, y) = \|\varphi_x\|^2 + \|\varphi_y\|^2 - 2\langle \varphi_x, \varphi_y \rangle + f(x) + \overline{f(y)}.$$

The other statements can be derived immediately. $\square$

The negative definite $2 \times 2$ matrix $\begin{pmatrix} 0 & i \\ -i & 0 \end{pmatrix}$ shows that not every negative definite kernel $\psi$ vanishing on the diagonal can be represented in the form

$$\psi(x, y) = \|\varphi_x\|^2 + \|\varphi_y\|^2 - 2\langle \varphi_x, \varphi_y \rangle$$

for some Hilbert space valued mapping $x \mapsto \varphi_x$, since this would imply $\operatorname{Re} \psi(x, y) = \|\varphi_x - \varphi_y\|^2 = 0$, hence $\varphi_x = \varphi_y$ and then $\psi(x, y) = 0$.

**3.3. Corollary.** *Let* $\varphi_p(x, y) = \exp(-|x - y|^p)$, $x, y \in \mathbb{R}$ *and* $0 < p < \infty$. *Then* $\varphi_p$ *is positive definite if and only if* $p \leqq 2$.

Proof. By 1.10 the kernel $\psi(x, y) = (x - y)^2$ is negative definite, hence so is $\psi^\alpha$ for $0 < \alpha \leqq 1$ by 2.10, and from Theorem 2.2 we get that $\varphi_p$ is positive definite for all $p \leqq 2$.

Suppose now there is some $p > 2$ such that $\varphi_p$ is positive definite. Then for any $t > 0$, $x_1, \ldots, x_n \in \mathbb{R}$ and $c_1, \ldots, c_n \in \mathbb{R}$ we have

$$\sum_{j,k=1}^{n} c_j c_k \exp(-t|x_j - x_k|^p) = \sum_{j,k=1}^{n} c_j c_k \exp(-|t^{1/p}x_j - t^{1/p}x_k|^p) \geqq 0,$$

so by Theorem 2.2 the kernel $|x - y|^p$ is negative definite, and by the proposition above $|x - y|^{p/2}$ is a metric on $\mathbb{R}$. However,

$$|0 - 1|^{p/2} = 1 = |1 - 2|^{p/2}, \qquad |0 - 2|^{p/2} = 2^{p/2} > 2$$

contradicting the triangle inequality. □

**3.4. Exercise.** Let the kernel $\varphi_p$ be defined on $\mathbb{R}_+ \times \mathbb{R}_+$ by $\varphi_p(x, y) = \exp[-(x + y)^p]$ where $0 < p < \infty$. Show that $\varphi_p$ is positive definite if and only if $p \leqq 1$.

**3.5. Exercise.** Let $H$ be the RKHS associated with the positive definite kernel $\varphi: X \times X \to \mathbb{C}$. Show that any closed subspace of $H$ is the RKHS for some positive definite kernel on $X \times X$.

**3.6. Exercise.** Let $\varphi, \psi: X \times X \to \mathbb{C}$ be two positive definite kernels. Then $\operatorname{RKHS}(\varphi) \subseteq \operatorname{RKHS}(\psi)$ if and only if $\lambda\psi - \varphi$ is positive definite for some $\lambda > 0$.

**3.7. Exercise.** Let $\varphi: X \times X \to \mathbb{C}$ be a positive definite kernel not identically zero. Show that the following conditions are equivalent:

(i) $\varphi$ generates an extreme ray in the convex cone $\mathscr{P}$ of positive definite kernels on $X \times X$ (i.e. $\{\lambda\varphi \mid \lambda \geqq 0\}$ is an extreme subset of $\mathscr{P}$).
(ii) $\operatorname{RKHS}(\varphi)$ is of dimension 1.
(iii) $\varphi(x, y) = f(x)\overline{f(y)}$ for some function $f: X \to \mathbb{C}$ not identically zero.

**3.8. Exercise.** Let $\varphi: X \times X \to \mathbb{C}$ be a positive definite kernel such that $\varphi(x, y) \neq 0$ for all $(x, y) \in X \times X$. Show that if also $1/\varphi$ is positive definite then $\varphi(x, y) = f(x)\overline{f(y)}$ for some function $f: X \to \mathbb{C}\setminus\{0\}$.

**3.9. Exercise.** Given a positive definite kernel $\varphi: X \times X \to \mathbb{C}$ and a non-empty subset $X' \subseteq X$, consider the restriction $\varphi' := \varphi | X' \times X'$ and the two RKHS's $H$ and $H'$ associated with $\varphi$ (resp. $\varphi'$). Show that $H' = \{f | X' \mid f \in H\}$. *Hint*: There is an isometry $U: H' \to H$ (not necessarily onto) mapping $\varphi'_{x'}$ to $\varphi_{x'}$ for each $x' \in X'$. Use $U$ as well as the adjoint operator $U^*$.

**3.10. Exercise.** If $p > 2$ then for no $\lambda > 0$ the kernel $\exp[-\lambda|x - y|^p]$ is positive definite on $\mathbb{R} \times \mathbb{R}$. Nevertheless there are mixtures

$$\varphi(x, y) = \int_0^\infty \exp[-\lambda|x - y|^p]\, d\mu(\lambda), \qquad \mu \in M_+^1(]0, \infty[),$$

which are positive definite.

**3.11. Exercise.** Let $\varphi$ be a positive definite kernel defined on $X \times X$ where $X$ is a topological space. Show that $\varphi$ is continuous if and only if Re $\varphi$ is continuous at each point of the diagonal.

## Notes and Remarks

Positive (semi-) definite matrices have a long history and we have not traced this history back to its origins. The first instance where a positive definite kernel on a nonfinite set has been considered, seems to be within the theory of integral equations, and the first systematic treatment in this connection was given by Mercer (1909). Mercer defines a continuous and symmetric real-valued function $\varphi$ on $[a, b] \times [a, b] \subseteq \mathbb{R}^2$ to be of *positive type* if and only if

$$\int_a^b \int_a^b c(x)c(y)\varphi(x, y)\, dx\, dy \geqq 0 \tag{1}$$

holds for all continuous functions $c: [a, b] \to \mathbb{R}$, and he shows that this condition is equivalent to $\varphi$ being a positive definite kernel in our terminology; cf. Exercise 1.24. Property (1) had already been singled out a few years earlier by Hilbert (1904) who called the function $\varphi$ *definite* in this case. Mercer also defined a function $\psi$ to be of *negative type* if and only if $-\psi$ is of positive type. The idea leading to the notion of a negative definite kernel (in our terminology) goes back to Schoenberg (1938b); however, he requires these kernels to vanish on the diagonal—a condition appearing natural in his context: imbedding of metric spaces in a Hilbert space.

The product stability of positive definiteness (Theorem 1.12) has been shown by Schur (1911, Satz VII). A remarkable result in this connection was found by Fitzgerald and Horn (1977): if $(a_{jk})$ is a positive definite $n \times n$ matrix with nonnegative entries then for all real $\alpha \geqq n - 2$ the matrix $(a_{jk}^{\alpha})$ is positive definite, too, and this lower bound for $\alpha$ is sharp.

The fundamental connection between positive and negative definite kernels expressed in Theorem 2.2 goes back to Schoenberg (1938b) in case the negative definite kernel vanishes on the diagonal. The general case seems to have been proved first in Herz (1962).

Infinitely divisible positive definite kernels appear at many places in analysis and—particularly important—in probability theory. Infinitely divisible complex-valued kernels are studied in some detail by Horn (1969a). A study of positive and negative definite kernels with invariance properties under a group action may be found in Parthasarathy and Schmidt (1972).

The possibility of representing a positive definite kernel $\varphi: X \times X \to \mathbb{C}$ as $\varphi(x, y) = \langle F(x), F(y) \rangle$ for some Hilbert space valued function $F$ on $X$—which may be looked at as some weak form of integral representation—has, of course, been well known for a long time in the case of a finite set $X$. For countable $X$ the result was shown by Kolmogorov (1941), and a few years later Aronszajn (1944, 1950) settled the question fully. His second named paper, an enlarged version of the first one, gives the first systematic treatment of the theory of reproducing kernels which has since found many applications in mathematical analysis; for example, in complex function theory, cf. Hille (1972), or in time series analysis, see Parzen (1971). See also Donoghue (1974).

Proposition 3.2 contains as a special case the statement that a real-valued negative definite kernel $\psi$ on $X \times X$ vanishing on the diagonal can be represented as

$$\psi(x, y) = \|F(x) - F(y)\|^2$$

for some Hilbert space valued mapping $F$ on $X$, a result due to Schoenberg (1938b).

CHAPTER 4

# Main Results on Positive and Negative Definite Functions on Semigroups

## §1. Definitions and Simple Properties

In the present book we will deal mainly with positive (and negative) definite functions on *abelian semigroups*, but nevertheless we will introduce the concepts for arbitrary semigroups with involution.

The subject of positive definite functions on locally compact groups splits into two completely different theories for abelian and nonabelian groups. At present there exists a rather satisfactory theory of positive definite functions on abelian semigroups whereas very little is known about the nonabelian case.

**1.1. Definition.** A *semigroup* $(S, \circ)$ is a nonempty set $S$ equipped with an *associative composition* $\circ$ and a *neutral element*† $e$.

A *semigroup with involution* or a *$*$-semigroup* $(S, \circ, *)$ is a semigroup $(S, \circ)$ together with a mapping $*: S \to S$, called *involution*, satisfying

(i) $(s \circ t)^* = t^* \circ s^*$ for $s, t \in S$;
(ii) $(s^*)^* = s$ for $s \in S$.

One should note that the axioms imply $e = e^*$, in fact: $e^* = e^* \circ e = (e^* \circ e)^* = (e^*)^* = e$.

For an abelian (= commutative) semigroup the composition and neutral element are always denoted $+$ and $0$, and the neutral element is called zero.

We now list some examples showing the great generality of the above concept.

† Many authors do not assume the existence of a neutral element in the definition of a semigroup. With the exception of Chapter 8 we always assume that semigroups have a neutral element.

**1.2. Examples.** (a) Any group $(G, \circ)$ is a semigroup with involution when we define $s^* = s^{-1}$.

(b) Any abelian semigroup $(S, +)$ is a semigroup with involution when we define $s^* = s$, the *identical involution.*

(c) Let $(T, +)$ be an abelian semigroup. The product semigroup $S = T \times T$ is a semigroup with involution if $(t_1, t_2)^* := (t_2, t_1)$.

(d) The closed unit disc $D := \{z \in \mathbb{C} \mid |z| \leqq 1\}$ with $z \circ w := zw$ and $z^* := \bar{z}$ is a semigroup with involution.

Concrete examples of semigroups occur throughout mathematics. We will study some in detail in §4.

**1.3.** Let $(S, \circ)$ denote a semigroup with neutral element $e$. A *subsemigroup* of $S$ is any subset $T \subseteq S$ such that $e \in T$, and which contains $s \circ t$ whenever $s, t \in T$. If $S$ has an involution, then $T \subseteq S$ is called a $*$-*subsemigroup*, if $T$ is a subsemigroup which contains $t^*$ whenever $t \in T$.

An element $\omega \in S$ is called *absorbing* if $s \circ \omega = \omega \circ s = \omega$ for all $s \in S$; obviously there can exist at most one absorbing element in $S$ and clearly $\omega^* = \omega$. In the additive semigroup $([0, \infty], +)$ we have $\omega = \infty$ and in the closed unit disc $(D, \cdot)$ we have $\omega = 0$. Each semigroup $S$ can be enlarged to a semigroup $T$ with absorbing element defining $T := S \cup \{\tilde{\omega}\}$ (where $\tilde{\omega}$ is some element not contained in $S$) and $\tilde{\omega} \circ s = s \circ \tilde{\omega} := \tilde{\omega} =: \tilde{\omega} \circ \tilde{\omega}$, $s \in S$. Note that if $S$ contains the absorbing element $\omega$ then $S \setminus \{\omega\}$ is not necessarily a subsemigroup of $S$.

Let $(S, \circ), (T, \circ)$ be semigroups with neutral elements $e_S$ and $e_T$. A mapping $f: S \to T$ is called a *homomorphism* if $f(e_S) = e_T$ and $f(x \circ y) = f(x) \circ f(y)$ for $x, y \in S$. If both $S$ and $T$ have involutions we add the requirement $f(s^*) = f(s)^*$ for $s \in S$ in order to call $f$ a homomorphism. Sometimes we use the word $*$-*homomorphism* in this case. If $f: S \to T$ is a homomorphism which is one-to-one and onto it is called an *isomorphism.*

**1.4.** Let $(S, \circ)$ be a semigroup. If $S$ is equipped with a topology $\mathcal{O}$ we call $(S, \circ, \mathcal{O})$ a *topological semigroup* provided the composition mapping $(s, t) \mapsto s \circ t$ is continuous from $S \times S$ into $S$. If $S$ in addition has an involution $*$ we require $s \mapsto s^*$ to be continuous, too.

In the rest of this section $S = (S, \circ, *)$ denotes a semigroup with involution.

**1.5. Definition.** A function $\varphi: S \to \mathbb{C}$ is called *positive definite* if $(s, t) \mapsto \varphi(s^* \circ t)$ is a positive definite kernel on $S \times S$, i.e. if

$$\sum_{j,k=1}^{n} c_j \bar{c}_k \varphi(s_j^* \circ s_k) \geqq 0$$

for all $n \in \mathbb{N}$, $\{s_1, \ldots, s_n\} \subseteq S$, $\{c_1, \ldots, c_n\} \subseteq \mathbb{C}$, and it is called *strictly positive definite* if the kernel $\varphi(s^* \circ t)$ is so.

The set of positive definite functions $\varphi: S \to \mathbb{C}$ is denoted $\mathscr{P}(S)$ and $\mathscr{P}_1(S) = \{\varphi \in \mathscr{P}(S) \mid \varphi(e) = 1\}$.

The above concept is not changed if the kernel $\varphi(s^* \circ t)$ is replaced by the kernel $\varphi(s \circ t^*)$.

On an abelian group $(G, +)$ we can consider two different involutions, the identical involution and the involution $x^* = -x$ for $x \in G$. Corresponding to these two involutions we have two different notions of positive definiteness for functions on $G$. We say that $\varphi: G \to \mathbb{C}$ is *positive definite in the group sense* if it is positive definite when $G$ carries the involution $x^* = -x$ for $x \in G$, and we say that $\varphi$ is *positive definite in the semigroup sense* if $G$ is equipped with the identical involution. Later on, in particular in Chapter 5, both of these notions of positive definiteness will occur simultaneously. Likewise for negative definite functions on $G$, to be introduced in Definition 1.8 below, we have the two notions of *negative definiteness in the group sense* and *in the semigroup sense.*

**1.6.** By 3.1.5 it follows that any positive definite function $\varphi$ is *hermitian*, i.e. $\varphi(s^*) = \overline{\varphi(s)}$ for $s \in S$, and by 3.1.8 we have

$$\varphi(s^* \circ s) \geqq 0 \qquad \text{and} \qquad |\varphi(s^* \circ t)|^2 \leqq \varphi(s^* \circ s)\varphi(t^* \circ t), \qquad s, t \in S,$$

in particular,

$$\varphi(e) \geqq 0 \qquad \text{and} \qquad |\varphi(s)|^2 \leqq \varphi(e)\varphi(s^* \circ s), \qquad s \in S.$$

The last inequality implies that $\varphi \equiv 0$ if $\varphi(e) = 0$.

The set $\mathscr{P}(S)$ is a convex cone in the vector space $\mathbb{C}^S$ of complex-valued functions on $S$. By 3.1.12 and 3.1.11 it follows that $\mathscr{P}(S)$ is stable under products and closed in the topology of pointwise convergence. The set $\mathscr{P}_1(S)$ is closed and convex and a *base* for $\mathscr{P}(S)$, i.e. for every $\varphi \in \mathscr{P}(S)\setminus\{0\}$ there exists a unique pair $(\lambda, \varphi_0)$ with $\lambda > 0$ and $\varphi_0 \in \mathscr{P}_1(S)$ such that $\varphi = \lambda\varphi_0$, namely $\lambda = \varphi(e)$ and $\varphi_0 = \lambda^{-1}\varphi$.

**1.7. Remark.** Let $H$ be a complex Hilbert space and $B(H)$ the set of bounded operators on $H$. Following Sz.-Nagy (1960) a function $\Phi: S \to B(H)$ is called of *positive type* if for all $n \in \mathbb{N}$, all $\{s_1, \ldots, s_n\} \subseteq S$ and all $\{\xi_1, \ldots, \xi_n\} \subseteq H$

$$\sum_{j,k=1}^{n} \langle \Phi(s_j^* \circ s_k)\xi_k, \xi_j \rangle \geqq 0.$$

If $\Phi$ is of positive type, then $\Phi$ is positive definite in the sense that all the scalar functions $\Phi_\xi(s) = \langle \Phi(s)\xi, \xi \rangle$, $\xi \in H$, are positive definite. Examples are known showing that the converse is not true, cf., e.g. Arveson (1969).

If $H = \mathbb{C}$ the functions of positive type are the same as positive definite functions. More generally if $\Phi(S)$ is contained in a commutative $C^*$-algebra of $B(H)$, it can be shown that the two notions coincide.

A $C^*$-algebra $S$ with unit can be considered as a semigroup with involution, the composition being the multiplication of the $C^*$-algebra. Then linear mappings $\Phi: S \to B(H)$ of positive type are widely studied under the name of completely positive mappings, whereas a linear $\Phi$ is positive definite if and only if it is positive, i.e. maps positive elements in $S$ into positive elements of $B(H)$.

There exists a vast literature about these operator-valued functions, see Arveson (1969), Evans and Lewis (1977) and Mlak (1978), and references therein, but a treatment of this subject falls outside the scope of this book.

**1.8. Definition.** A function $\psi: S \to \mathbb{C}$ is called *negative definite* if $(s, t) \mapsto \psi(s^* \circ t)$ is a negative definite kernel on $S \times S$, i.e. if $\psi$ is *hermitian* and

$$\sum_{j,k=1}^{n} c_j \overline{c_k} \psi(s_j^* \circ s_k) \leqq 0$$

for all $n \geqq 2$, $\{s_1, \ldots, s_n\} \subseteq S$ and $\{c_1, \ldots, c_n\} \subseteq \mathbb{C}$ with $\sum_{j=1}^{n} c_j = 0$. The set of negative definite functions $\psi: S \to \mathbb{C}$ is denoted $\mathcal{N}(S)$.

The set $\mathcal{N}(S)$ is a closed convex cone in $\mathbb{C}^S$ containing the real constants. For $\psi \in \mathcal{N}(S)$ we have by 3.1.7 that

$$2 \operatorname{Re} \psi(s^* \circ t) \geqq \psi(s^* \circ s) + \psi(t^* \circ t), \qquad s, t \in S,$$

in particular

$$2 \operatorname{Re} \psi(t) \geqq \psi(e) + \psi(t^* \circ t), \qquad t \in S. \tag{1}$$

As an application of Lemma 3.2.1 we get

**1.9. Proposition.** *Let $\psi: S \to \mathbb{C}$ be a hermitian function satisfying $\psi(e) \geqq 0$. Then $\psi$ is negative definite if and only if the kernel $(s, t) \mapsto \overline{\psi(s)} + \psi(t) - \psi(s^* \circ t)$ is positive definite.*

In the further development of the theory we need to consider certain boundedness conditions for positive definite functions.

**1.10. Definition.** A function $\alpha: S \to \mathbb{R}_+$ is called an *absolute value* if

(i) $\alpha(e) = 1$;
(ii) $\alpha(s \circ t) \leqq \alpha(s)\alpha(t)$ for $s, t \in S$;
(iii) $\alpha(s^*) = \alpha(s)$ for $s \in S$.

We will later see many examples of absolute values. At present we note that the constant function $s \mapsto 1$ is an absolute value. If $\alpha$, $\beta$ are absolute values and $k > 0$ then $\alpha\beta$, $\max(\alpha, \beta)$ and $\alpha^k$ are again absolute values.

**1.11. Definition.** A function $f: S \to \mathbb{C}$ is called *bounded* with respect to an absolute value $\alpha$ (shortly: *$\alpha$-bounded*) if there exists a constant $C > 0$ such that

$$|f(s)| \leqq C\alpha(s) \qquad \text{for} \quad s \in S,$$

and $f$ is called *exponentially bounded* if there exists an absolute value with respect to which $f$ is bounded.

The set of exponentially bounded functions is an algebra.

**1.12. Proposition.** *Let $\varphi \in \mathscr{P}(S)$ be bounded with respect to an absolute value $\alpha$. Then*

$$|\varphi(s)| \leqq \varphi(e)\alpha(s) \qquad \textit{for} \quad s \in S.$$

PROOF. Without loss of generality we may assume that $\varphi(e) = 1$, so that $|\varphi(s)|^2 \leqq \varphi(s^* \circ s)$. By iteration we get for $n \in \mathbb{N}$

$$|\varphi(s)|^{2^n} \leqq \varphi((s^* \circ s)^{2^{n-1}}),$$

and using $|\varphi(t)| \leqq C\alpha(t)$ for some constant $C > 0$ we get

$$|\varphi(s)|^{2^n} \leqq C\alpha((s^* \circ s)^{2^{n-1}}) \leqq C\alpha(s)^{2^n},$$

hence

$$|\varphi(s)| \leqq \alpha(s) \lim_{n \to \infty} C^{2^{-n}} = \alpha(s). \qquad \square$$

**1.13.** Let $\varphi \in \mathscr{P}(S)$ and let $H_0$ be the linear subspace of $\mathbb{C}^S$ generated by the functions $\{\varphi_s | s \in S\}$, where $\varphi_s(t) = \varphi(s^* \circ t)$. Then $H_0$ is equipped with a scalar product $\langle \cdot, \cdot \rangle$ such that

$$\langle \varphi_s, \varphi_t \rangle = \varphi(s^* \circ t), \qquad s, t \in S,$$

and the completion $H$ of $H_0$ (realized in $\mathbb{C}^S$) is the RKHS associated with $\varphi$, cf. 3.3.1.

For each $s \in S$ there exists a linear transformation $\pi(s): H_0 \to H_0$ such that

$$\pi(s)\varphi_t = \varphi_{s \circ t}, \qquad s, t \in S,$$

and $\pi$ is a *representation* of $S$ in the space $\mathrm{Hom}(H_0)$ of linear transformations of $H_0$, i.e.

$$\pi(s \circ t) = \pi(s)\pi(t), \qquad \pi(e) = I, \qquad \pi(s^*) = \pi(s)^*,$$

where the last equation means that

$$\langle \pi(s)f, g \rangle = \langle f, \pi(s^*)g \rangle \qquad \text{for} \quad f, g \in H_0.$$

We see that

$$\varphi(s) = \langle \varphi, \pi(s)\varphi \rangle, \qquad s \in S.$$

Conversely, if $K_0$ is a pre-Hilbert space, $\xi \in K_0$ and $\pi: S \to \mathrm{Hom}(K_0)$ is a representation, then it is easily seen that $\varphi(s) := \langle \xi, \pi(s)\xi \rangle$ is positive definite.

We shall now characterize the positive definite functions $\varphi$ for which the operators $\pi(s)$ are bounded on $H_0$. If this is the case, then $\pi(s)$ can be uniquely extended to a bounded operator $\bar{\pi}(s)$ on the Hilbert space $H$, and $\bar{\pi}$ is a representation of $S$ in $B(H)$, the $C^*$-algebra of bounded operators on $H$.

**1.14. Theorem.** *Let $\varphi \in \mathscr{P}(S)$, and let $\pi: S \to \mathrm{Hom}(H_0)$ be the representation as above. Then $\varphi$ is exponentially bounded if and only if $\pi(s)$ is a bounded operator on $H_0$ for each $s$.*

PROOF. Suppose first that $\pi(s)$ is bounded on $H_0$ for each $s \in S$. Then $\alpha(s) := \|\pi(s)\|$ is an absolute value on $S$ and $|\varphi(s)| = |\langle \varphi, \pi(s)\varphi \rangle| \leqq \|\varphi\|^2 \|\pi(s)\|$, thus showing that $\varphi$ is bounded with respect to $\alpha$.

Conversely, suppose that $|\varphi(s)| \leqq \varphi(e)\alpha(s)$ for some absolute value $\alpha$. Let $f = \sum_{j=1}^n c_j \varphi_{x_j} \in H_0$ and define

$$g(s) = \sum_{j,k=1}^{n} c_j \bar{c}_k \varphi(x_j^* \circ s \circ x_k), \qquad s \in S.$$

Then $g \in \mathscr{P}(S)$, for if $\{y_1, \ldots, y_m\} \subseteq S$, $\{d_1, \ldots, d_m\} \subseteq \mathbb{C}$ we have

$$\sum_{p,q=1}^{m} d_p \bar{d}_q g(y_p^* \circ y_q) = \sum_{p,q=1}^{m} \sum_{j,k=1}^{n} (d_p c_j)\overline{(d_q c_k)}\varphi((y_p \circ x_j)^* \circ (y_q \circ x_k)) \geqq 0.$$

Furthermore, $g$ is bounded with respect to $\alpha$ since

$$|g(s)| \leqq \left( \sum_{j=1}^{n} |c_j| \alpha(x_j) \right)^2 \varphi(e)\alpha(s).$$

By Proposition 1.12 it follows that

$$\|\pi(s)f\|^2 = g(s^* \circ s) \leqq g(e)\alpha(s)^2 = \|f\|^2 \alpha(s)^2,$$

which shows that $\pi(s)$ is a bounded operator on $H_0$ of norm $\leqq \alpha(s)$. □

**1.15. Exercise.** Let $E_s$ denote the shift operator on $\mathbb{C}^S$ defined by $E_s f(t) = f(s \circ t)$, $s, t \in S, f \in \mathbb{C}^S$. Let $\varphi \in \mathscr{P}(S)$ and let $H_0 \subseteq H \subseteq \mathbb{C}^S$ be the associated pre-Hilbert space and Hilbert space. Use the closed graph theorem to prove that $E_{s*}(H) \subseteq H$ if and only if $\pi(s)$ is a bounded operator on $H_0$. If $\pi(s)$ is a bounded operator on $H_0$, then the unique continuous extension $\bar{\pi}(s)$ of $\pi(s)$ to $H$ is given by $\bar{\pi}(s)f = E_{s*} f$ for $f \in H$.

**1.16. Exercise.** Suppose that $S$ is a group with $s^* = s^{-1}$. Show that every $\varphi \in \mathscr{P}(S)$ satisfies $|\varphi(s)| \leqq \varphi(e)$, $s \in S$.

**1.17. Exercise.** Let $l^1(S)$ denote the set of functions $f: S \to \mathbb{C}$ for which

$$\|f\| = \sum_{s \in S} |f(s)| < \infty,$$

and define

$$f * g(s) = \sum_{\substack{(a,b) \in S^2 \\ a \circ b = s}} f(a)g(b), \qquad f^*(s) = \overline{f(s^*)}, \qquad s \in S.$$

Show that $l^1(S)$ is a Banach algebra with involution and unit.

Let $\varphi: S \to \mathbb{C}$ be a bounded function and let $L_\varphi: l^1(S) \to \mathbb{C}$ be defined by $L_\varphi(f) = \sum_{s \in S} f(s)\overline{\varphi(s)}$. Show that $\varphi \in \mathscr{P}(S)$ if and only if $L_\varphi(f^* * f) \geqq 0$ for all $f \in l^1(S)$.

**1.18. Exercise.** Let $(S, \circ, *)$ denote a semigroup with involution and let $H \subseteq S$ be a nonempty subset of $S$. Then if $1_H \in \mathscr{P}(S)$ it follows that $H$ is a $*$-subsemigroup of $S$. In case $S$ is an abelian group and $s^* = -s$ the converse also holds, i.e. $1_H$ is positive definite in the group sense if and only if $H$ is a subgroup of $S$ (cf. Exercise 3.1.23). Try to find a counterexample for the converse statement on a semigroup with identical involution.

**1.19. Exercise.** Let $(S, \circ, *)$ be a semigroup with involution and add an absorbing element $\tilde{\omega}$ to $S$ (cf. 1.3). If $\varphi \in \mathscr{P}(S)$ then $\tilde{\varphi}: S \cup \{\tilde{\omega}\} \to \mathbb{C}$ defined by $\tilde{\varphi}|S = \varphi$ and $\tilde{\varphi}(\tilde{\omega}) = 0$ is positive definite on $S \cup \{\tilde{\omega}\}$. For any positive definite extension $\tilde{\tilde{\varphi}}$ of $\varphi$ we have $0 \leqq \tilde{\tilde{\varphi}}(\tilde{\omega}) \leqq \inf_{s \in S} \varphi(s^* \circ s)$.

## §2. Exponentially Bounded Positive Definite Functions on Abelian Semigroups

Throughout this section, $S = (S, +, *)$ is an abelian semigroup with involution, which may or may not be the identity.

**2.1. Definition.** A function $\rho: S \to \mathbb{C}$ is called a *semicharacter* if

(i) $\rho(0) = 1$;
(ii) $\rho(s + t) = \rho(s)\rho(t)$ for $s, t \in S$;
(iii) $\rho(s^*) = \overline{\rho(s)}$ for $s \in S$.

Note that if (ii) holds, then (i) is equivalent with $\rho \not\equiv 0$. A semicharacter is a homomorphism of $(S, +, *)$ into the semigroup $(\mathbb{C}, \cdot, -)$, where the composition and involution are multiplication and conjugation.

If the involution $*$ is the identity, a semicharacter is automatically real-valued. If $(S, +)$ is an abelian group and $s^* = -s$, a semicharacter has its values in the circle group $\mathbb{T} = \{z \in \mathbb{C} \,|\, |z| = 1\}$ and is a group character.

**2.2.** The set of semicharacters on $S$ is denoted $S^*$. We equip $S^*$ with the topology inherited from $\mathbb{C}^S$, having the topology of pointwise convergence. In particular $S^*$ is a Hausdorff space. (Since $\mathbb{C}^S$ is completely regular and $S^*$ is a (closed) subset of $\mathbb{C}^S$ we have in fact that $S^*$ is a completely regular space.)

Note that $S^*$ is a topological semigroup under pointwise multiplication, the mapping $\rho \mapsto \bar{\rho}$ is an involution and the function 1 is the neutral element. We call $S^*$ the *dual semigroup* of $S$.

**2.3.** The semigroup structure of $S^*$ leads to a *convolution* in $M_+^b(S^*)$, cf. 2.1.16. Let $\pi\colon S^* \times S^* \to S^*$ be the continuous mapping given by $\pi(\rho_1, \rho_2) = \rho_1\rho_2$. For $\mu, \nu \in M_+^b(S^*)$ the convolution $\mu * \nu \in M_+^b(S^*)$ is by definition the image measure of the Radon product measure $\mu \otimes \nu$ under the mapping $\pi$, i.e.

$$\mu * \nu(B) = \mu \otimes \nu(\pi^{-1}(B)) \qquad \text{for} \quad B \in \mathscr{B}(S^*).$$

Note that by 2.1.24 $\operatorname{supp}(\mu * \nu) = \overline{\operatorname{supp}(\mu) \cdot \operatorname{supp}(\nu)}$.

**2.4.** Every semicharacter is positive definite. Our main concern is whether every positive definite function can be represented as an integral of semicharacters. We shall return to this question in Chapter 6, where it will be shown that the answer is in general negative. However, exponentially bounded positive definite functions admit such an integral representation as we shall see next.

For $\mu \in M_+^c(S^*)$, the set of Radon measures on $S^*$ with compact support, the function

$$\varphi(s) = \int_{S^*} \rho(s)\, d\mu(\rho), \qquad s \in S$$

is positive definite. In fact, for $\{s_1, \ldots, s_n\} \subseteq S$, $\{c_1, \ldots, c_n\} \subseteq \mathbb{C}$ we find

$$\sum_{j,k=1}^{n} c_j \bar{c}_k \varphi(s_j + s_k^*) = \int \left| \sum_{j=1}^{n} c_j \rho(s_j) \right|^2 d\mu(\rho) \geqq 0.$$

Furthermore, $\varphi$ is exponentially bounded. To see this we define an absolute value $\alpha_K$ associated with a compact subset $K \subseteq S^*$ by

$$\alpha_K(s) = \sup\{|\rho(s)| \mid \rho \in K\}, \qquad s \in S.$$

In the special case of $K = \operatorname{supp}(\mu)$ we find

$$|\varphi(s)| \leqq \mu(K)\alpha_K(s) = \varphi(0)\alpha_K(s) \qquad \text{for} \quad s \in S,$$

showing that $\varphi$ is bounded with respect to $\alpha_K$. This establishes the "only if" part of the following result:

**2.5. Theorem.** *A function $\varphi\colon S \to \mathbb{C}$ has an integral representation of the form*

$$\varphi(s) = \int_{S^*} \rho(s)\, d\mu(\rho), \tag{1}$$

*with $\mu \in M_+^c(S^*)$ if and only if it is positive definite and exponentially bounded. If $\varphi$ has these properties there is exactly one measure $\mu \in M_+(S^*)$ such that* (1) *holds.*

As a preparation to the proof we fix an absolute value $\alpha$ and consider the set $\mathscr{P}^\alpha(S)$ of $\alpha$-bounded functions $\varphi \in \mathscr{P}(S)$, and the subset $\mathscr{P}_1^\alpha(S)$ of functions $\varphi \in \mathscr{P}^\alpha(S)$ satisfying $\varphi(0) = 1$. Clearly $\mathscr{P}^\alpha(S)$ is a closed convex subcone of $\mathscr{P}(S)$ and $\mathscr{P}_1^\alpha(S)$ is a closed convex set, which is a base for $\mathscr{P}^\alpha(S)$.

We remark that the set $\mathscr{P}^e(S)$ of exponentially bounded positive definite functions is given as

$$\mathscr{P}^e(S) = \bigcup_\alpha \mathscr{P}^\alpha(S),$$

where the union is taken over the set of absolute values, so $\mathscr{P}^e(S)$ is a convex cone, stable under products, cf. 1.10.

**2.6. Theorem.** *Let $\alpha$ be an absolute value on $S$. Then $\mathscr{P}_1^\alpha(S)$ is a compact convex set whose extreme points are precisely the $\alpha$-bounded semicharacters.*

PROOF. Since the functions in $\mathscr{P}_1^\alpha(S)$ are pointwise bounded, Tychonoff's theorem implies that $\mathscr{P}_1^\alpha(S)$ is compact. A semicharacter $\rho$ is $\alpha$-bounded if and only if $|\rho(s)| \leqq \alpha(s)$ for $s \in S$.

We show below that $\operatorname{ex}(\mathscr{P}_1^\alpha(S)) \subseteq S^*$, and defining $\gamma: S \to S$ by $\gamma(s) = s^* + s$ we conclude from Corollary 2.5.12 that $\operatorname{ex}(\mathscr{P}_1^\alpha(S))$ is precisely the set of $\alpha$-bounded semicharacters.

In order to establish $\operatorname{ex}(\mathscr{P}_1^\alpha(S)) \subseteq S^*$, we define for $\varphi \in \mathscr{P}^\alpha(S)$, $a \in S$ and $\sigma \in \mathbb{C}$ with $|\sigma| \leqq 1$ the function $T_{a,\sigma}\varphi$ by

$$T_{a,\sigma}\varphi(s) = \alpha(a)\varphi(s) + \frac{\sigma}{2}\varphi(s+a) + \frac{\bar{\sigma}}{2}\varphi(s+a^*), \qquad s \in S.$$

We claim that $T_{a,\sigma}\varphi \in \mathscr{P}^\alpha(S)$. For $\{s_1, \ldots, s_n\} \subseteq S$, $\{c_1, \ldots, c_n\} \subseteq \mathbb{C}$ we have

$$\begin{aligned}\sum_{j,k=1}^n c_j\overline{c_k}T_{a,\sigma}\varphi(s_j^* + s_k) &= \alpha(a)\sum_{j,k=1}^n c_j\overline{c_k}\varphi(s_j^* + s_k) \\ &\quad + \operatorname{Re}\left\{\sigma\sum_{j,k=1}^n c_j\overline{c_k}\varphi(s_j^* + s_k + a)\right\} \\ &= \alpha(a)g(0) + \operatorname{Re}\{\sigma g(a)\},\end{aligned}$$

where

$$g(s) := \sum_{j,k=1}^n c_j\overline{c_k}\varphi(s_j^* + s_k + s), \qquad s \in S.$$

According to the proof of Theorem 1.14 we have $g \in \mathscr{P}^\alpha(S)$, so the inequality $|g(a)| \leqq g(0)\alpha(a)$ (cf. 1.12) implies that $T_{a,\sigma}\varphi \in \mathscr{P}(S)$ and $T_{a,\sigma}\varphi$ is clearly $\alpha$-bounded.

Let $\varphi \in \operatorname{ex}(\mathscr{P}_1^\alpha(S))$. We shall show that $\varphi(s+a) = \varphi(s)\varphi(a)$ for $s, a \in S$. This is clear if $\alpha(a) = 0$ because $|\varphi(s+a)| \leqq \alpha(s+a) \leqq \alpha(s)\alpha(a) = 0$ and $|\varphi(a)| \leqq \alpha(a) = 0$. Suppose $\alpha(a) > 0$. Since

$$T_{a,1}\varphi + T_{a,-1}\varphi = T_{a,i}\varphi + T_{a,-i}\varphi = 2\alpha(a)\varphi,$$

all the functions $T_{a,\sigma}\varphi$ are proportional to $\varphi$ for $\sigma \in \{\pm 1, \pm i\}$, and it follows that $\varphi(s + a) = k(a)\varphi(s)$, $s \in S$, for a certain proportionality factor $k(a)$. Putting $s = 0$ we get $k(a) = \varphi(a)$. □

Proof of Theorem 2.5. If $\varphi$ is positive definite, exponentially bounded and not identically zero, there exists an absolute value $\alpha$ such that $\varphi(0)^{-1}\varphi \in \mathscr{P}_1^\alpha(S)$. By Theorem 2.5.6 there exists a Radon probability measure $\tau$ on the compact set of $\alpha$-bounded semicharacters having $\varphi(0)^{-1}\varphi$ as barycentre. Using the continuous linear functionals $f \mapsto f(s)$ on $\mathbb{C}^S$ we get

$$\varphi(s) = \int_{S^*} \rho(s)\, d\mu(\rho), \qquad s \in S,$$

where $\mu = \varphi(0)\tau$ belongs to $M_+^c(S^*)$.

To prove the uniqueness assertion assume that

$$\varphi(s) = \int_{S^*} \rho(s)\, d\mu(\rho) = \int_{S^*} \rho(s)\, d\nu(\rho), \qquad s \in S,$$

where $\mu \in M_+^c(S^*)$ and $\nu \in M_+(S^*)$ is such that $\rho \mapsto \rho(s)$ is $\nu$-integrable for all $s \in S$. The set

$$\mathscr{A} = \left\{\rho \mapsto \sum_{j=1}^n c_j\rho(s_j) \,|\, c_1, \ldots, c_n \in \mathbb{C}, s_1, \ldots, s_n \in S, n \in \mathbb{N}\right\}$$

of continuous complex-valued functions on $S^*$ is an algebra stable under conjugation, containing the constant functions and separating the points of $S^*$. It follows by the Stone–Weierstrass theorem that the set of restrictions to a compact subset $K \subseteq S^*$ of the functions in $\mathscr{A}$ is a dense subset of $C(K)$ in the uniform norm. All functions $\Phi \in \mathscr{A}$ are $\nu$-integrable and $\int \Phi\, d\mu = \int \Phi\, d\nu$.

We claim that $\operatorname{supp}(\nu) \subseteq \operatorname{supp}(\mu)$. If not, there exists a compact set $L \subseteq S^*$ disjoint from $\operatorname{supp}(\mu)$ such that $\nu(L) > 0$. Choose $\varepsilon > 0$ such that $\varepsilon^2\varphi(0) < \nu(L)$. By the Stone–Weierstrass theorem there exists a real-valued function $\Phi \in \mathscr{A}$ such that $\Phi \geqq 1$ on $L$ and $0 \leqq \Phi \leqq \varepsilon$ on $\operatorname{supp}(\mu)$. Then $\Phi^2 \in \mathscr{A}$, $\Phi^2 \geqq 0$ and

$$\nu(L) \leqq \int_L \Phi^2\, d\nu \leqq \int \Phi^2\, d\nu = \int \Phi^2\, d\mu \leqq \varepsilon^2\mu(S^*) = \varepsilon^2\varphi(0),$$

which is a contradiction. Finally, the set of restrictions to $\operatorname{supp}(\mu)$ of the functions in $\mathscr{A}$ is dense in $C(\operatorname{supp}(\mu))$, so by the Riesz representation theorem $\mu = \nu$. □

The uniqueness statement in Theorem 2.5 combined with Theorem 2.6 can be expressed in the terminology of the theory of integral representations.

**2.7. Corollary.** *Let $\alpha$ be an absolute value on $S$. Then $\mathscr{P}_1^\alpha(S)$ is a Bauer simplex.*

As a special case we consider the absolute value $b(s) = 1$, $s \in S$. The cone $\mathscr{P}^b(S)$ is equal to the cone of bounded positive definite functions, and we define

$$\hat{S} := S^* \cap \mathscr{P}^b(S) = \{\rho \in S^* \,|\, |\rho(s)| \leqq 1 \text{ for } s \in S\},$$

which is a compact subsemigroup of $S^*$ called the *restricted dual semigroup.* It is easy to see that $\hat{S}$ is the set of bounded semicharacters.

For this special case Theorem 2.6 and Corollary 2.7 lead to the following result:

**2.8. Theorem.** *The set $\mathscr{P}_1^b(S)$ of bounded positive definite functions $\varphi$ with $\varphi(0) = 1$ is a Bauer simplex and its set of extreme points is $\hat{S}$. A bounded positive definite function $\varphi$ has an integral representation*

$$\varphi(s) = \int_{\hat{S}} \rho(s)\, d\mu(\rho), \qquad s \in S,$$

*where $\mu \in M_+(\hat{S})$ is uniquely determined.*

In the special case of $S$ being an abelian group with $s^* = -s$, Theorem 2.8 reduces to Bochner's theorem for discrete abelian groups, cf. Rudin (1962):

**2.9. Theorem.** *Let $G$ be a discrete abelian group. A function $\varphi: G \to \mathbb{C}$ is positive definite in the group sense if and only if it is the Fourier transform of a (nonnegative) Radon measure on the compact dual group $\hat{G}$.*

**2.10.** The set of signed measures of the form $\mu_1 - \mu_2 + i(\mu_3 - \mu_4)$ with $\mu_j \in M_+^c(S^*)$, $j = 1, \ldots, 4$ will be denoted $M^c(S^*)$. Extending the convolution of measures in $M_+^c(S^*)$ to $M^c(S^*)$ by bilinearity, $M^c(S^*)$ becomes an algebra.

For $\mu \in M^c(S^*)$ we denote by $\hat{\mu}$ the function

$$\hat{\mu}(s) = \int_{S^*} \rho(s)\, d\mu(\rho), \qquad s \in S$$

and $\hat{\mu}$ is called the *generalized Laplace transform* of $\mu$. The mapping $\mu \mapsto \hat{\mu}$ of $M^c(S^*)$ into $\mathbb{C}^S$, the *generalized Laplace transformation*, has the following properties, where $\alpha, \beta \in \mathbb{C}$, $\mu, \nu \in M^c(S^*)$:

(i) $(\alpha\mu + \beta\nu)\hat{} = \alpha\hat{\mu} + \beta\hat{\nu}$;
(ii) $(\mu * \nu)\hat{} = \hat{\mu} \cdot \hat{\nu}$;
(iii) $\hat{\mu} = 0 \Rightarrow \mu = 0$;

i.e. $\mu \mapsto \hat{\mu}$ is an injective algebra homomorphism. Since $M_+^c(S^*)$ is mapped onto the cone $\mathscr{P}^e(S)$, $M^c(S^*)$ is mapped onto the subspace of $\mathbb{C}^S$ spanned by $\mathscr{P}^e(S)$. Among the above properties (i) and (ii) are straightforward to establish, and (iii) follows from the proof in Theorem 2.5. Indeed, if $\mu = (\mu_1 - \mu_2) + i(\mu_3 - \mu_4)$, with $\mu_j \in M_+^c(S^*)$, and $K \subseteq S^*$ is compact such that $\text{supp}(\mu_j) \subseteq K$, $j = 1, \ldots, 4$, then $\int \Phi\, d\mu = 0$ for all $\Phi \in \mathscr{A}$, and the restrictions $\Phi|K$ form a dense subspace of $C(K)$. By Theorem 2.2.4 it follows that $\mu = 0$.

We identify $M_+(\hat{S})$ with the set of $\mu \in M_+^c(S^*)$ for which $\operatorname{supp}(\mu) \subseteq \hat{S}$. For $\mu, \nu \in M_+^c(S^*)$ with support in $\hat{S}$, $\mu * \nu$ also has its support in $\hat{S}$, so we can consider the convolution as a composition in $M_+(\hat{S})$.

In the following result $M_+(\hat{S})$ is equipped with the weak topology, which is equal to the vague topology. The result is analogous to the Lévy continuity theorem in probability theory, but simpler.

**2.11. Theorem.** *The transformation $\mu \mapsto \hat{\mu}$ is a homeomorphism of $M_+(\hat{S})$ onto $\mathscr{P}^b(S)$.*

PROOF. Only the continuity of the inverse mapping $\hat{\mu} \mapsto \mu$ needs some explanation. Let $(\mu_\alpha)$ be a net in $M_+(\hat{S})$, let $\mu \in M_+(\hat{S})$ and assume $\hat{\mu}_\alpha(s) \to \hat{\mu}(s)$ for each $s \in S$. In particular there exists $\alpha_0$ and $C > 0$ such that $\mu_\alpha(\hat{S}) = \hat{\mu}_\alpha(0) \leqq C$ for $\alpha \geqq \alpha_0$, showing that $(\mu_\alpha)$ is eventually in the compact set of measures in $M_+(\hat{S})$ with total mass $\leqq C$. It therefore suffices to show that $\mu$ is the only accumulation point for $(\mu_\alpha)$. Indeed, if $\sigma$ is an accumulation point for $(\mu_\alpha)$, then $\hat{\mu} = \hat{\sigma}$ by the continuity of $\hat{}$, hence $\mu = \sigma$. □

**2.12. Exercise.** Let $S$ be an abelian semigroup with involution. Show that the Banach algebra $l^1(S)$ from Exercise 1.17 is abelian. Let $\Delta$ be the Gelfand spectrum of nonzero multiplicative linear functionals (Rudin 1973, p. 265), and let $\Delta_h$ denote the set of hermitian elements, i.e. the set of $L \in \Delta$ such that $L(f^*) = \overline{L(f)}$ for $f \in l^1(S)$. Let $\Gamma$ denote the set of nonzero bounded functions $\gamma: S \to \mathbb{C}$ satisfying $\gamma(s + t) = \gamma(s)\gamma(t)$ for $s, t \in S$ with the topology of pointwise convergence. For $\gamma \in \Gamma$ let $L_\gamma: l^1(S) \to \mathbb{C}$ be defined by

$$L_\gamma(f) = \sum_{s \in S} f(s)\gamma(s).$$

Show that $\gamma \mapsto L_\gamma$ is a homeomorphism of $\Gamma$ onto $\Delta$ which maps $\hat{S}$ onto $\Delta_h$. It can be proved that $l^1(S)$ is semisimple if and only if $\Gamma$ separates the points of $S$, cf. Hewitt and Zuckermann (1956).

**2.13. Exercise.** Let $(S, +, *)$ and $(T, +, *)$ be abelian semigroups with involution, and let $(S \times T, +, *)$ be the product semigroup defined by $(s, t) + (u, v) = (s + u, t + v)$, $(s, t)^* = (s^*, t^*)$. Show that there exists a topological semigroup isomorphism of $S^* \times T^*$ onto $(S \times T)^*$, which maps $\hat{S} \times \hat{T}$ onto $\widehat{S \times T}$.

**2.14. Exercise.** Let $(S, +, *)$ and $(T, +, *)$ be abelian semigroups with involutions and let $h: S \to T$ be a homomorphism. The *dual map* $\tilde{h}: \mathbb{C}^T \to \mathbb{C}^S$ is defined by $\tilde{h}(f) = f \circ h$. Show that $\tilde{h}(\mathscr{P}(T)) \subseteq \mathscr{P}(S)$, $\tilde{h}(\mathscr{P}^e(T)) \subseteq \mathscr{P}^e(S)$, $\tilde{h}(T^*) \subseteq S^*$, $\tilde{h}(\hat{T}) \subseteq \hat{S}$ and that $h^* := \tilde{h}|T^*$ is a continuous homomorphism of $T^*$ into $S^*$ which, furthermore, is one-to-one if $h$ is onto.

Let $\varphi \in \mathscr{P}^e(S)$ have the representing measure $\mu \in M_+^c(S^*)$. Show that $\varphi = f \circ h$ for some $f \in \mathscr{P}^e(T)$ if and only if $\mu = \nu^{h^*}$ for some $\nu \in M_+^c(T^*)$.

**2.15. Exercise.** Let $\alpha$ be an absolute value on $S$ and let $\varphi \in \mathscr{P}_1^\alpha(S)$ satisfy $|\varphi(s)|^2 = \varphi(s + s^*)$ for all $s \in S$. Show that $\varphi \in S^*$. Show also that if $\varphi \in \mathscr{P}_1^\alpha(S)$ satisfies $|\varphi(s)| = \alpha(s)$ for all $s \in S$, then $\varphi \in S^*$.

**2.16. Exercise.** Let $(S, +, *)$ be an abelian semigroup with involution. Show that $S^*$ is a linearly independent subset of $\mathbb{C}^S$ and conclude that if $S$ is finite then card $S^* \leqq$ card $S$. Give an example where card $S^* <$ card $S < \infty$.

**2.17. Exercise.** Show that $([-1, 1], \cdot)\hat{}$ can be described as $T \cup \text{sgn} \cdot T \cup \{\text{sgn}^2\}$ where $T := \{t \mapsto |t|^\alpha, \alpha \in [0, \infty]\}$, $\text{sgn} := 1_{]0,1]} - 1_{[-1,0[}$, and $|t|^0 := 1$, $|t|^\infty := 1_{\{-1,1\}}(t)$ for $t \in [-1, 1]$.

**2.18. Exercise.** Let $\varphi \in \mathscr{P}^\alpha(S)$. Then $\{s \in S \,|\, \varphi(s) = \alpha(s)\}$ as well as $\{s \in S \,|\, |\varphi(s)| = \alpha(s)\}$ are $*$-subsemigroups of $S$.

**2.19. Exercise.** Let $(S, +, *)$ be an abelian semigroup with involution and let $T := \{\rho \in \hat{S} \,|\, |\rho| \equiv 1\}$. If $\varphi \in \mathscr{P}^b(S)$ has the representing measure $\mu$ then $\mu(T) = \inf_{s \in S} \varphi(s + s^*)$.

**2.20. Exercise.** Let $(S, +, *)$ be an abelian semigroup and consider $S \cup \{\tilde{\omega}\}$ as in Exercise 1.19. Let $\varphi \in \mathscr{P}^b(S)$ have the representing measure $\mu$. Then an extension $\tilde{\varphi}$ of $\varphi$ to $S \cup \{\tilde{\omega}\}$ is positive definite if and only if $0 \leqq \tilde{\varphi}(\tilde{\omega}) \leqq \mu(\{1_S\})$.

**2.21. Exercise.** Let $S$ be an abelian semigroup with involution, let $n \in \mathbb{N}$ and let $\widehat{S^n}$ be identified with $(\hat{S})^n$ in accordance with Exercise 2.13. Show that if $\varphi \in \mathscr{P}^b(S^n)$ has the representing measure $\mu \in M_+(\hat{S}^n)$, and if $\sigma$ is a permutation of $n$ elements, then $\varphi \circ \sigma$ has the representing measure $\mu^{\sigma^{-1}}$. Conclude that $\varphi \in \mathscr{P}^b(S^n)$ is symmetric, i.e. $\varphi \circ \sigma = \varphi$ for all permutations $\sigma$, if and only if $\mu$ is symmetric.

**2.22. Exercise.** Let $S = [0, 1]$ and define $s \circ t = (s + t - 1)^+$ for $s, t \in S$. Show that $(S, \circ)$ is an abelian semigroup with neutral element 1 and absorbing element 0. Show further that $S$ is 2-divisible (cf. 6.8) and that $S^* = \{1, 1_{\{1\}}\}$.

# §3. Negative Definite Functions on Abelian Semigroups

In most of this section, we consider negative definite functions on an abelian semigroup with involution $(S, +, *)$. From 3.20 onwards, we will specialize to semigroups with the identical involution.

For $\psi \in \mathcal{N}(S)$ we know by Theorem 3.2.2 that $e^{-t\psi} \in \mathcal{P}(S)$ for $t > 0$. Since we only have an integral representation for exponentially bounded positive definite functions and, in particular, for bounded positive definite functions, the following simple result is of some interest.

**3.1. Proposition.** *Let $\psi \in \mathcal{N}(S)$. Then $e^{-t\psi} \in \mathcal{P}^e(S)$ (resp. $\in \mathcal{P}^b(S)$) for all $t > 0$ if and only if there exists an absolute value $\alpha \geqq 1$ and a constant $C \in \mathbb{R}$ such that*

$$\operatorname{Re} \psi(s) \geqq C - \log \alpha(s) \qquad \textit{for} \quad s \in S,$$

$$(\textit{resp.}\ \operatorname{Re} \psi(s) \geqq C \qquad \textit{for} \quad s \in S).$$

*If this is the case then $C = \psi(0)$ can be used.*

PROOF. If $e^{-\psi}$ is exponentially bounded (resp. bounded) there exists an absolute value $\alpha$, where we may assume $\alpha \geqq 1$ (resp. $\alpha = 1$), such that

$$|e^{-\psi(s)}| = e^{-\operatorname{Re}\psi(s)} \leqq e^{-\psi(0)}\alpha(s), \qquad s \in S,$$

and the stated inequalities follow. Conversely, if the first inequality of the proposition holds, we get for $t > 0$

$$|e^{-t\psi(s)}| \leqq e^{-tC}\alpha(s)^t, \qquad s \in S,$$

which shows that $e^{-t\psi} \in \mathcal{P}^e(S)$ (resp. $\in \mathcal{P}^b(S)$ when $\alpha = 1$) since $\alpha^t$ is an absolute value. □

The set of negative definite functions $\psi$ for which $\operatorname{Re} \psi$ is bounded below will be denoted $\mathcal{N}^l(S)$.

**3.2. Corollary.** *For $\psi \in \mathcal{N}^l(S)$ we have*

$$\operatorname{Re} \psi(s) \geqq \psi(0), \qquad s \in S.$$

**3.3. Proposition.** *Let $\psi \in \mathcal{N}^l(S)$ satisfy $\psi(0) \geqq 0$. Then*

$$\sqrt{|\psi(s+t)|} \leqq \sqrt{|\psi(s)|} + \sqrt{|\psi(t)|}, \qquad s, t \in S.$$

PROOF. By Proposition 1.9 the matrix

$$\begin{pmatrix} 2\operatorname{Re}\psi(s) - \psi(s^* + s) & \overline{\psi(s)} + \psi(t) - \psi(s^* + t) \\ \psi(s) + \overline{\psi(t)} - \psi(s + t^*) & 2\operatorname{Re}\psi(t) - \psi(t^* + t) \end{pmatrix}$$

is positive definite, hence, using inequality (1) of §1,

$$\begin{aligned} |\overline{\psi(s)} + \psi(t) - \psi(s^* + t)|^2 &\leqq (2\operatorname{Re}\psi(s) - \psi(s^* + s)) \\ &\quad \times (2\operatorname{Re}\psi(t) - \psi(t^* + t)) \\ &\leqq 4|\psi(s)||\psi(t)|. \end{aligned}$$

Replacing $s$ by $s^*$ and extracting the square root gives

$$|\psi(s+t)| \leqq |\psi(s)| + |\psi(t)| + 2\sqrt{|\psi(s)|}\sqrt{|\psi(t)|},$$

and the inequality follows. □

In some later applications we shall need the following boundedness result for negative definite functions on semigroups with absorbing element (cf. 1.3).

**3.4. Proposition.** *Let $S$ contain the absorbing element $\omega$. Then the functions in $\mathcal{N}^l(S)$ are automatically bounded. More precisely if $\psi \in \mathcal{N}^l(S)$ and $\psi(0) \geqq 0$ then $\|\psi\|_\infty \leqq \sqrt{5}\psi(\omega)$.*

Proof. For fixed $s \in S$ the $3 \times 3$ matrix

$$\begin{pmatrix} \psi(0) & \psi(s^*) & \psi(\omega) \\ \psi(s) & \psi(s+s^*) & \psi(\omega) \\ \psi(\omega) & \psi(\omega) & \psi(\omega) \end{pmatrix} = (a_{jk})_{j,k=1,2,3}$$

is negative definite. Writing down explicitly the inequality $\sum c_j \bar{c}_k a_{jk} \leqq 0$ for $c_1 = c_2 = 1$, $c_3 = -2$ gives

$$\psi(0) + 2 \operatorname{Re} \psi(s) + \psi(s + s^*) \leqq 4\psi(\omega),$$

and since $\psi(s + s^*) = \operatorname{Re} \psi(s + s^*) \geqq 0$ by Corollary 3.2 we get $0 \leqq \operatorname{Re} \psi(s) \leqq 2\psi(\omega)$. Similarly, the choice of $c_1 = 1$, $c_2 = i$, $c_3 = -1 - i$ gives $-\operatorname{Im} \psi(s) \leqq \psi(\omega)$, and the coefficients $c_1 = 1$, $c_2 = -i$, $c_3 = i - 1$ yield $\operatorname{Im} \psi(s) \leqq \psi(\omega)$. Hence $|\operatorname{Im} \psi(s)| \leqq \psi(\omega)$ and therefore $\|\psi\|_\infty \leqq \sqrt{5}\psi(\omega)$. □

We shall next establish an important relationship between functions $\psi \in \mathcal{N}^l(S)$ and convolution semigroups of Radon measures on the compact semigroup $\hat{S}$, which is the restricted dual semigroup of $S$.

**3.5. Definition.** Let $T$ be a Hausdorff topological semigroup with neutral element $e$. A *convolution semigroup* on $T$ is a family $(\mu_t)_{t \geqq 0}$ from $M_+^b(T)$ such that $t \mapsto \mu_t$ is a continuous homomorphism from the semigroup $(\mathbb{R}_+, +)$ into the semigroup $(M_+^b(T), *)$, i.e. such that:

(i) $\mu_0 = \varepsilon_e$;
(ii) $\mu_t * \mu_r = \mu_{t+r}$ for $t, r \in \mathbb{R}_+$;
(iii) $t \mapsto \mu_t$ is weakly continuous.

In the case of $T = \hat{S}$ condition (iii) can be replaced by a seemingly weaker one as the following lemma shows.

**3.6. Lemma.** *Let $t \mapsto \mu_t$ be a homomorphism from $(\mathbb{R}_+, +)$ into $M_+(\hat{S})$ such that* $\lim_{t \to 0} \mu_t = \varepsilon_1$ *weakly, then $(\mu_t)_{t \geqq 0}$ is a convolution semigroup on $\hat{S}$.*

Proof. By Theorem 2.11 it suffices to prove that $j_s(t) := \hat{\mu}_t(s)$ is a continuous mapping from $\mathbb{R}_+$ into $\mathbb{C}$ for each $s \in S$. By assumption we have $j_s(t + r) = j_s(t) j_s(r)$ and $\lim_{t \to 0} j_s(t) = 1$, so there exists $A > 0$ such that $|j_s(t)| \leqq 2$ for $t \in [0, A]$. It follows that $|j_s(t)| \leqq 2^n$ for $t \in [0, nA]$, in particular $j_s$ is locally bounded. This combined with the equations $(t, r, t - r > 0)$

$$|j_s(t + r) - j_s(t)| = |j_s(t)||j_s(r) - 1|,$$
$$|j_s(t - r) - j_s(t)| = |j_s(t - r)||j_s(r) - 1|$$

implies that $j_s$ is continuous. □

**3.7. Theorem.** *There is a one-to-one correspondence between convolution semigroups $(\mu_t)_{t \geqq 0}$ on $\hat{S}$ and negative definite functions $\psi \in \mathcal{N}^l(S)$ established via the formula*

$$\hat{\mu}_t(s) = e^{-t\psi(s)} \qquad \text{for} \quad t \geqq 0, \quad s \in S. \tag{1}$$

PROOF. Let $(\mu_t)_{t \geqq 0}$ be a convolution semigroup. For $s \in S$ the mapping $j_s: t \mapsto \hat{\mu}_t(s)$ is continuous by Theorem 2.11 and satisfies $j_s(t + r) = j_s(t)j_s(r)$, hence of the form $j_s(t) = e^{-t\psi(s)}$ for a uniquely determined complex number $\psi(s)$. Since $e^{-t\psi} \in \mathcal{P}^b(S)$ for each $t > 0$ it follows by Theorem 3.2.2 and Proposition 3.1 that $\psi \in \mathcal{N}(S)$ and that $\operatorname{Re} \psi \geqq \psi(0)$. Conversely, if $\psi$ has these properties there exists a uniquely determined family $(\mu_t)_{t \geqq 0}$ from $M_+(\hat{S})$ such that (1) holds. Clearly

$$\hat{\mu}_t(s)\hat{\mu}_r(s) = \hat{\mu}_{t+r}(s) \qquad \text{and} \qquad \lim_{t \to 0} \hat{\mu}_t(s) = 1 = \hat{\varepsilon}_1(s)$$

for $s \in S$, so $(\mu_t)_{t \geqq 0}$ is a convolution semigroup by 2.10, 2.11 and 3.6. □

We will now consider two special types of functions $\psi \in \mathcal{N}^l(S)$, namely, purely imaginary homomorphisms and nonnegative quadratic forms.

**3.8. Definition.** A function $\psi: S \to \mathbb{C}$ is called a *purely imaginary homomorphism* if it has the form $\psi = il$, where $l: S \to \mathbb{R}$ is $*$-additive, i.e. $l(s + t) = l(s) + l(t)$ and $l(s^*) = -l(s)$ for $s, t \in S$.

A function $q: S \to \mathbb{R}$ is called a *quadratic form* if

$$2q(s) + 2q(t) = q(s + t) + q(s + t^*) \qquad \text{for} \quad s, t \in S.$$

If the involution on $S$ is the identical, every purely imaginary homomorphism is identically zero, and a quadratic form is a homomorphism of $(S, +)$ into $(\mathbb{R}, +)$.

If $S$ is an abelian group, with the involution being $s^* = -s$ for $s \in S$, the functional equation for a quadratic form $q$ is

$$2q(s) + 2q(t) = q(s + t) + q(s - t) \qquad \text{for} \quad s, t \in S.$$

**3.9. Theorem.** *Purely imaginary homomorphisms and nonnegative quadratic forms belong to $\mathcal{N}^l(S)$.*

PROOF. Let $\psi: S \to \mathbb{C}$ be a purely imaginary homomorphism, let $\{s_1, \ldots, s_n\} \subseteq S$ and $\{c_1, \ldots, c_n\} \subseteq \mathbb{C}$ with $\sum_{j=1}^n c_j = 0$. Then

$$\sum_{j,k=1}^n c_j \overline{c_k} \psi(s_j^* + s_k) = \sum_{j,k=1}^n c_j \overline{c_k} (\overline{\psi(s_j)} + \psi(s_k)) = 0,$$

hence $\psi \in \mathcal{N}^l(S)$.

Let $q: S \to \mathbb{R}$ be a quadratic form. Note that $q(0) = 0$ and $q(s^*) = q(s)$. We define $B: S \times S \to \mathbb{R}$ by

$$B(s, t) = q(s) + q(t) - q(s + t^*), \qquad s, t \in S,$$

and claim that $B$ has the following properties for $s, t, r \in S$:

(i) $B(s, t) = B(t, s)$;
(ii) $B(s^*, t) = B(s, t^*) = -B(s, t)$;
(iii) $B(s + r, t) = B(s, t) + B(r, t)$;
(iv) $\frac{1}{2}B(s, s) = \lim_{n\to\infty}(q(ns)/n^2)$.

Here (i) is clear, and

$$\begin{aligned} B(s^*, t) = B(s, t^*) &= q(s) + q(t) - q(s + t) \\ &= q(s + t^*) - q(s) - q(t) = -B(s, t). \end{aligned}$$

To see (iii), we first remark that

$$q(s) + q(t + t^*) = q(s + t + t^*) \qquad \text{for} \quad s, t \in S,$$

and therefore we have

$$\begin{aligned} B(s, t) + B(r, t) &= q(s) + q(t) - q(s + t^*) + q(r) + q(t) - q(r + t^*) \\ &= q(s) + q(r) + 2q(t) - \tfrac{1}{2}[q(s + t^* + r + t^*) \\ &\quad + q(s + t^* + t + r^*)] \\ &= q(s) + q(r) + 2q(t) - \tfrac{1}{2}q(s + r^*) - \tfrac{1}{2}q(t + t^*) \\ &\quad - \tfrac{1}{2}q(s + r + t^* + t^*) \\ &= \tfrac{1}{2}q(s + r) + 2q(t) - \tfrac{1}{2}q(t + t^*) \\ &\quad - \tfrac{1}{2}\{2q(s + r + t^*) + 2q(t) - q(s + r + t + t^*)\} \\ &= q(s + r) + q(t) - q(s + r + t^*) = B(s + r, t). \end{aligned}$$

To see (iv) we remark that for $n \in \mathbb{N}$

$$q(ns) = n^2 q(s) - \frac{n(n-1)}{2} q(s + s^*).$$

This formula is clearly true for $n = 1, 2$, and assuming it is correct for $n - 1$, $n$ we get

$$\begin{aligned} q((n + 1)s) &= 2q(ns) + 2q(s) - q((n - 1)s + (s + s^*)) \\ &= (2n^2 + 2)q(s) - n(n - 1)q(s + s^*) - q((n - 1)s) - q(s + s^*) \\ &= (n + 1)^2 q(s) - \frac{(n + 1)n}{2} q(s + s^*). \end{aligned}$$

It follows that

$$\lim_{n\to\infty} \frac{q(ns)}{n^2} = q(s) - \tfrac{1}{2}q(s + s^*) = \tfrac{1}{2}B(s, s).$$

If $q$ is a nonnegative quadratic form it follows by (iv) that $B(s, s) \geqq 0$ for all $s \in S$. For $\{s_1, \ldots, s_n\} \subseteq S$, $\{c_1, \ldots, c_n\} \subseteq \mathbb{Z}$ we now get

$$\sum_{j,k=1}^{n} c_j c_k(q(s_j) + q(s_k) - q(s_j^* + s_k)) = \sum_{j,k=1}^{n} c_j c_k B(s_j, s_k)$$
$$= B(t, t) \geqq 0,$$

where

$$t = \sum_{j, c_j \geqq 0} c_j s_j + \sum_{j, c_j < 0} (-c_j) s_j^*.$$

We conclude from Proposition 1.9 and Exercise 3.1.21 that $q$ is negative definite. □

**3.10. Remark.** If a quadratic form $q$ is bounded below then it is nonnegative and negative definite. In fact, the proof shows that $B(s, s) \geqq 0$, hence $q \in \mathcal{N}^l(S)$ and, finally, $q(s) \geqq q(0) = 0$ by Corollary 3.2.

For $f \in \mathbb{C}^S$ and $a \in S$, we define $\Gamma_a f \in \mathbb{C}^S$ by

$$\Gamma_a f(s) = \tfrac{1}{2}(f(s + a) + f(s + a^*)), \qquad s \in S.$$

**3.11. Proposition.** *Let $\psi \in \mathcal{N}^l(S)$ and let $(\mu_t)_{t \geqq 0}$ be the corresponding convolution semigroup on $\hat{S}$. For $a \in S$ we have $\Gamma_a\psi - \psi \in \mathcal{P}^b(S)$, and if $\sigma_a \in M_+(\hat{S})$ is the unique Radon measure such that $\Gamma_a \psi - \psi = \hat{\sigma}_a$ then*

$$\sigma_a = \lim_{t \to 0} (1 - \operatorname{Re} \rho(a)) \frac{1}{t} \mu_t$$

*weakly.*

Proof. If $I$ is the identity operator on $\mathbb{C}^S$ we have

$$\frac{1}{t}(\Gamma_a - I)(1 - e^{-t\psi})(s) = \frac{1}{t}(I - \Gamma_a)(e^{-t\psi})(s) = \frac{1}{t}\int_{\hat{S}} \rho(s)(1 - \operatorname{Re} \rho(a))\, d\mu_t(\rho),$$

showing that $(\Gamma_a - I)((1/t)(1 - e^{-t\psi})) \in \mathcal{P}^b(S)$. For $t \to 0$ this function tends pointwise to $\Gamma_a\psi - \psi$, which then necessarily belongs to $\mathcal{P}^b(S)$. The assertion about the measure follows immediately from the Lévy continuity theorem (2.11). □

We now consider an arbitrary hermitian function $\psi: S \to \mathbb{C}$ with the property that $\Gamma_a\psi - \psi \in \mathcal{P}^b(S)$ for all $a \in S$, and we denote by $\sigma_a$ the unique Radon measure on $\hat{S}$ such that $\Gamma_a\psi - \psi = \hat{\sigma}_a$.

**3.12. Lemma.** *Let $\psi: S \to \mathbb{C}$ be a hermitian function such that $\Gamma_a\psi - \psi \in \mathcal{P}^b(S)$ for each $a \in S$. There exists a unique Radon measure $\mu \in M_+(\hat{S} \setminus \{1\})$ such that*

$$(1 - \operatorname{Re} \rho(a))\mu = \sigma_a | (\hat{S} \setminus \{1\}) \qquad \textit{for} \quad a \in S. \tag{2}$$

*If* $\psi \in \mathcal{N}^l(S)$ *and* $(\mu_t)_{t \geqq 0}$ *is the corresponding convolution semigroup then* $\mu = \lim_{t \to 0}(1/t)\mu_t | (\hat{S} \setminus \{1\})$ *vaguely.*

PROOF. For $a \in S$ we let

$$\mathcal{O}_a := \{\rho \in \hat{S} \mid \operatorname{Re} \rho(a) < 1\},$$

which is an open subset of $\hat{S}$. Since $\mathcal{O}_a = \{\rho \in \hat{S} \mid \rho(a) \neq 1\}$ the family $(\mathcal{O}_a)_{a \in S}$ is an open covering of the locally compact space $\hat{S} \setminus \{1\}$. On each $\mathcal{O}_a$ we consider the Radon measure

$$\tau_a := (1 - \operatorname{Re} \rho(a))^{-1}(\sigma_a | \mathcal{O}_a).$$

Since

$$\begin{aligned} -(\Gamma_a - I)(\Gamma_b - I)\psi(s) &= -(\Gamma_b - I)(\Gamma_a - I)\psi(s) \\ &= \int_{\hat{S}} \rho(s)(1 - \operatorname{Re} \rho(a))\, d\sigma_b(\rho) \\ &= \int_{\hat{S}} \rho(s)(1 - \operatorname{Re} \rho(b))\, d\sigma_a(\rho), \end{aligned}$$

we get by the unicity assertion of Theorem 2.8 that

$$(1 - \operatorname{Re} \rho(a))\sigma_b = (1 - \operatorname{Re} \rho(b))\sigma_a, \qquad a, b \in S,$$

so the compatibility condition of Theorem 2.1.18 is satisfied for the measures $(\tau_a)_{a \in S}$. Consequently, there exists a uniquely determined Radon measure $\mu$ on $\hat{S} \setminus \{1\}$ such that $\mu | \mathcal{O}_a = \tau_a$ for $a \in S$, and it is clear that $\mu$ is the only Radon measure satisfying (2).

To finish the proof it is enough by Theorem 2.4.9 to prove that for each $a \in S$

$$\lim_{t \to 0} \frac{1}{t} \mu_t | \mathcal{O}_a = \tau_a \qquad \text{vaguely on } \mathcal{O}_a.$$

Let $f \in C_+^c(\mathcal{O}_a)$ be given. Then

$$g(\rho) := \begin{cases} f(\rho)(1 - \operatorname{Re} \rho(a))^{-1}, & \rho \in \mathcal{O}_a, \\ 0, & \rho \in \hat{S} \setminus \mathcal{O}_a \end{cases}$$

is a bounded continuous function on $\hat{S}$, so by 3.11,

$$\lim_{t \to 0} \frac{1}{t} \int f\, d(\mu_t | \mathcal{O}_a) = \int g\, d\sigma_a = \int f\, d\tau_a. \qquad \square$$

**3.13. Definition.** Let $\psi$ be as in Lemma 3.12. The measure $\mu$ is called the *Lévy measure* for $\psi$ or for the corresponding convolution semigroup $(\mu_t)_{t \geqq 0}$ in case $\psi \in \mathcal{N}^l(S)$.

The Lévy measure might have infinite total mass, but (2) implies that

$$\int_{\hat{S}\setminus\{1\}} (1 - \operatorname{Re} \rho(a))\, d\mu(\rho) < \infty \qquad \text{for all} \quad a \in S. \tag{3}$$

Let $\mathfrak{L}$ denote the set of $\mu \in M_+(\hat{S}\setminus\{1\})$ for which (3) holds.

With the same assumptions as in Lemma 3.12 we have:

**3.14. Lemma.** *The following conditions are equivalent:*

(i) $\mu = 0$;
(ii) $\Gamma_a\psi - \psi$ *is constant for each* $a \in S$;
(iii) $\psi = \psi(0) + q + il$, *where* $q$ *is a nonnegative quadratic form, and* $l$ *is* $*$*-additive.*

Proof. The equivalence "(i) $\Leftrightarrow$ (ii)" is obvious. Suppose that (ii) holds. Defining $q = \operatorname{Re}(\psi - \psi(0))$, $l = \operatorname{Im}(\psi - \psi(0))$, we have

$$\Gamma_a q - q = \sigma_a(\{1\}) \geqq 0, \qquad \Gamma_a l - l = 0$$

and

$$q(a^*) = q(a), \qquad l(a^*) = -l(a)$$

for all $a \in S$, and it follows easily that $q$ is a nonnegative quadratic form and $l$ is $*$-additive. In fact

$$\Gamma_a q(s) - q(s) = \Gamma_a q(0) - q(0) = q(a), \qquad s \in S,$$

which is the functional equation for a quadratic form. Furthermore,

$$\Gamma_a l(s) - l(s) = \tfrac{1}{2}[l(s + a) + l(s + a^*)] - l(s) = 0, \qquad s \in S,$$

so by interchanging the roles of $a$ and $s$ we have

$$l(s + a) + l(s + a^*) = 2l(s),$$

$$l(a + s) + l(a + s^*) = 2l(a),$$

and using $l(s + a^*) = -l(a + s^*)$ we find

$$l(a + s) = l(a) + l(s).$$

Conversely, if (iii) holds, it is easy to see that $\Gamma_a\psi - \psi = q(a)$. □

As another use of the Lévy measure we can completely characterize the set $\mathcal{N}^b(S)$ of bounded negative definite functions.

**3.15. Proposition.** *Let* $\psi \in \mathcal{N}^b(S)$. *Then the Lévy measure* $\mu$ *for* $\psi$ *satisfies* $\mu(\hat{S}\setminus\{1\}) \leqq \|\psi\|_\infty$ *and*

$$\psi(s) = \psi(0) + \int_{\hat{S}\setminus\{1\}} (1 - \rho(s))\, d\mu(\rho), \qquad s \in S.$$

*In particular* $\psi$ *has the form* $c - \varphi$ *where* $c \in \mathbb{R}$ *and* $\varphi \in \mathcal{P}^b(S)$.

PROOF. Without loss of generality we may assume $\psi(0) = 0$. The corresponding convolution semigroup $(\mu_t)_{t \geqq 0}$ consists of probability measures, and we have for $t > 0$

$$1 - e^{-t\psi(s)} = \int_{\hat{S}} (1 - \rho(s))\, d\mu_t(\rho) = \int_{\hat{S}\setminus\{1\}} (1 - \rho(s))\, d\mu_t(\rho).$$

Using $\operatorname{Re} \psi(s) \geqq 0$ (cf. 3.2) and $|1 - e^{-z}| \leqq |z|$ for $\operatorname{Re} z \geqq 0$, we find

$$0 \leqq \int_{\hat{S}\setminus\{1\}} (1 - \operatorname{Re} \rho(s))\, d\mu_t(\rho) \leqq t|\psi(s)| \leqq t\, \|\psi\|_\infty.$$

For any molecular measure $\alpha$ on $S$ we define

$$\hat{\alpha}(\rho) = \int_S \rho(s)\, d\alpha(s), \qquad \rho \in \hat{S}.$$

If $\alpha$ is a molecular probability measure on $S$ we have

$$\begin{aligned} 0 &\leqq \int_S \left( \int_{\hat{S}\setminus\{1\}} (1 - \operatorname{Re} \rho(s))\, d\mu_t(\rho) \right) d\alpha(s) \\ &= \int_{\hat{S}\setminus\{1\}} (1 - \operatorname{Re} \hat{\alpha}(\rho))\, d\mu_t(\rho) \leqq t\|\psi\|_\infty. \end{aligned}$$

Let $C \subseteq \hat{S}\setminus\{1\}$ be a compact set. For any $\rho_0 \in C$ there exists $a \in S$ such that $\operatorname{Re} \rho_0(a) < 1$, and there exists an open neighbourhood $U$ of $\rho_0$ in $\hat{S}$ such that $\operatorname{Re} \rho(a) < 1$ for all $\rho \in U$. By compactness of $C$ there exist finitely many points $a_1, \ldots, a_n \in S$ such that

$$\frac{1}{n} \sum_{i=1}^{n} \operatorname{Re} \rho(a_i) < 1 \qquad \text{for all} \quad \rho \in C.$$

If we define the molecular probability $\alpha$ by

$$\alpha = \sum_{i=1}^{n} \frac{1}{2n} (\varepsilon_{a_i} + \varepsilon_{a_i^*}),$$

then

$$\hat{\alpha}(\rho) = \frac{1}{n} \sum_{i=1}^{n} \operatorname{Re} \rho(a_i) \in [-1, 1], \qquad \rho \in \hat{S}$$

and $\hat{\alpha}(\rho) < 1$ for $\rho \in C$. For every natural number $p$ the convolution power $\alpha^{*p}$ is again a molecular probability measure and $(\alpha^{*p})\hat{}\,(\rho) = (\hat{\alpha}(\rho))^p$, $\rho \in \hat{S}$. Applying Fatou's lemma gives

$$\begin{aligned} \mu_t(C) &\leqq \int_{\hat{S}\setminus\{1\}} \liminf_{p\to\infty} (1 - (\hat{\alpha}(\rho))^{2p+1})\, d\mu_t(\rho) \\ &\leqq \liminf_{p\to\infty} \int (1 - (\hat{\alpha}(\rho))^{2p+1})\, d\mu_t(\rho) \leqq t\|\psi\|_\infty, \end{aligned}$$

and $C$ being arbitrary, we get

$$\mu_t(\hat{S}\setminus\{1\}) \leqq t\|\psi\|_\infty.$$

Since $(1/t)\mu_t|(\hat{S}\setminus\{1\}) \to \mu$ vaguely for $t \to 0$ we get by Proposition 2.4.4 that $\mu(\hat{S}\setminus\{1\}) \leqq \|\psi\|_\infty$ and that

$$\begin{aligned}\psi(s) &= \lim_{t\to 0} \frac{1}{t}(1 - e^{-t\psi(s)}) \\ &= \lim_{t\to 0} \int_{\hat{S}\setminus\{1\}} (1 - \rho(s)) \frac{1}{t}\, d\mu_t(\rho) \\ &= \int_{\hat{S}\setminus\{1\}} (1 - \rho(s))\, d\mu(\rho),\end{aligned}$$

because $\rho \mapsto 1 - \rho(s)$ belongs to $C^0(\hat{S}\setminus\{1\})$. □

**3.16. Corollary.** *Let $S$ contain an absorbing element $\omega$. Then the transformation*

$$\mathscr{P}_1^b(S) \to \{\psi \in \mathscr{N}^l(S) \mid \psi(0) = 0, \psi(\omega) \leqq 1\},$$
$$\varphi \mapsto 1 - \varphi$$

*is onto, i.e. is an affine homeomorphism.*

Proof. Let $\psi \in \mathscr{N}^l(S)$ fulfil $\psi(0) = 0$ and $\psi(\omega) \leqq 1$. By Proposition 3.4 $\psi$ is bounded and therefore $\psi$ has the representation

$$\psi(s) = \int_{\hat{S}\setminus\{1\}} (1 - \rho(s))\, d\mu(\rho), \qquad s \in S,$$

where $\psi(\omega) = \mu(\hat{S}\setminus\{1\}) \leqq 1$. Put $\alpha := 1 - \mu(\hat{S}\setminus\{1\})$ and $\mu' := \mu + \alpha\varepsilon_1 \in M_+^1(\hat{S})$, then

$$1 - \psi(s) = \alpha + \int_{\hat{S}\setminus\{1\}} \rho(s)\, d\mu(\rho) = \int_{\hat{S}} \rho(s)\, d\mu'(\rho)$$

showing that $1 - \psi \in \mathscr{P}_1^b(S)$. □

Our goal is to extend the integral representation in Proposition 3.15 to functions $\psi \in \mathscr{N}^l(S)$, thus looking for a representation analogous to the classical Lévy–Khinchin formula in probability theory, cf., e.g. Loève (1963). If $\mu \in M_+(\hat{S}\setminus\{1\})$ is such that

$$\int |1 - \rho(s)|\, d\mu(\rho) < \infty \qquad \text{for} \quad s \in S, \tag{4}$$

then

$$\psi_\mu(s) := \int_{\hat{S}\setminus\{1\}} (1 - \rho(s))\, d\mu(\rho), \qquad s \in S \tag{5}$$

belongs to $\mathcal{N}^l(S)$, as is easily seen. However (4) need not hold for all Lévy measures, and therefore it is necessary to introduce a compensating term in the integral in (5). This is made precise in the following.

**3.17. Definition.** A function $L\colon S \times \hat{S} \to \mathbb{R}$ is called a *Lévy function for S* if

(i) $L(\cdot, \rho)$ is a $*$-additive function from $S$ into $\mathbb{R}$ for each $\rho \in \hat{S}$.
(ii) $L(s, \cdot)$ is Borel measurable on $\hat{S}$ for each $s \in S$.
(iii) $\int |1 - \rho(s) + iL(s, \rho)|\, d\mu(\rho) < \infty$ for all $s \in S$ and all measures $\mu \in \mathfrak{L}$ (i.e. such that (3) holds).

We do not know if every semigroup with involution has a Lévy function, and it is probably a difficult question. If $S$ is an abelian group and $s^* = -s$ there exists a Lévy function. This may be seen from Parthasarathy *et al.* (1963) and depends on the structure theory for abelian groups. See also Forst (1976). If $S$ is an abelian semigroup with $s^* = s$ the function $L \equiv 0$ is a Lévy function. This also holds (cf., 3.4 and 3.15) if $S$ contains an absorbing element. We shall later give other examples of Lévy functions.

**3.18. Proposition.** *Suppose S has a Lévy function L. For any measure $\mu \in \mathfrak{L}$ the function*

$$\psi_\mu(s) := \int_{\hat{S}\setminus\{1\}} (1 - \rho(s) + iL(s, \rho))\, d\mu(\rho), \qquad s \in S$$

*belongs to $\mathcal{N}^l(S)$ and the Lévy measure for $\psi_\mu$ is $\mu$.*

PROOF. The function $\psi_\mu$ is easily seen to be negative definite and $\operatorname{Re} \psi \geqq 0$. Furthermore,

$$(\Gamma_a - I)\psi_\mu(s) = \int_{\hat{S}\setminus\{1\}} \rho(s)(1 - \operatorname{Re} \rho(a))\, d\mu(\rho), \qquad a, s \in S,$$

which shows that $\sigma_a' \in M_+(\hat{S})$, determined such that $\hat{\sigma}_a' = (\Gamma_a - I)\psi_\mu$, is concentrated on $\hat{S}\setminus\{1\}$ and equal to $(1 - \operatorname{Re} \rho(a))\mu$. By Lemma 3.12 it follows that $\mu$ is the Lévy measure for $\psi_\mu$. □

We can now state and prove the "Lévy–Khinchin" integral representation for $\mathcal{N}^l(S)$.

**3.19. Theorem.** *Suppose S has Lévy function L. The following conditions are equivalent for a function $\psi\colon S \to \mathbb{C}$:*

(i) $\psi \in \mathcal{N}^l(S)$.
(ii) *$\psi$ is hermitian and $\Gamma_a\psi - \psi \in \mathcal{P}^b(S)$ for each $a \in S$.*
(iii) *There exists a triple $(l, q, \mu)$ where $l\colon S \to \mathbb{R}$ is $*$-additive, $q$ is a non-negative quadratic form and $\mu \in \mathfrak{L}$ such that*

$$\psi(s) = \psi(0) + il(s) + q(s) + \int_{\hat{S}\setminus\{1\}} (1 - \rho(s) + iL(s, \rho))\, d\mu(\rho)$$

*for all $s \in S$.*

*The triple $(l, q, \mu)$ is uniquely determined by $\psi \in \mathcal{N}^l(S)$, $\mu$ being the Lévy measure for $\psi$ and*

$$q(s) = \lim_{n\to\infty} \frac{\operatorname{Re}\psi(ns)}{n^2} + \lim_{n\to\infty} \frac{\psi(n(s+s^*))}{2n}, \qquad s \in S.$$

PROOF. "(ii) ⇒ (iii)" Let $\mu$ be the Lévy measure for $\psi$ and let $\psi_\mu$ be defined as in Proposition 3.18. The function

$$r := \psi - \psi(0) - \psi_\mu$$

is hermitian and $\Gamma_a r - r = \hat{\sigma}_a - (\sigma_a|(\hat{S}\setminus\{1\}))^\wedge = \sigma_a(\{1\})$ for $a \in S$, hence positive definite and constant. By Lemma 3.14 it follows that $\psi$ has the representation stated in (iii). Finally, if (iii) holds, it is clear by 3.9, 3.14 and 3.18 that $\psi \in \mathcal{N}^l(S)$, and that the Lévy measure for $\psi$ is $\mu$. This shows that the triple $(l, q, \mu)$ is uniquely determined by $\psi$, and we find

$$\frac{\operatorname{Re}\psi(ns)}{n^2} = \frac{\psi(0)}{n^2} + \frac{q(ns)}{n^2} + \int_{\hat{S}\setminus\{1\}} \frac{1}{n^2}(1 - \operatorname{Re}(\rho(s)^n))\, d\mu(\rho),$$

$$\frac{\psi(n(s+s^*))}{n} = \frac{\psi(0)}{n} + \frac{q(n(s+s^*))}{n} + \int_{\hat{S}\setminus\{1\}} \frac{1}{n}(1 - |\rho(s)|^{2n})\, d\mu(\rho).$$

By the proof of Theorem 3.9 we have

$$\lim_{n\to\infty} \frac{q(ns)}{n^2} = q(s) - \tfrac{1}{2}q(s+s^*),$$

and by the functional equation for quadratic forms

$$q(n(s+s^*)) = nq(s+s^*),$$

which gives the formula for $q(s)$ provided the integrals tend to zero. That this is true follows from the dominated convergence theorem because of the inequalities

$$\frac{1}{n^2}(1 - \operatorname{Re}(\rho(s)^n)) \leqq \frac{\pi^2}{4}(1 - \operatorname{Re}\rho(s)),$$

$$\frac{1}{n}(1 - |\rho(s)|^{2n}) \leqq 1 - \operatorname{Re}\rho(s+s^*).$$

Putting $\rho(s) = re^{i\theta}$, $r \in [0, 1]$, $\theta \in ]-\pi, \pi]$, these inequalities become

$$\frac{1}{n^2}(1 - r^n\cos(n\theta)) \leqq \frac{\pi^2}{4}(1 - r\cos\theta), \qquad \frac{1}{n}(1 - r^{2n}) \leqq 1 - r^2, \qquad n \in \mathbb{N}.$$

Clearly $1 - r^n \leqq n(1 - r)$ for $r \in [0, 1]$, $n \in \mathbb{N}$, and using the inequalities

$$\left|\frac{\sin x}{x}\right| \leqq 1, \quad x \in \mathbb{R}, \qquad \frac{\sin x}{x} \geqq \frac{2}{\pi} \quad \text{for} \quad |x| \leqq \frac{\pi}{2},$$

we find for $\theta \in ]-\pi, \pi]$

$$1 - \cos(n\theta) = 2\sin^2\frac{n\theta}{2} = \left(\frac{\sin\frac{n\theta}{2}}{\frac{n\theta}{2}}\right)^2 \left(\frac{\frac{\theta}{2}}{\sin\frac{\theta}{2}}\right)^2 2n^2 \sin^2\frac{\theta}{2}$$

$$\leqq \frac{\pi^2}{4} n^2(1 - \cos\theta),$$

hence

$$\frac{1}{n^2}(1 - r^n\cos(n\theta)) \leqq \frac{1}{n^2}(1 - r^n) + \frac{r^n}{n^2}(1 - \cos n\theta)$$

$$\leqq \frac{1}{n}(1 - r) + \frac{\pi^2}{4} r^n(1 - \cos\theta) \leqq \frac{\pi^2}{4}(1 - r\cos\theta). \quad \square$$

**Remark.** In the above representation $q$ and $\mu$ are independent of the Lévy function $L$, whereas $l$ depends on the choice of $L$.

In the rest of this section, we assume that $S$ is an abelian semigroup with the identical involution. Then negative definite functions are real-valued and $\mathcal{N}^l(S)$ is the set of negative definite functions $\psi$ which are bounded below or equivalently satisfy $\psi(s) \geqq \psi(0)$.

For $f \in \mathbb{R}^S$ and $a \in S$, we define $\Delta_a f \in \mathbb{R}^S$ by

$$\Delta_a f(s) = f(s + a) - f(s), \qquad s \in S,$$

hence $\Delta_a f = (\Gamma_a - I)f$ with the previous notation.

The function $L \equiv 0$ is a Lévy function for $S$, and from the previous results it is easy to get the following:

**3.20. Theorem.** *Let $S$ have the identical involution. The following conditions are equivalent for a function $\psi: S \to \mathbb{R}$:*

(i) $\psi \in \mathcal{N}^l(S)$.
(ii) $\Delta_a \psi \in \mathcal{P}^b(S)$ *for each* $a \in S$.
(iii) *There exist an additive function $q: S \to [0, \infty[$ and a Radon measure $\mu \in M_+(\hat{S}\setminus\{1\})$ such that*

$$\psi(s) = \psi(0) + q(s) + \int_{\hat{S}\setminus\{1\}} (1 - \rho(s))\, d\mu(\rho).$$

*The decomposition in* (iii) *is uniquely determined, $\mu$ being the Lévy measure for $\psi$ and*

$$q(s) = \lim_{n\to\infty} \frac{\psi(ns)}{n}, \qquad s \in S.$$

*If $q \equiv 0$ then $\psi$ is bounded if and only if $\mu(\hat{S}\setminus\{1\}) < \infty$.*

**3.21. Corollary.** *Every $\psi \in \mathcal{N}_+(S)$ satisfies the inequalities*

$$|\psi(s) - \psi(t)| \leqq \psi(s+t) \leqq \psi(s) + \psi(t) \qquad \text{for} \quad s, t \in S.$$

Proof. The set of functions $\psi: S \to [0, \infty[$ satisfying the above inequalities is a closed convex cone $C$ in $\mathbb{R}^S$ containing the nonnegative constants, the additive functions $q: S \to [0, \infty[$ and the functions $1 - \rho$, $\rho \in \hat{S}$. By 3.20 it follows that $\mathcal{N}_+(S) \subseteq C$. □

The following result is inspired by a remark of van Harn and Steutel (1980).

**3.22. Proposition.** *Let $\psi: S \to \mathbb{R}$ and define $\psi_a(s) := \psi(a) + \psi(s) - \psi(a+s)$ for $a \in S$. If $\psi \in \mathcal{N}^l(S)$ then $\psi_a \in \mathcal{N}^b(S)$ for $a \in S$. Conversely, if $\psi_a \in \mathcal{N}^b(S)$ for all $a \in S$ then $\psi \in \mathcal{N}(S)$.*

Proof. If $\psi \in \mathcal{N}^l(S)$ has the representation of (iii) in 3.20 we get

$$\psi_a(s) = \psi(0) + \int_{\hat{S}\setminus\{1\}} (1 - \rho(s))(1 - \rho(a))\, d\mu(\rho),$$

and since $(1 - \rho(a))\mu$ is a finite measure, $\psi_a \in \mathcal{N}^b(S)$.

Conversely, if $\psi_a \in \mathcal{N}^b(S)$ with Lévy measure $\mu_a$, we have by 3.15 that

$$\psi_a(s) = \psi(0) + \int_{\hat{S}\setminus\{1\}} (1 - \rho(s))\, d\mu_a(\rho).$$

Now

$$\Delta_b \psi_a(s) = \Delta_a \psi_b(s) = \psi(a+s) + \psi(b+s) - \psi(s) - \psi(a+b+s)$$

and

$$\Delta_b \psi_a(s) = \int_{\hat{S}\setminus\{1\}} \rho(s)(1 - \rho(b))\, d\mu_a(\rho),$$

$$\Delta_a \psi_b(s) = \int_{\hat{S}\setminus\{1\}} \rho(s)(1 - \rho(a))\, d\mu_b(\rho).$$

By uniqueness of representing measures we get

$$(1 - \rho(b))\mu_a = (1 - \rho(a))\mu_b, \qquad a, b \in S.$$

As in the proof of Lemma 3.12, there exists a unique Radon measure $\mu$ on $\hat{S}\setminus\{1\}$ such that

$$(1 - \rho(a))\mu = \mu_a \qquad \text{for} \quad a \in S,$$

and since $\mu_a$ is finite we have $\mu \in \mathfrak{L}$. The function

$$h(s) := \psi(s) - \psi(0) - \int (1 - \rho(s))\, d\mu(\rho), \qquad s \in S$$

is real-valued, and we get

$$h(a) + h(s) - h(a + s) = \psi_a(s) - \psi(0) - \int (1 - \rho(s))(1 - \rho(a))\, d\mu(\rho)$$

$$= \psi_a(s) - \psi(0) - \int (1 - \rho(s))\, d\mu_a(\rho) = 0,$$

which shows that $h$ is additive, hence $h \in \mathcal{N}(S)$.

The formula

$$\psi(s) = \psi(0) + h(s) + \int (1 - \rho(s))\, d\mu(\rho), \qquad s \in S,$$

shows that $\psi$ is negative definite. □

**3.23. Exercise.** Let $S$ be an abelian semigroup with involution and $l\colon S \to \mathbb{R}$ a function such that $\psi(s) = il(s)$ is negative definite. Show that $l$ is $*$-additive.

**3.24. Exercise.** Let $S$, $T$ be abelian semigroups with involution having Lévy functions. Construct a Lévy function for $S \times T$.

**3.25. Exercise.** Let $S$ be an abelian semigroup with involution and Lévy function. Show that if $\psi \in \mathcal{N}^l(S)$ then $\operatorname{Re} \psi(a) - (\Gamma_a - I)\psi \in \mathcal{N}^b(S)$ for all $a \in S$. Let $\psi\colon S \to \mathbb{C}$ be a hermitian function such that $\operatorname{Re} \psi(a) - (\Gamma_a - I)\psi \in \mathcal{N}^b(S)$ for all $a \in S$. Show that $\psi$ is sum of a quadratic form and a function in $\mathcal{N}^l(S)$. Show finally that $\psi \in \mathcal{N}^l(S)$ if $\operatorname{Re} \psi$ in addition is bounded below.

**3.26. Exercise.** Let $S$ have the identical involution and let $\varphi \in \mathcal{P}_1^b(S)$ be infinitely divisible (cf. 3.2.6). Show the inequalities:

(a) $\varphi(s)\varphi(t) \leqq \varphi(s + t) \leqq \sqrt{\varphi(2s)}\sqrt{\varphi(2t)}$;
(b) $\varphi(s + t)\varphi(s) \leqq \varphi(t)$.

Show that on $S = (\mathbb{R}_+, +)$ inequality (a) holds even without the assumption of $\varphi$ being infinitely divisible.

**3.27. Exercise.** Let $S$ have an absorbing element $\omega$. If $\psi \in \mathcal{N}^l(S)$ and $\psi(0) = 0$, then $\|\psi\|_\infty \leqq 2\psi(\omega)$, i.e. the constant $\sqrt{5}$ in Proposition 3.4 can be amended to 2 in case $\psi(0) = 0$.

**3.28. Exercise.** Let $(S, +, *)$ be an abelian semigroup with involution and let $G \subseteq S$ be a generator set (see 5.7). Show that if $\mu \in M_+(\hat{S}\setminus\{1\})$ satisfies

$$\int_{\hat{S}\setminus\{1\}} (1 - \operatorname{Re} \rho(a))\, d\mu(\rho) < \infty \qquad \text{for all} \quad a \in G$$

then $\mu \in \mathfrak{L}$ (see 3.13).

**3.29. Exercise.** Let $S$ be an abelian semigroup with the identical involution and let $G \subseteq S$ be a generator set. Show the following generalization of Theorem 3.20:

If $\psi: S \to \mathbb{R}$ satisfies $\Delta_a \psi \in \mathscr{P}^b(S)$ for all $a \in G$ then $\psi \in \mathscr{N}^l(S)$.

# §4. Examples of Positive and Negative Definite Functions

**4.1.** $S = (\mathbb{R}_+, +)$. The closed half-line $\mathbb{R}_+$ is an abelian semigroup. As involution we use the identical involution $x = x^*$. Since $\mathbb{R}_+$ is 2-*divisible* (i.e. every $x \in \mathbb{R}_+$ can be written $x = 2y$ for $y \in \mathbb{R}_+$), we see that positive definite functions on $\mathbb{R}_+$ are nonnegative; if they are bounded they are even completely monotone, cf. Corollary 6.8.

For $a \in [0, \infty]$ we define $\rho_a: \mathbb{R}_+ \to \mathbb{R}$ by $\rho_a(s) = e^{-as}$ with the convention that $0 \cdot \infty = 0$, $a \cdot \infty = \infty$ for $a > 0$, i.e. $\rho_\infty = 1_{\{0\}}$. Then $\rho_a$ is a bounded semicharacter, continuous when $a \in [0, \infty[$ and discontinuous when $a = \infty$. Conversely, if $\rho$ is a bounded semicharacter, then $0 \leqq \rho(s) \leqq 1$, which implies that $\rho$ is decreasing. Let $a \in [0, \infty]$ be determined such that $\rho(1) = e^{-a}$. Then it is easy to see that $\rho = \rho_a$, so $a \mapsto \rho_a$ is a bijection of $[0, \infty]$ onto $\hat{S}$, and it is continuous when $[0, \infty]$ is considered as the one-point compactification of $[0, \infty[$. It follows by a well-known theorem from topology that $a \mapsto \rho_a$ is a homeomorphism.

It is not possible to give a similar simple representation of $S^*$. However, if $h: \mathbb{R} \to \mathbb{R}$ is any solution of the functional equation $h(x + y) = h(x) + h(y)$, $x, y \in \mathbb{R}$, then $\exp(h)|\mathbb{R}_+$ is a semicharacter, and all $\rho \in S^* \setminus \{\rho_\infty\}$ are given in this way. By the axiom of choice there exist nonmeasurable solutions $h$ of the functional equation.

An application of Theorem 2.8 leads to the following result:

**4.2. Proposition.** *A function $\varphi: \mathbb{R}_+ \to \mathbb{R}$ is positive definite and bounded if and only if it has the form*

$$\varphi(s) = \int_0^\infty e^{-as}\, d\mu(a) + b\rho_\infty(s), \qquad s \geqq 0,$$

*where $\mu \in M_+^b(\mathbb{R}_+)$ and $b \geqq 0$. The pair $(\mu, b)$ is uniquely determined by $\varphi$.*

An additive function $q: \mathbb{R}_+ \to [0, \infty[$ is necessarily of the form $q(s) = cs$ with $c = q(1) \geqq 0$. By 3.20 we then have:

**4.3. Proposition.** *A function $\psi: \mathbb{R}_+ \to \mathbb{R}$ belongs to $\mathscr{N}^l(\mathbb{R}_+)$ if and only if it has the form*

$$\psi(s) = \psi(0) + cs + b1_{]0, \infty[}(s) + \int_0^\infty (1 - e^{-as})\, d\mu(a), \qquad s \geqq 0,$$

*where $b, c \geqq 0$ and $\mu \in M_+(]0, \infty[)$ are uniquely determined by $\psi$.*

**Remark.** The measures $\mu \in M_+(]0, \infty[)$ which occur in the above formula can be characterized by the single integrability condition

$$\int_0^\infty \frac{x}{1+x}\, d\mu(x) < \infty.$$

The functions of the form

$$f(s) = \gamma + cs + \int_0^\infty (1 - e^{-as})\, d\mu(a), \qquad s \geqq 0,$$

where $c, \gamma \geqq 0$ and $\mu \in M_+(]0, \infty[)$ such that $\int_0^\infty [x/(1+x)]\, d\mu(x) < \infty$, are called *Bernstein functions* by some authors, cf. Berg and Forst (1975). By Proposition 4.3 we see that $f\colon \mathbb{R}_+ \to \mathbb{R}$ is a Bernstein function if and only if it is continuous and belongs to $\mathcal{N}_+(\mathbb{R}_+)$. The expression for a Bernstein function is well defined in the right half-plane $\{z \in \mathbb{C} \mid \operatorname{Re} z \geqq 0\}$. It is continuous there and holomorphic in the interior. Bernstein functions $f$ operate on negative definite kernels $\psi\colon X \times X \to \mathbb{C}$ with $\psi|\Delta \geqq 0$ in the sense that $f \circ \psi$ is again a kernel of this type, cf. Proposition 3.2.9.

**4.4.** The function

$$s \mapsto \int_0^\infty e^{-sa}\, d\mu(a)$$

is called the *Laplace transform* of $\mu$ and is denoted $\mathscr{L}\mu$. This function is actually well defined in the right half-plane $\operatorname{Re} z \geqq 0$ by the formula

$$\mathscr{L}\mu(z) = \int_0^\infty e^{-za}\, d\mu(a),$$

and it is easily seen to be continuous, and holomorphic in the open half-plane $\operatorname{Re} z > 0$. Furthermore,

$$(\mathscr{L}\mu)^{(n)}(z) = \int_0^\infty (-a)^n e^{-za}\, d\mu(a) \qquad \text{for} \quad \operatorname{Re} z > 0, \quad n \geqq 0,$$

so in particular

$$(-1)^n (\mathscr{L}\mu)^{(n)}(x) \geqq 0 \qquad \text{for} \quad x > 0.$$

We have the following corollary of Proposition 4.2.

**4.5. Corollary.** *A function $\varphi\colon \mathbb{R}_+ \to \mathbb{R}$ is continuous, positive definite and bounded if and only if it is the Laplace transform of a measure in $M_+^b(\mathbb{R}_+)$.*

In Theorem 6.13 we will give a completely different characterization of the functions in Corollary 4.5.

**4.6.** $S = (\mathbb{R}_+^k, +)$. For $s = (s_1, \ldots, s_k) \in \mathbb{R}_+^k$ and $a = (a_1, \ldots, a_k) \in [0, \infty]^k$ we put $\langle s, a \rangle = \sum_{j=1}^k s_j a_j$ and define $\rho_a\colon \mathbb{R}_+^k \to \mathbb{R}$ by $\rho_a(s) = e^{-\langle s, a \rangle}$. Then the mapping $a \mapsto \rho_a$ is a homeomorphism and an isomorphism of $([0, \infty]^k, +)$ onto $(\hat{S}, \cdot)$. The semicharacter $\rho_a$ is continuous on $\mathbb{R}_+^k$ when $a \in \mathbb{R}_+^k$ and discontinuous when $a \in [0, \infty]^k \setminus \mathbb{R}_+^k$. Clearly we have results analogous to 4.2 and 4.3. For $\mu \in M_+^b(\mathbb{R}_+^k)$ the function

$$\mathscr{L}\mu(s) = \int_{\mathbb{R}_+^k} e^{-\langle s, a \rangle}\, d\mu(a), \qquad s \in \mathbb{R}_+^k$$

is the ($k$-dimensional) *Laplace transform* of $\mu$. This formula makes sense for $s \in \mathbb{C}_+^k$ and defines a continuous function which is holomorphic in $\{z \in \mathbb{C} \mid \operatorname{Re} z > 0\}^k$.

A $k$-dimensional version of Corollary 4.5 is

**4.7. Proposition.** *A function $\varphi$: $\mathbb{R}_+^k \to \mathbb{R}$ is continuous, positive definite and bounded if and only if it is the Laplace transform of a measure in $M_+^b(\mathbb{R}_+^k)$.*

PROOF. It is enough to prove that a continuous, bounded and positive definite function $\varphi$ is a Laplace transform of a measure in $M_+^b(\mathbb{R}_+^k)$, the converse being obvious. There exists $\mu \in M_+([0, \infty]^k)$ such that

$$\varphi(s) = \int_{[0, \infty]^k} e^{-\langle s, a \rangle}\, d\mu(a), \qquad s \in \mathbb{R}_+^k.$$

If $\pi_j$: $[0, \infty]^k \to [0, \infty]$ denotes the projection onto the $j$th coordinate, we have

$$\varphi(0, \ldots, 0, s_j, 0, \ldots, 0) = \mu(\{a \in [0, \infty]^k \mid a_j = \infty\}) 1_{\{0\}}(s_j) + \int_0^\infty e^{-s_j t}\, d\sigma_j(t),$$

where $\sigma_j$ is the restriction of the image measure $\mu^{\pi_j}$ to $\mathbb{R}_+$, and it follows that $\mu(\{a \in [0, \infty]^k \mid a_j = \infty\}) = 0$ for $j = 1, \ldots, k$, hence $\mu([0, \infty]^k \setminus \mathbb{R}_+^k) = 0$, and $\varphi$ is the Laplace transform of $\mu$. □

**4.8.** $S = (\mathbb{N}_0^k, +)$. Here $\mathbb{N}_0 = \{0, 1, 2, \ldots\}$ and $k \geqq 1$. The involution is the identical mapping. For $x = (x_1, \ldots, x_k) \in \mathbb{R}^k$ the function $\rho_x$: $\mathbb{N}_0^k \to \mathbb{R}$ given by

$$\rho_x(n) = x^n = x_1^{n_1} \ldots x_k^{n_k}, \qquad n = (n_1, \ldots, n_k) \in \mathbb{N}_0^k,$$

is a semicharacter, and it is easy to see that the mapping $x \mapsto \rho_x$ is a topological semigroup isomorphism of $(\mathbb{R}^k, \cdot)$ onto $(\mathbb{N}_0^k)^*$, where the composition in $\mathbb{R}^k$ is $x \cdot y = (x_1 y_1, \ldots, x_k y_k)$. Under this isomorphism $[-1, 1]^k$ corresponds to $\hat{\mathbb{N}}_0^k$.

A measure $\mu \in M_+(\mathbb{R}^k)$ such that $x^n \in \mathscr{L}^1(\mu)$ for all $n \in \mathbb{N}_0^k$ is said to have moments of all orders. The function

$$\varphi(n) = \int x^n\, d\mu(x), \qquad n \in \mathbb{N}_0^k$$

is called the *moment function* of $\mu$, and the number $\varphi(n)$ is called the $n$th *moment* of $\mu$.

The classical moment problem consists in characterizing the set of moment functions. We return to this in Chapter 6. Using that $n \mapsto a^{n_1+\cdots+n_k}$ is an absolute value on $\mathbb{N}_0^k$, for each $a > 0$, we get the following partial solution of the moment problem from Theorem 2.6.

**4.9. Proposition.** *A function $\varphi\colon \mathbb{N}_0^k \to \mathbb{R}$ is positive definite and verifies*

$$|\varphi(n)| \leqq Ca^{n_1+\cdots+n_k} \qquad \text{for} \quad n \in \mathbb{N}_0^k,$$

*where $C$, $a > 0$, if and only if $\varphi$ is a moment function of a measure $\mu \in M_+([-a, a]^k)$, i.e. is of the form*

$$\varphi(n) = \int_{[-a,a]^k} x^n \, d\mu(x) \qquad \text{for} \quad n \in \mathbb{N}_0^k.$$

*In particular $\varphi$ is positive definite and bounded if and only if $\varphi$ is a moment function of a measure $\mu \in M_+([-1, 1]^k)$.*

From Theorem 3.20 we get the following description of $\mathcal{N}^l(\mathbb{N}_0^k)$:

**4.10. Proposition.** *A function $\psi\colon \mathbb{N}_0^k \to \mathbb{R}$ is negative definite and bounded below if and only if*

$$\psi(n) = \psi(0) + \langle n, b\rangle + \int_{[-1,1]^k\setminus\{\mathbf{1}\}} (1 - x^n)\, d\mu(x), \qquad n \in \mathbb{N}_0^k$$

*with $b = (b_1, \ldots, b_k) \in \mathbb{R}_+^k$ and $\mu \in M_+([-1, 1]^k\setminus\{\mathbf{1}\})$, where $\mathbf{1} = (1, \ldots, 1)$.*

**Remark.** The measures $\mu \in M_+([-1, 1]^k\setminus\{\mathbf{1}\})$ which can occur in the above formula are characterized by

$$\int_{[-1,1]^k\setminus\{\mathbf{1}\}} (k - (x_1 + \cdots + x_k))\, d\mu(x) < \infty.$$

In fact, this condition is equivalent with the $k$ conditions

$$\int_{[-1,1]^k\setminus\{\mathbf{1}\}} (1 - x_j)\, d\mu(x) < \infty, \qquad j = 1, \ldots, k,$$

which are clearly necessary. They are also sufficient (cf. 3.28), as is seen from the inequalities

$$0 \leqq 1 - x_1^{n_1} \ldots x_k^{n_k} \leqq \sum_{j=1}^{k} n_j(1 - x_j), \qquad x \in [-1, 1]^k.$$

**4.11.** $S = (\mathbb{N}_0^2, +, *)$. In this case we equip the additive semigroup $(\mathbb{N}_0^2, +)$ with the involution $(n, m)^* = (m, n)$.

For $z \in \mathbb{C}$, the function $\rho_z\colon \mathbb{N}_0^2 \to \mathbb{C}$ given by

$$\rho_z(n, m) = z^n\bar{z}^m \qquad \text{for} \quad (n, m) \in \mathbb{N}_0^2$$

is a semicharacter, and it is easy to see that the mapping $z \mapsto \rho_z$ is a topological semigroup isomorphism of $(\mathbb{C}, \cdot)$ onto $S^*$. Under this isomorphism the unit disc $D = \{z \in \mathbb{C} \mid |z| \leq 1\}$ corresponds to $\hat{S}$.

For a measure $\mu \in M_+(\mathbb{C})$ such that $z^n \bar{z}^m \in \mathscr{L}^1(\mu)$ for all $(n, m) \in \mathbb{N}_0^2$ the function

$$\varphi(n, m) = \int_{\mathbb{C}} z^n \bar{z}^m \, d\mu(z), \qquad (n, m) \in \mathbb{N}_0^2$$

is called the *complex moment function* of $\mu$, and the number $\varphi(n, m)$ is called the $(n, m)$th *complex moment* of $\mu$.

The complex moment problem consists in characterizing the set of complex moment functions. We return to this in Chapter 6, cf. 6.3.5. Using that $(n, m) \mapsto a^{n+m}$ is an absolute value on $\mathbb{N}_0^2$ for each $a > 0$, we get the following partial solution of the complex moment problem from Theorem 2.6.

**4.12. Proposition.** *A complex-valued function $\varphi$ is positive definite on $(\mathbb{N}_0^2, +, *)$ and verifies*

$$|\varphi(n, m)| \leq C a^{n+m} \qquad \textit{for} \quad (n, m) \in \mathbb{N}_0^2,$$

*where $C, a > 0$, if and only if $\varphi$ is a complex moment function of a measure $\mu \in M_+(\{z \in \mathbb{C} \mid |z| \leq a\})$, i.e. is of the form*

$$\varphi(n, m) = \int_{|z| \leq a} z^n \bar{z}^m \, d\mu(z), \qquad (n, m) \in \mathbb{N}_0^2.$$

*In particular $\varphi$ is positive definite and bounded if and only if $\varphi$ is a complex moment function of a measure $\mu \in M_+(D)$.*

In preparation for the Lévy–Khinchin representation in this case we need the following two lemmas.

**4.13. Lemma.** *The $*$-homomorphisms $l\colon (\mathbb{N}_0^2, +, *) \to \mathbb{R}$ are given by*

$$l(n, m) = \alpha(n - m), \qquad \alpha \in \mathbb{R},$$

*and the quadratic forms $q\colon (\mathbb{N}_0^2, +, *) \to \mathbb{R}$ are given by*

$$q(n, m) = \alpha(n + m) + \beta(n - m)^2, \qquad \alpha, \beta \in \mathbb{R}.$$

*Here $q \geq 0$ if and only if $\alpha, \beta \geq 0$.*

PROOF. The assertion about the $*$-homomorphisms is easily established. Let $q$ be a quadratic form and put $a = q(1, 0)$, $b = q(1, 1)$. Since $q(s + t) = q(s) + q(t)$ if $t = t^*$ we have

$$q(n, m) = q(n - m, 0) + mb \qquad \text{if} \quad n \geq m,$$

and from the proof of Theorem 3.9 we know that

$$q(n, 0) = n^2 a - \frac{n(n-1)}{2} b, \qquad n \geqq 0.$$

For $n \geqq m$ we therefore have

$$q(n, m) = (n-m)^2\left(a - \frac{b}{2}\right) + (n+m)\frac{b}{2},$$

and since $q(n, m) = q(m, n)$ this holds for all $(n, m) \in \mathbb{N}_0^2$.

Conversely, it is clear that any function of the form $q(n, m) = \alpha(n + m) + \beta(n - m)^2$, $\alpha, \beta \in \mathbb{R}$, is a quadratic form, which is nonnegative if $\alpha, \beta \geqq 0$; and if $q \geqq 0$ then $q(1, 1) = 2\alpha \geqq 0$ and $\lim_{n\to\infty}(1/n^2)q(n, 0) = \beta \geqq 0$. □

**4.14. Lemma.** *The function* $L\colon \mathbb{N}_0^2 \times D \to \mathbb{R}$ *given by*

$$L((n, m), z) = y(n - m), \qquad z = x + iy$$

*is a Lévy function for* $(\mathbb{N}_0^2, +, *)$. *A measure* $\mu \in M_+(D\setminus\{1\})$ *satisfies* (3) *of* 3.13 *if and only if*

$$\int_{D\setminus\{1\}} (1 - x)\, d\mu(z) < \infty.$$

Proof. The condition given on $\mu$ is clearly necessary. Suppose next that $\mu$ satisfies the given condition. For $(n, m) \in \mathbb{N}_0^2$ such that $n + m > 0$ and $z = x + iy \in D$ we have

$$1 - \operatorname{Re}(z^n \bar{z}^m) = 1 - \sum \binom{n}{p}\binom{m}{q} x^{n+m-p-q}(-1)^q(-1)^{(p+q)/2} y^{p+q},$$

where the sum is over $(p, q)$ such that $p \leqq n$, $q \leqq m$ and $p + q$ is even. For $p = q = 0$ we have the term $x^{n+m}$. In all other terms we have a factor $y^{p+q}$, where $p + q$ is even $\geqq 2$, so $y^{p+q} \leqq (1 - x^2)^{(p+q)/2}$. Therefore

$$0 \leqq 1 - \operatorname{Re}(z^n \bar{z}^m) \leqq 1 - x^{n+m} + \sum \binom{n}{p}\binom{m}{q}(1 - x^2)^{(p+q)/2},$$

which shows that there exists a constant $K > 0$ (depending on $n, m$) such that

$$1 - \operatorname{Re}(z^n \bar{z}^m) \leqq K(1 - x)$$

so (3) of 3.13 holds. (This result also follows from Exercise 3.28.)

The function $L$ clearly satisfies (i) and (ii) of 3.17, and (iii) follows if we establish

$$|1 - z^n \bar{z}^m + i(n - m)y| \leqq K(1 - x) \qquad \text{for} \quad z = x + iy \in D$$

and a suitable constant $K$ depending on $(n, m)$. This is easily seen for $n + m \leqq 1$ and for $n + m \geqq 2$ we find

$$\begin{aligned} 1 - z^n \bar{z}^m + i(n - m)y = 1 - x^{n+m} + iy(n - m)(1 - x^{n+m-1}) \\ + y^2 P(x, y), \end{aligned}$$

where $P$ is a polynomial, hence

$$|1 - z^n\bar{z}^m + i(n-m)y| \leqq (1 - x^{n+m}) + |n-m|(1 - x^{n+m-1}) + (1 - x^2)\max_D |P(x, y)|$$

and the inequality follows. □

We can now give the Lévy–Khinchin representation.

**4.15. Proposition.** *A function* $\psi\colon \mathbb{N}_0^2 \to \mathbb{C}$ *belongs to* $\mathcal{N}^l(\mathbb{N}_0^2)$ *if and only if it has a representation*

$$\psi(n, m) = \psi(0, 0) + ia(n-m) + b(n+m) + c(n-m)^2 + \int_{D\setminus\{1\}} (1 - z^n\bar{z}^m + iy(n-m))\, d\mu(z),$$

*where* $a \in \mathbb{R}$, $b, c \geqq 0$ *and* $\mu \in M_+(D\setminus\{1\})$ *satisfies*

$$\int_{D\setminus\{1\}} (1 - x)\, d\mu(z) < \infty.$$

**4.16. Definition.** A commutative semigroup $(S, +)$ is called *idempotent* if $s + s = s$, for every $s \in S$.

We shall always equip an idempotent semigroup with the identical involution.

Let $(S, +)$ be an idempotent semigroup. We define an ordering $\leqq$ on $S$ by $s \leqq t$ if $s + t = t$. It is easy to see that $\leqq$ is reflexive, transitive and antisymmetric, and that $0 \leqq s$, $s + t = \sup(s, t)$ for $s, t \in S$. If $S$ has a greatest element then it is absorbing and vice versa.

A semicharacter $\rho$ on $S$ is 0–1-valued so $\hat{S} = S^*$. Furthermore, the set $I = \rho^{-1}(\{1\})$ is a subsemigroup which is *hereditary on the left*, i.e. for $s, t \in S$ with $s \leqq t$ and $t \in I$ we have $s \in I$. Conversely, it is easy to see that if $I \subseteq S$ is a subsemigroup and hereditary on the left then $1_I \in S^*$. Therefore $S^*$ is isomorphic with the set $\mathcal{S}$ of subsemigroups which are hereditary on the left, considered as a semigroup under intersection. The neutral element is $S$ and $\{0\}$ is an absorbing element. The topology from $S^*$ transported to $\mathcal{S}$ is the coarsest topology under which the mappings $(\chi_s)_{s\in S}$ from $\mathcal{S}$ to $\{0, 1\}$ are continuous, where

$$\chi_s(I) = \begin{cases} 1 & \text{if } s \in I, \\ 0 & \text{if } s \notin I, \end{cases} \qquad \text{for} \quad I \in \mathcal{S}.$$

A function $\varphi \in \mathcal{P}(S)$ is nonnegative, decreasing and bounded, so, in particular, $\mathcal{P}(S) = \mathcal{P}^b(S)$. In fact $\varphi(s) = \varphi(s + s) \geqq 0$, and if $s \leqq t$ then $\varphi(t)^2 = \varphi(s + t)^2 \leqq \varphi(s + s)\varphi(t + t) = \varphi(s)\varphi(t)$ so $\varphi(t) \leqq \varphi(s)$. In particular, $\varphi(t) \leqq \varphi(0)$. Similarly, $\psi \in \mathcal{N}(S)$ is increasing and satisfies $\psi(0) \leqq \psi(s)$ so $\mathcal{N}(S) = \mathcal{N}^l(S)$.

By specialization of the Theorems 2.8 and 3.20 we get:

**4.17. Proposition.** *Let* $(S, +)$ *be an idempotent semigroup. For* $\varphi \in \mathcal{P}(S)$ *there is a unique* $\mu \in M_+(\mathcal{S})$ *such that*

$$\varphi(s) = \mu(\{I \in \mathcal{S} \mid s \in I\}), \qquad s \in S,$$

*and for* $\psi \in \mathcal{N}(S)$ *there is a unique* $\mu \in M_+(\mathcal{S}\setminus\{S\})$ *such that*

$$\psi(s) = \psi(0) + \mu(\{I \in \mathcal{S}\setminus\{S\} \mid s \notin I\}), \qquad s \in S.$$

A function $\varphi: S \to [0, \infty[$ which is decreasing and bounded need not be positive definite, cf. Exercise 4.25. However, if $S$ has the further property of the order being *total*, i.e. for any $s, t \in S$ either $s \leqq t$ or $t \leqq s$, then we have

**4.18. Proposition.** *Let* $(S, +)$ *be an idempotent semigroup for which the ordering is total. Then* $\varphi: S \to \mathbb{R}$ *is positive (resp. negative) definite if and only if* $\varphi$ *is nonnegative and decreasing (resp. increasing).*

Proof. Let $\varphi: S \to [0, \infty[$ be decreasing and let $s_1, \ldots, s_n \in S$, $c_1, \ldots, c_n \in \mathbb{R}$. By 3.1.2 it is no restriction to assume $s_1 \geqq s_2 \geqq \cdots \geqq s_n$. From probability theory we know the existence of $n$ independent normally distributed random variables $X_1, \ldots, X_n$ with mean zero and variances $\varphi(s_1), \varphi(s_2) - \varphi(s_1), \ldots, \varphi(s_n) - \varphi(s_{n-1})$. (If one of the variances is zero the corresponding normal distribution is degenerate.)

Put $Y_k = \sum_{j=1}^k X_j$, $k = 1, \ldots, n$ and denote the expectation by $\mathbb{E}$; then if $j \leqq k$,

$$\begin{aligned} \mathbb{E}(Y_j Y_k) &= \mathbb{E}(Y_j^2) = \varphi(s_1) + \varphi(s_2) - \varphi(s_1) + \cdots + \varphi(s_j) - \varphi(s_{j-1}) \\ &= \varphi(s_j) = \varphi(s_j + s_k) \end{aligned}$$

implying

$$\sum_{j,k=1}^n c_j c_k \varphi(s_j + s_k) = \mathbb{E}\left\{\left(\sum_{j=1}^n c_j Y_j\right)^2\right\} \geqq 0.$$

If $\varphi: S \to \mathbb{R}$ is increasing then $e^{-t\varphi}$ is nonnegative and decreasing for each $t > 0$, and it follows by Theorem 3.2.2 that $\varphi$ is negative definite. □

**4.19.** As an example of an idempotent semigroup whose ordering is total we consider $S = [0, 1]$, the composition being maximum. The neutral element is zero and 1 is an absorbing element. The ordering of 4.16 is the usual ordering. The set $\mathcal{S}$ of Proposition 4.17 has the following elements: $\{0\}$, $[0, a]$, $[0, a[$ where $a \in ]0, 1]$, and we will now describe the topology on $\mathcal{S}$ induced by the isomorphism between $S^*$ and $\mathcal{S}$.

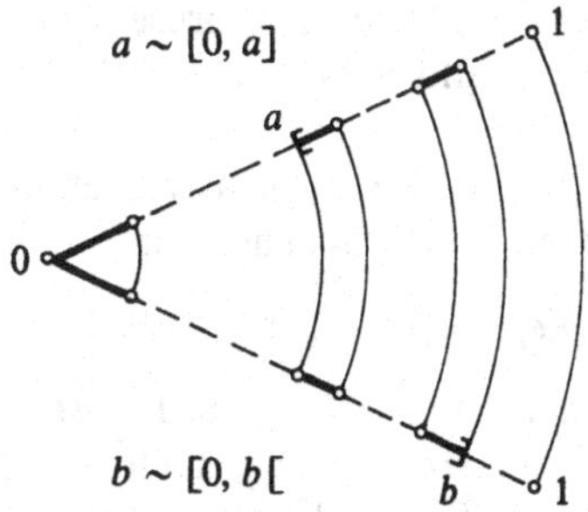


Figure 1

A base of neighbourhoods of $\{0\}$, $[0, a]$ and $[0, b[$, where $a, b \in ]0, 1]$, is given, respectively, by

(i) $\{\{0\}\} \cup \{[0, s] \mid s \in ]0, \varepsilon[\} \cup \{[0, s[ \mid s \in ]0, \varepsilon[\}$ for $0 < \varepsilon < 1$;
(ii) $\{[0, s] \mid s \in [a, a + \varepsilon[ \cap [0, 1]\} \cup \{[0, s[ \mid s \in ]a, a + \varepsilon[ \cap [0, 1]\}$ for $0 < \varepsilon < 1$;
(iii) $\{[0, s] \mid s \in ]b - \varepsilon, b[ \cap [0, 1]\} \cup \{[0, s[ \mid s \in ]b - \varepsilon, b] \cap ]0, 1]\}$ for $0 < \varepsilon < 1$.

We remark that $\mathscr{S}$ is totally ordered under inclusion and that the corresponding order topology on $\mathscr{S}$ is the topology just described. It is useful to think of $\mathscr{S}$ as two copies of $[0, 1]$ glued together in the point 0 as Figure 1 illustrates. On the upper segment $a \in ]0, 1]$ represents $[0, a]$ and on the lower segment $b \in ]0, 1]$ represents $[0, b[$.

We know that $\mathscr{S}$ is a compact Hausdorff space and see that $[0, 1]$ is an isolated point. Furthermore $\mathscr{S}$ is one of those strange nonmetrizable spaces which are first countable (i.e. every point has a countable neighbourhood base). In fact, if $\mathscr{S}$ was metrizable, then a classical theorem from topology (cf. Bauer (1978, p. 217)) implies that the Banach space $C(\mathscr{S})$ of continuous functions on $\mathscr{S}$ is separable, which is not true: For $a \in [0, 1]$ let $\delta_a \in C(\mathscr{S})$ be the function $I \mapsto 1_I(a)$. Then $\|\delta_a - \delta_b\| = 1$ for $a \neq b$, so the open balls $B(\delta_a, \frac{1}{2})$ in $C(\mathscr{S})$, $a \in [0, 1]$ form a noncountable family of disjoint open sets contradicting the separability.

As an application of 4.17 and 4.18 we get

**4.20. Proposition.** *To every decreasing function* $\varphi\colon [0, 1] \to [0, \infty[$ *there exists a unique Radon measure* $\mu$ *on* $\mathscr{S}$ *such that*

$$\varphi(s) = \mu(\{I \in \mathscr{S} \mid s \in I\}), \qquad s \in [0, 1].$$

**4.21. Exercise.** Let $(\mathbb{N}, \cdot)$ be the multiplicative semigroup of natural numbers. Show that the dual semigroup and the restricted dual semigroup are isomorphic to $\mathbb{R}^{\mathbb{P}}$ and $[-1, 1]^{\mathbb{P}}$ where $\mathbb{P} = \{2, 3, 5, \ldots\}$ denotes the set of primes.

**4.22. Exercise.** Let $\varphi \in \mathscr{P}^b(\mathbb{N}_0)$ have the representing measure $\mu$ on $[-1, 1]$. Show that $\sum_0^\infty |\varphi(n)| < \infty$ if and only if $(1 - |t|)^{-1} \in \mathscr{L}^1(\mu)$.

**4.23. Exercise.** Let $A = (a_{ij})$ be a real positive definite $k \times k$ matrix. Show that there exists a function $\varphi \in \mathscr{P}^b(\mathbb{N}_0^k)$ such that

$$a_{ij} = \varphi(e_i + e_j) \qquad \text{for} \quad i, j = 1, \ldots, k,$$

where $e_1 = (1, 0, \ldots, 0), \ldots, e_k = (0, \ldots, 0, 1)$. *Hint*: Assume $|a_{ij}| \leqq 1$. There exists a real $k \times k$ matrix $B = (b_{ij})$ such that $A = B^*B$. Let $x_1, \ldots, x_k \in \mathbb{R}^k$ be the row-vectors of $B$, and define a measure $\mu \in M_+([-1, 1]^k)$ by putting mass 1 at each of the points $x_1, \ldots, x_k$.

**4.24. Exercise.** Let $\varphi\colon [0, 1] \to [0, \infty[$ be decreasing and let $\mu$ be the representing measure on $\mathscr{S}$ according to Proposition 4.20. Show that

$$\varphi(a) - \varphi(a + 0) = \mu(\{[0, a]\}), \qquad 0 \leqq a < 1,$$

$$\varphi(a - 0) - \varphi(a) = \mu(\{[0, a[\}), \qquad 0 < a \leqq 1.$$

Show further that

$$\varphi(s) = \varphi(1) + \sum_{a \in A} \mu(\{[0, a]\}) 1_{[0, a]}(s) + \sum_{b \in B} \mu(\{[0, b[\}) 1_{[0, b[}(s)$$

$$+ \mu_c(\{I \in \mathscr{S} \mid s \in I\}),$$

where $A$ is the at most countable set of $a \in [0, 1[$ such that $\mu(\{[0, a]\}) > 0$, $B$ is the at most countable set of $b \in ]0, 1]$ such that $\mu(\{[0, b[\}) > 0$ and $\mu_c$ is the continuous part of $\mu$. The function $s \mapsto \mu_c(\{I \in \mathscr{S} \mid s \in I\})$ is continuous and decreasing.

**4.25. Exercise.** Let $X$ be a nonempty set. The set $\mathbb{P}(X)$ of subsets of $X$ is an idempotent semigroup under intersection. Describe the semicharacters of $\mathbb{P}(X)$ explicitly in the case where $X$ has two elements, and show that $\varphi = 1 - 1_{\{\varnothing\}}$ is decreasing but not positive definite.

**4.26. Exercise.** On $\mathbb{R}_+$ we define $x \circ y = x + y + xy$. Show that $(\mathbb{R}_+, \circ)$ is an abelian semigroup and that $x \mapsto \log(1 + x)$ is an isomorphism of $(\mathbb{R}_+, \circ)$ onto $(\mathbb{R}_+, +)$. Show that the bounded semicharacters on $(\mathbb{R}_+, \circ)$ are given by

$$x \mapsto \left(\frac{1}{1 + x}\right)^t \qquad \text{for} \quad 0 \leqq t \leqq \infty.$$

**4.27. Exercise.** An infinite matrix $(a_{jk})$ of the form $a_{jk} = \varphi(j + k)$ with $\varphi\colon \mathbb{N}_0 \to \mathbb{R}$ is called a *Hankel matrix*. It is positive (resp. negative) definite if and only if $\varphi$ is positive (resp. negative) definite on the semigroup $(\mathbb{N}_0, +)$, and it is infinitely divisible if and only if $\varphi^t$ is positive definite for all $t > 0$. Show that the so-called Hilbert matrix $(1/(j + k + 1))_{j,k \geqq 0}$ is infinitely divisible.

**4.28. Exercise.** Let $\psi: \mathbb{R}_+ \to \mathbb{R}_+$ be given. Show that $\psi \in \mathcal{N}(\mathbb{R}_+)$ if and only if $\varphi \circ \psi \in \mathcal{P}^b(\mathbb{R}_+)$ for all $\varphi \in \mathcal{P}^b(\mathbb{R}_+)$.

**4.29. Exercise.** Show that if $\psi_1, \psi_2 \in \mathcal{N}_+(\mathbb{R}_+)$ then $\psi_1 \circ \psi_2 \in \mathcal{N}_+(\mathbb{R}_+)$.

**4.30. Exercise.** For $M \subseteq M_+^1(\mathbb{R}_+^k)$ the following three conditions are equivalent:

(i) $M$ is relatively compact with respect to the weak topology.
(ii) $\{\mathcal{L}\mu \mid \mu \in M\}$ is uniformly equicontinuous on $\mathbb{R}_+^k$.
(iii) $\{\mathcal{L}\mu \mid \mu \in M\}$ is equicontinuous in 0.

*Hint*: Use Prohorov's compactness criterion mentioned in the Notes and Remarks to Chapter 2.

**4.31. Exercise.** (Continuity Theorem for Laplace Transforms). Let a sequence $\mu_1, \mu_2, \ldots \in M_+^1(\mathbb{R}_+^k)$ be given such that $\varphi(t) = \lim_{n\to\infty}(\mathcal{L}\mu_n)(t)$ exists for all $t \in \mathbb{R}_+^k$ and assume that $\varphi$ is continuous at 0. Then $\varphi = \mathcal{L}\mu$ for some $\mu \in M_+^1(\mathbb{R}_+^k)$ and $\mu_n \to \mu$ weakly.

**4.32. Exercise.** If $M_+^1(\mathbb{R}_+^k)$ is equipped with the weak topology and $\mathcal{P}_1^b(\mathbb{R}_+^k) \cap C(\mathbb{R}_+^k)$ with the topology of uniform convergence on compact subsets of $\mathbb{R}_+^k$, the Laplace transformation $\mathcal{L}: M_+^1(\mathbb{R}_+^k) \to \mathcal{P}_1^b(\mathbb{R}_+^k) \cap C(\mathbb{R}_+^k)$ is a homeomorphism.

## §5. τ-Positive Functions

In this section, we shall present an approach due to Maserick (1977) to the integral representation of positive definite functions.

Let $A$ be a real or complex commutative algebra with identity $e$ and involution $*$. If $A$ is a real algebra we always assume $a^* = a$ for all $a \in A$.

The basic idea is to find an integral representation of certain "positive" linear functionals on $A$, the representing measure being concentrated on a set of multiplicative linear functionals. The connection to semigroups is obtained if $A$ is chosen to be the algebra of shift operators generated by the shifts $E_s$, $s \in S$, where $(E_s f)(t) = f(s + t)$ for $f \in \mathbb{C}^S$. By the relation $L(E_s) = f(s)$ functions $f$ on $S$ are in one-to-one correspondence with linear functionals $L$ on $A$.

The positivity concept for linear functionals depends on the notion of an admissible subset $\tau$ of $A$.

**5.1. Definition.** A nonempty subset $\tau \subseteq A$ is called *admissible* if

(i) $a^* = a$ for all $a \in \tau$;
(ii) $e - a \in \operatorname{alg\,span}^+(\tau)$ for all $a \in \tau$;
(iii) $\operatorname{alg\,span}(\tau) = A$.

Here alg span($\tau$) (resp. alg span$^+(\tau)$) is the set of linear combinations $\sum_{i=1}^n \lambda_i x_i$, where each $\lambda_i$ is a scalar (resp. $\geqq 0$) and each $x_i$ is a finite product of elements from $\tau$. Note that alg span$^+(\tau)$ is the smallest convex cone in $A$, stable under multiplication and containing $\tau$. Furthermore $e \in$ alg span$^+(\tau)$ because of (ii).

**5.2. Lemma.** *Let $\tau$ be an admissible subset of $A$. Then:*

(i) $e - x_1 \dots x_n \in$ alg span$^+(\tau)$ *for* $x_1, \dots, x_n \in \tau$;
(ii) *for* $x \in$ alg span$^+(\tau)$ *there exists* $\varepsilon(x) > 0$ *such that* $e - \varepsilon x \in$ alg span$^+(\tau)$ *for* $\varepsilon \in [0, \varepsilon(x)]$.

Proof. The assertion (i) follows from the algebraic identity

$$e - \prod_{j=1}^n x_j = \sum_{\sigma \neq 1} \prod_{j=1}^n x_j^{\sigma_j}(e - x_j)^{1-\sigma_j}, \tag{1}$$

where the summation is taken over all $\sigma = (\sigma_1, \dots, \sigma_n) \in \{0, 1\}^n \setminus \{(1, \dots, 1)\}$ (note that $x^0 := e$), and this identity follows in turn by the formula

$$e = \prod_{j=1}^n (x_j + (e - x_j)).$$

If $x = \sum_{j=1}^n \lambda_i y_j$, where $\lambda_j > 0$ and $y_j$ is a product of elements from $\tau$, we put $\varepsilon(x) = (\sum_{j=1}^n \lambda_j)^{-1}$. For $\varepsilon \in [0, \varepsilon(x)]$, we find

$$e - \varepsilon x = \left(1 - \frac{\varepsilon}{\varepsilon(x)}\right)e + \varepsilon \sum_{j=1}^n \lambda_j(e - y_j) \in \text{alg span}^+(\tau)$$

because of (i). □

A linear functional $L: A \to \mathbb{C}$ is called *$\tau$-positive*, where $\tau$ is admissible, if

$$L(a) \geqq 0 \qquad \text{for all} \quad a \in \text{alg span}^+(\tau).$$

This holds if and only if

$$L(a_1 \dots a_n) \geqq 0 \qquad \text{for all finite sets} \quad \{a_1, \dots, a_n\} \subseteq \tau.$$

The set $A_\tau^*$ of $\tau$-positive linear functionals on $A$ is a convex cone in the algebraic dual space $A^*$, closed in the topology $\sigma(A^*, A)$. Note that $L(x^*) = \overline{L(x)}$ for $x \in A$ when $L \in A_\tau^*$. By Lemma 5.2 it follows that $L(e) > 0$ for $L \in A_\tau^* \setminus \{0\}$, so

$$B := \{L \in A_\tau^* \mid L(e) = 1\}$$

is a base for $A_\tau^*$.

**5.3. Lemma.** *The base $B$ is compact.*

Proof. For $x \in A$ there exists a constant $K_x > 0$ such that

$$|L(x)| \leqq K_x L(e) \qquad \text{for} \quad L \in A_\tau^*.$$

In fact, if $x \in \operatorname{alg\,span}^+(\tau)$ we can use $K_x = \varepsilon(x)^{-1}$ with $\varepsilon(x)$ from 5.2(ii), and every $x \in A$ can be written $x = x_1 - x_2 + i(x_3 - x_4)$ with $x_j \in \operatorname{alg\,span}^+(\tau)$, $j = 1, \ldots, 4$. If $(L_\alpha)$ is a universal net on $B$ the above inequality shows that $L(x) := \lim_\alpha L_\alpha(x)$ exists for every $x \in A$, and this implies the compactness of $B$. □

Let $\Delta$ denote the set of $\tau$-positive, multiplicative linear functionals on $A$, which are not identically zero. Clearly $\Delta$ is a compact subset of $B$.

**5.4. Theorem.** *A linear functional $L$ on $A$ is $\tau$-positive if and only if there exists a (necessarily unique) Radon measure $\mu \in M_+(\Delta)$ such that*

$$L(x) = \int_\Delta \delta(x)\, d\mu(\delta) \qquad \text{for} \quad x \in A. \tag{2}$$

*The base $B$ for $A_\tau^*$ is a Bauer simplex with* $\operatorname{ex}(B) = \Delta$.

Proof. A functional of the form (2) is clearly $\tau$-positive. Given $L \in A_\tau^*$, the existence of $\mu$ such that (2) holds follows from Theorem 2.5.6, once we have shown that $\operatorname{ex}(B) \subseteq \Delta$. Suppose $L \in \operatorname{ex}(B)$ and define for $a \in A$ the translate $L_a$ of $L$ by $L_a(x) = L(ax)$. For $a \in \operatorname{alg\,span}^+(\tau)$ we have $L_a \in A_\tau^*$. Since $L = L_a + L_{e-a}$ and $L_a, L_{e-a} \in A_\tau^*$ for $a \in \tau$, $L_a$ is proportional to $L$. The set of elements $a \in A$ for which $L_a$ is proportional to $L$ is a subalgebra of $A$ containing $\tau$, hence $L_a = \lambda_a L$ for all $a \in A$. Evaluating at $x = e$ shows $\lambda_a = L(a)$, hence $L \in \Delta$.

Defining $\gamma: A \to A$ by $\gamma(x) = xx^*$ an application of Corollary 2.5.12 yields $\operatorname{ex}(B) = \Delta$.

The set of functions $\delta \mapsto \delta(x)$, $x \in A$, is a point-separating $*$-subalgebra of $C(\Delta, \mathbb{C})$ containing the constant functions, hence uniformly dense in $C(\Delta, \mathbb{C})$ by the Stone–Weierstrass theorem. This implies that the representing measure is uniquely determined. □

From the integral representation in Theorem 5.4 we immediately get:

**5.5. Corollary.** *Any functional $L \in A_\tau^*$ is positive in the sense that $L(xx^*) \geqq 0$ for all $x \in A$.*

It is interesting to note that the corollary can be proved directly without use of the integral representation. This was done by Maserick and Szafraniec (1984) as follows:

For a function $f$: $[0, 1] \to \mathbb{C}$ and $x \in A$ we define

$$B_n(x; f) = \sum_{k=0}^{n} f\left(\frac{k}{n}\right)\binom{n}{k} x^k (e - x)^{n-k},$$

and for $f(t) = 1, t, t^2$ we find (cf. Davis (1963, p. 109))

$$B_n(x; 1) = e, \qquad B_n(x; t) = x, \qquad B_n(x; t^2) = \frac{1}{n}x + \frac{n-1}{n}x^2.$$

Similarly, for a function $f\colon [0,1]^2 \to \mathbb{C}$ and $x_1, x_2 \in A$ we define

$$B_n(x_1, x_2; f) = \sum_{j,k=0}^{n} f\left(\frac{j}{n}, \frac{k}{n}\right)\binom{n}{j}\binom{n}{k} x_1^j(e - x_1)^{n-j} x_2^k (e - x_2)^{n-k},$$

and for $f(s,t) = (s-t)^2$, we find

$$\begin{aligned} B_n(x_1, x_2; (s-t)^2) &= B_n(x_1, x_2; s^2) + B_n(x_1, x_2; -2st) + B_n(x_1, x_2; t^2) \\ &= B_n(x_1; s^2) - 2B_n(x_1; s)B_n(x_2; t) + B_n(x_2; t^2) \\ &= (x_1 - x_2)^2 + \frac{1}{n}(x_1 + x_2 - x_1^2 - x_2^2). \end{aligned}$$

If $x_1, x_2 \in A$ such that $x_1, x_2, e - x_1, e - x_2 \in \operatorname{alg\,span}^+(\tau)$, we have for $L \in A_\tau^*$

$$0 \leqq L(B_n(x_1, x_2; (s-t)^2)) = L((x_1 - x_2)^2) + \frac{1}{n} L(x_1 + x_2 - x_1^2 - x_2^2),$$

hence for $n \to \infty$: $L((x_1 - x_2)^2) \geqq 0$. For $x_1, x_2 \in \operatorname{alg\,span}^+(\tau)$ there exists $\varepsilon > 0$ such that $\varepsilon x_1, \varepsilon x_2, e - \varepsilon x_1, e - \varepsilon x_2 \in \operatorname{alg\,span}^+(\tau)$, cf. 5.2, and therefore

$$L(\varepsilon^2(x_1 - x_2)^2) \geqq 0,$$

so $L((x_1 - x_2)^2) \geqq 0$. An element $x \in A$ can be written $x = x_1 - x_2 + i(x_3 - x_4)$ with $x_j \in \operatorname{alg\,span}^+(\tau)$, $j = 1, \ldots, 4$, so $xx^* = (x_1 - x_2)^2 + (x_3 - x_4)^2$ and finally $L(xx^*) \geqq 0$.

Let $S$ be a commutative semigroup with involution, and let

$$A = \left\{ \sum_{j=1}^{n} c_j E_{s_j} \,\middle|\, n \in \mathbb{N},\ c_j \in \mathbb{C},\ s_j \in S \right\}$$

be the algebra of shift operators on $\mathbb{C}^S$. We recall that $E_s f(t) = f(s+t)$ for $s, t \in S$, $f \in \mathbb{C}^S$. Clearly, $A$ is a commutative algebra with unit $I = E_0$, and defining

$$\left(\sum_{j=1}^{n} c_j E_{s_j}\right)^* = \sum_{j=1}^{n} \overline{c_j} E_{s_j^*}$$

we get an involution on $A$. Since $(E_s)_{s \in S}$ is a basis for $A$, functions $f\colon S \to \mathbb{C}$ and linear functionals $L\colon A \to \mathbb{C}$ are in one-to-one correspondence via the formula $L(E_s) = f(s)$. If $f$ and $L$ correspond to each other and

$$T = \sum_{j=1}^{n} c_j E_{s_j} \in A$$

then

$$L(T) = \sum_{j=1}^{n} c_j f(s_j) = Tf(0).$$

It follows that

$$L(TT^*) = \sum_{j,k=1}^{n} c_j \overline{c_k} f(s_j + s_k^*),$$

so $L$ is positive if and only if $f \in \mathscr{P}(S)$.

If $\tau \subseteq A$ is admissible, we call $f$ *τ-positive* if the corresponding linear functional $L$ is τ-positive, i.e. if and only if

$$Tf(0) \geqq 0 \qquad \text{for all} \quad T \in \text{alg span}^+(\tau),$$

or if and only if

$$T_1 \ldots T_n f(0) \geqq 0 \qquad \text{for all} \quad T_1, \ldots, T_n \in \tau.$$

Note that a semicharacter $\rho$ is τ-positive if and only if $T\rho(0) \geqq 0$ for all $T \in \tau$, since $T_1 \ldots T_n \rho(0) = T_1\rho(0) \ldots T_n\rho(0)$.

**5.6. Theorem.** *Let $\tau$ be an admissible subset of the algebra $A$ of shift operators. Every τ-positive function $f: S \to \mathbb{C}$ is positive definite, exponentially bounded and has an integral representation*

$$f(s) = \int_{S^*} \rho(s)\, d\mu(\rho),$$

*where $\mu \in M_+(S^*)$ is concentrated on the compact set of τ-positive semicharacters.*

**Proof.** By Theorem 5.4 the linear functional $L$ corresponding to the τ-positive function $f$ has a representation

$$L(T) = \int_\Delta \delta(T)\, d\tilde{\mu}(\delta), \qquad T \in A,$$

where $\tilde{\mu} \in M_+(\Delta)$. For $\delta \in \Delta$ the function $s \mapsto \delta(E_s)$ is a τ-positive semicharacter, and the mapping $j: \Delta \to S^*$ given by $j(\delta)(s) = \delta(E_s)$ is a homeomorphism of $\Delta$ onto the compact set $j(\Delta)$ of τ-positive semicharacters. The image measure $\mu := \tilde{\mu}^j$ of $\tilde{\mu}$ under $j$ is a Radon measure on $S^*$ with compact support contained in $j(\Delta)$, and replacing $T$ by $E_s$ we get

$$f(s) = \int_{S^*} \rho(s)\, d\mu(\rho), \qquad s \in S.$$

This shows that $f$ is exponentially bounded and positive definite. □

**5.7.** A subset $G \subseteq S$ is called a *generator set* for $S$, if every element in $S \setminus \{0\}$ is a finite sum of elements from $G \cup \{a^* \mid a \in G\}$.

Let $\alpha: S \to \mathbb{R}_+$ be an absolute value such that $\alpha(a) > 0$ for all $a \in S$. For $\sigma \in \mathbb{C}$, $a \in S$, we define

$$\Omega_{\sigma,a} = \tfrac{1}{2}\left(I + \frac{\sigma}{2\alpha(a)} E_a + \frac{\bar{\sigma}}{2\alpha(a^*)} E_{a^*}\right).$$

The family

$$\tau_G = \{\Omega_{\sigma,a} | \sigma \in \{\pm 1, \pm i\}, a \in G\}$$

is easily seen to be admissible if $G$ is a generator set.

The following converse of Theorem 5.6 holds:

*If $f$ is positive definite and exponentially bounded there exists an admissible $\tau$ such that $f$ is $\tau$-positive.*

In fact, there exists an absolute value $\alpha$ such that $f$ is $\alpha$-bounded, and we may assume $\alpha(a) > 0$ for all $a \in S$. Then

$$f(s) = \int \rho(s)\, d\mu(\rho), \qquad s \in S,$$

where $\mu \in M_+(S^*)$ is supported by the compact set of $\alpha$-bounded semicharacters. With $\tau = \tau_G$ as above, where $G$ is a generator set, we get

$$\Omega_{\sigma,a} f(s) = \int \rho(s)\tfrac{1}{2}[1 + \alpha(a)^{-1} \operatorname{Re}(\sigma\rho(a))]\, d\mu(\rho),$$

and since $|\rho(a)| \leqq \alpha(a)$ for $\rho \in \operatorname{supp}(\mu)$ it follows that $\Omega_{\sigma,a} f \in \mathscr{P}^\alpha(S)$. Hence $\Omega_{\sigma_1,a_1} \cdots \Omega_{\sigma_n,a_n} f \in \mathscr{P}^\alpha(S)$ for $\sigma_j \in \{\pm 1, \pm i\}$, $a_j \in G$, in particular,

$$\Omega_{\sigma_1,a_1} \cdots \Omega_{\sigma_n,a_n} f(0) \geqq 0$$

showing that $f$ is $\tau_G$-positive.

Using the absolute value $\alpha \equiv 1$ we get "(i) $\Rightarrow$ (ii)" of the following result:

**5.8. Proposition.** *For a function $f: S \to \mathbb{C}$ the following conditions are equivalent*:

(i) $f \in \mathscr{P}^b(S)$;
(ii) *$f$ is $\tau$-positive, where* $\tau = \{\Omega_{\sigma,a} | \sigma \in \{\pm 1, \pm i\}, a \in S\}$ *and*

$$\Omega_{\sigma,a} = \tfrac{1}{2}\left(I + \frac{\sigma}{2} E_a + \frac{\bar{\sigma}}{2} E_{a^*}\right).$$

Proof. To see "(ii) $\Rightarrow$ (i)" it suffices by Theorem 5.6 to verify that a $\tau$-positive semicharacter is bounded. But $\rho \in S^*$ is $\tau$-positive if and only if $1 + \operatorname{Re}(\sigma\rho(s)) \geqq 0$ for $s \in S$, $\sigma \in \{\pm 1, \pm i\}$, i.e. if and only if $\operatorname{Re} \rho(s), \operatorname{Im} \rho(s) \in [-1, 1]$ for all $s \in S$, which is equivalent with $\rho \in \hat{S}$. □

**5.9. Remark.** In the above proposition it is not possible to replace $\tau$ by

$$\tau_G = \{\Omega_{\sigma,a} | \sigma \in \{\pm 1, \pm i\}, a \in G\},$$

where $G$ is a generator set for $S$. The implication "(i) $\Rightarrow$ (ii)" is, of course, still true, but "(ii) $\Rightarrow$ (i)" might fail to hold. In fact, if $S = (\mathbb{N}_0^2, +, *)$ is the semigroup studied in 4.11, then $G = \{(1, 0)\}$ is a generator set and the

$\tau_G$-positive functions are given by

$$f(n, m) = \int_{[-1,1]^2} z^n \bar{z}^m \, d\mu(z), \qquad (n, m) \in \mathbb{N}_0^2,$$

which is a larger class than $\mathscr{P}^b(S)$.

In the case where the involution is the identical this remark does not apply:

**5.10. Proposition.** *Assume $S$ has the identical involution and let $G$ be a generator set for $S$. Then the following conditions are equivalent:*

(i) $f \in \mathscr{P}^b(S)$;
(ii) $(I \pm E_{a_1}) \ldots (I \pm E_{a_n}) f(0) \geqq 0$ *for all* $a_1, \ldots, a_n \in G$, $n \in \mathbb{N}$.

Proof. It is clear that "(i) $\Rightarrow$ (ii)" holds, and (ii) is equivalent with $f$ being $\tau$-positive, where

$$\tau = \{\tfrac{1}{2}(I \pm E_a) \mid a \in G\}$$

is admissible. Finally every $\tau$-positive semicharacter $\rho$ is bounded, because $|\rho(a)| \leqq 1$ for $a \in G$ implies $|\rho(s)| \leqq 1$ for all $s \in S$.

Note that $\Omega_{\pm 1, a} = \frac{1}{2}(I \pm E_a)$ and $\Omega_{\pm i, a} = \frac{1}{2}I$, and the families

$$\{\tfrac{1}{2}(I \pm E_a) \mid a \in G\} \qquad \text{and} \qquad \{\Omega_{\sigma, a} \mid \sigma \in \{\pm 1, \pm i\}, a \in G\}$$

lead to the same $\tau$-positive functions. □

**5.11. Exercise.** Let $S = (\mathbb{N}_0^k, +)$ and let $A^{(k)}$ be the algebra of real polynomials in $k$ variables $x_1, \ldots, x_k$. Show that the set

$$\tau = \{x_1, \ldots, x_k, 1 - x_1 - \cdots - x_k\}$$

is admissible. There is a one-to-one correspondence between functions $f: \mathbb{N}_0^k \to \mathbb{R}$ and linear functionals $L: A^{(k)} \to \mathbb{R}$ established via $L(x^n) = f(n)$, $n \in \mathbb{N}_0^k$. Show that $f$ is $\tau$-positive (in the sense that the corresponding linear functional $L$ is $\tau$-positive) if and only if there exists $\mu \in M_+(K)$ such that

$$f(n) = \int_K x^n \, d\mu(x), \qquad n \in \mathbb{N}_0^k,$$

where $K = \{(x_1, \ldots, x_k)\} \in \mathbb{R}^k \mid x_1 \geqq 0, \ldots, x_k \geqq 0, x_1 + \cdots + x_k \leqq 1\}$.

## §6. Completely Monotone and Alternating Functions

In this section, $S$ is an abelian semigroup with the identical involution. We have already introduced the shift operator $E_a: \mathbb{R}^S \to \mathbb{R}^S$ defined by $E_a f(s) = f(s + a)$ and $\Delta_a = E_a - I$, where $I$ denoted the identity operator. Since $\{E_a \mid a \in S\}$ is a commuting family of operators, the algebra generated by this family is commutative, too; in particular, we have

$$\Delta_a \Delta_b = \Delta_b \Delta_a \qquad \text{for} \quad a, b \in S.$$

Sometimes it is convenient to use also the operators

$$\nabla_a := I - E_a = -\Delta_a, \qquad a \in S,$$

which are, of course, again commuting; note that $\nabla_a \nabla_b = \Delta_a \Delta_b$, and that

$$\nabla_{a_1} \ldots \nabla_{a_n} \rho(s) = \rho(s) \prod_{i=1}^{n} [1 - \rho(a_i)] \tag{1}$$

for $\rho \in S^*$.

In his fundamental work on capacities Choquet (1954) studied functions on $S$ which he called *monotone* (resp. *alternating*) *of infinite order*. Here we shall call these functions *completely monotone* (resp. *completely alternating*).

**6.1. Definition.** A function $\varphi: S \to \mathbb{R}$ is called *completely monotone* if it is nonnegative and if for all finite sets $\{a_1, \ldots, a_n\} \subseteq S$ and $s \in S$

$$\nabla_{a_1} \ldots \nabla_{a_n} \varphi(s) \geqq 0.$$

A function $\psi: S \to \mathbb{R}$ is called *completely alternating* if for all $\{a_1, \ldots, a_n\} \subseteq S$ and $s \in S$

$$\nabla_{a_1} \ldots \nabla_{a_n} \psi(s) \leqq 0.$$

The set of completely monotone (resp. alternating) functions is denoted $\mathscr{M}(S)$ (resp. $\mathscr{A}(S)$). It is clear that $\mathscr{M}(S)$ and $\mathscr{A}(S)$ are closed convex cones in $\mathbb{R}^S$, and the nonnegative (resp. real) constant functions are contained in $\mathscr{M}(S)$ (resp. $\mathscr{A}(S)$).

If $\varphi \in \mathscr{M}(S)$ then $E_a \varphi \in \mathscr{M}(S)$ and, similarly, if $\psi \in \mathscr{A}(S)$ then $E_a \psi \in \mathscr{A}(S)$ for all $a \in S$, since $\nabla_{a_1} \ldots \nabla_{a_n}(E_a f) = E_a \nabla_{a_1} \ldots \nabla_{a_n} f$ for $f \in \mathbb{R}^S$ and since $E_a$ is a positive operator. Notice that $\varphi \in \mathscr{M}(S)$ and $\psi \in \mathscr{A}(S)$ satisfy $0 \leqq \varphi \leqq \varphi(0)$ and $\psi \geqq \psi(0)$.

**6.2. Remark.** The following terminology is frequently used for $s, a_1, \ldots, a_n \in S$:

$$\nabla_1 f(s; a_1) = f(s) - f(s + a_1),$$

and inductively for $n \geqq 2$

$$\nabla_n f(s; a_1, \ldots, a_n) = \nabla_{n-1} f(s; a_1, \ldots, a_{n-1}) - \nabla_{n-1} f(s + a_n; a_1, \ldots, a_{n-1}).$$

Clearly, $\nabla_1 f(\cdot\,; a_1) = \nabla_{a_1} f$ and

$$\nabla_n f(\cdot\,; a_1, \ldots, a_n) = \nabla_{a_1} \ldots \nabla_{a_n} f.$$

The following result connects the two concepts of completely monotone and alternating functions, and it is easily established.

**6.3. Lemma.** *A function $\psi: S \to \mathbb{R}$ belongs to $\mathscr{A}(S)$ if and only if for every $a \in S$ the function $\Delta_a \psi$ belongs to $\mathscr{M}(S)$.*

PROOF. If $\psi \in \mathscr{A}(S)$ then $\Delta_a \psi = -\nabla_a \psi \geqq 0$ and

$$\nabla_{a_1} \dots \nabla_{a_n}(\Delta_a \psi) = \Delta_a \nabla_{a_1} \dots \nabla_{a_n} \psi = -\nabla_a \nabla_{a_1} \dots \nabla_{a_n} \psi \geqq 0.$$

On the other hand, if $\Delta_a \psi \in \mathscr{M}(S)$ for all $a \in S$, then

$$\nabla_{a_1} \dots \nabla_{a_n} \psi = -\nabla_{a_1} \dots \nabla_{a_{n-1}}(\Delta_{a_n} \psi) \leqq 0. \qquad \square$$

The relation of completely monotone functions to $\tau$-positive functions is described in the following:

**6.4. Theorem.** *Let $G \subseteq S$ be a generator set. The family $\tau = \{E_a, I - E_a | a \in G\}$ is admissible, and for a function $\varphi: S \to \mathbb{R}$ the following conditions are equivalent:*

(i) *$\varphi$ is completely monotone.*
(ii) *$\varphi$ is $\tau$-positive.*
(iii) *There exists a measure $\mu \in M_+(\hat{S}_+)$ such that*

$$\varphi(s) = \int_{\hat{S}_+} \rho(s)\, d\mu(\rho), \qquad s \in S.$$

PROOF. For $a_1, \dots, a_n, s_1, \dots, s_m \in G$ we have

$$(I - E_{a_1}) \dots (I - E_{a_n}) E_{s_1} \dots E_{s_m} \varphi(0) = \nabla_{a_1} \dots \nabla_{a_n} \varphi(s_1 + \dots + s_m)$$

so (i) $\Rightarrow$ (ii). The implication "(ii) $\Rightarrow$ (iii)" follows from Theorem 5.6 since $\rho \in S^*$ is $\tau$-positive if and only if $0 \leqq \rho(a) \leqq 1$ for $a \in G$, but this is equivalent with $\rho \in \hat{S}_+$ because every $s \in S \setminus \{0\}$ is a finite sum of elements from $G$.

Finally, if (iii) holds, then $\varphi(s) \geqq 0$ and by (1)

$$\nabla_{a_1} \dots \nabla_{a_n} \varphi(s) = \int_{\hat{S}_+} \rho(s) \prod_{j=1}^{n} (1 - \rho(a_j))\, d\mu(\rho) \geqq 0,$$

so $\varphi \in \mathscr{M}(S)$. $\square$

A function $\varphi \in \mathscr{M}(S)$ is bounded by $\varphi(0)$. The set $\mathscr{M}^1(S)$ of functions $\varphi \in \mathscr{M}(S)$ such that $\varphi(0) = 1$ is therefore a compact convex base for the cone $\mathscr{M}(S)$.

**6.5. Theorem.** *The cone $\mathscr{M}(S)$ is an extreme subset of $\mathscr{P}^b(S)$ and $\mathscr{M}^1(S)$ is a Bauer simplex with $\operatorname{ex}(\mathscr{M}^1(S)) = \hat{S}_+$. For $\varphi_1, \varphi_2 \in \mathscr{M}(S)$ also $\varphi_1 \cdot \varphi_2 \in \mathscr{M}(S)$. A function $\varphi \in \mathscr{P}^b(S)$ is completely monotone if and only if the representing measure $\mu$ is concentrated on $\hat{S}_+$.*

PROOF. The previous theorem shows that $\mathscr{M}(S) \subseteq \mathscr{P}^b(S)$. If $\varphi = \varphi_1 + \varphi_2 \in \mathscr{M}(S)$ and $\varphi_i \in \mathscr{P}^b(S)$ has representing measure $\mu_i \in M_+(\hat{S})$, $i = 1, 2$, then $\mu = \mu_1 + \mu_2$, where $\mu \in M_+(\hat{S}_+)$ is the representing measure for $\varphi$. It follows that $\mu_1, \mu_2$ are concentrated on $\hat{S}_+$ so $\varphi_1, \varphi_2 \in \mathscr{M}(S)$, and we have shown that $\mathscr{M}(S)$ is an extreme subset of $\mathscr{P}^b(S)$.

By transitivity of extremality an extreme point of $\mathscr{M}^1(S)$ is also an extreme point of $\mathscr{P}_1^b(S)$, hence $\text{ex}(\mathscr{M}^1(S)) \subseteq \hat{S}_+$, and in fact there is equality since $\hat{S}_+ \subseteq \mathscr{M}^1(S) \cap \text{ex}(\mathscr{P}_1^b(S))$. For $\mu, \nu \in M_+(\hat{S}_+)$ we have $\text{supp}(\mu * \nu) \subseteq \hat{S}_+$, cf. 2.3, and it follows that $\mathscr{M}(S)$ is stable under multiplication.

By unicity of the representing measure for $\varphi \in \mathscr{P}^b(S)$ it follows that $\mathscr{M}^1(S)$ is a simplex, and that the representing measure for $\varphi$ is concentrated on $\hat{S}_+$ if $\varphi \in \mathscr{P}^b(S)$ is completely monotone. □

In analogy with Theorem 6.4 it suffices to check the defining conditions for a completely alternating function for elements in a generator set.

**6.6. Proposition.** *Let $G$ be a generator set for $S$. A function $\psi: S \to \mathbb{R}$ is completely alternating if and only if*

$$\nabla_{a_1} \dots \nabla_{a_n} \psi(s) \leqq 0 \qquad \textit{for} \quad s \in S \quad \textit{and} \quad a_1, \dots, a_n \in G, \quad n \geqq 1.$$

Proof. Suppose the conditions of the proposition hold. For $s \in S \setminus \{0\}$ there exist $a_1, \dots, a_n \in G$ with $s = a_1 + \dots + a_n$, and the identity (1) in the proof of Lemma 5.2 gives in this case

$$I - E_s = \sum_{\sigma \neq 1} \prod_{j=1}^{n} E_{a_j}^{\sigma_j} (I - E_{a_j})^{1-\sigma_j}.$$

Applying the above procedure to elements $s_1, \dots, s_k \in S \setminus \{0\}$ we find

$$(I - E_{s_1}) \dots (I - E_{s_k})\psi = \nabla_{s_1} \dots \nabla_{s_k} \psi \leqq 0,$$

hence $\psi \in \mathscr{A}(S)$. □

**6.7. Theorem.** *The cone $\mathscr{A}(S)$ is an extreme subset of $\mathscr{N}^l(S)$. A function $\psi \in \mathscr{N}^l(S)$ is completely alternating if and only if the Lévy measure $\mu$ is concentrated on $\hat{S}_+ \setminus \{1\}$.*

Proof. If $\psi \in \mathscr{A}(S)$ we have by 6.3 and 6.5 that $\Delta_a \psi \in \mathscr{P}^b(S)$ for $a \in S$, hence $\psi \in \mathscr{N}^l(S)$ by Theorem 3.20. If $\psi \in \mathscr{A}(S)$ and $\psi = \psi_1 + \psi_2$, where $\psi_1, \psi_2 \in \mathscr{N}^l(S)$, then $\Delta_a \psi = \Delta_a \psi_1 + \Delta_a \psi_2 \in \mathscr{M}(S)$ and $\Delta_a \psi_1, \Delta_a \psi_2 \in \mathscr{P}^b(S)$ for $a \in S$. As $\mathscr{M}(S)$ is extreme in $\mathscr{P}^b(S)$ we get $\Delta_a \psi_1, \Delta_a \psi_2 \in \mathscr{M}(S)$, and by Lemma 6.3 we conclude that $\psi_1, \psi_2 \in \mathscr{A}(S)$, and have thereby shown that $\mathscr{A}(S)$ is extreme in $\mathscr{N}^l(S)$.

Let $\psi \in \mathscr{N}^l(S)$ have the Lévy–Khinchin representation

$$\psi(s) = \psi(0) + q(s) + \int_{\hat{S} \setminus \{1\}} (1 - \rho(s))\, d\mu(\rho).$$

It is easy to see that $\psi(0), q \in \mathscr{A}(S)$ and that $1 - \rho \in \mathscr{A}(S)$ when $\rho \in \hat{S}_+ \setminus \{1\}$, so if $\mu$ is concentrated on $\hat{S}_+ \setminus \{1\}$ then $\psi \in \mathscr{A}(S)$. Let us next suppose that $\psi \in \mathscr{A}(S)$. Since $\mathscr{A}(S)$ is extreme in $\mathscr{N}^l(S)$ we conclude that

$$\psi_a(s) := \int_{T_a} (1 - \rho(s))\, d\mu(\rho)$$

belongs to $\mathscr{A}(S)$, too, where $T_a = \{\rho \in \hat{S} \mid \rho(a) < 0\}$. By Lemma 6.3 we have $\Delta_a \psi_a \in \mathscr{M}(S)$, hence

$$0 \leqq \Delta_a \psi_a(a) = \int_{T_a} \rho(a)(1 - \rho(a))\, d\mu(\rho).$$

But the integrand is strictly negative on $T_a$ and therefore $\mu(T_a) = 0$. Using that $T_a$ is open for every $a \in S$ and $\bigcup_{a \in S} T_a = \hat{S} \setminus \hat{S}_+$ we get $\mu(\hat{S} \setminus \hat{S}_+) = 0$. □

**6.8. Corollary.** *Suppose that S is 2-divisible, i.e. every $s \in S$ is of the form 2t for some $t \in S$. Then $\mathscr{M}(S) = \mathscr{P}^b(S)$ and $\mathscr{A}(S) = \mathscr{N}^l(S)$.*

Proof. If $S$ is 2-divisible every semicharacter is nonnegative, so the assertions follow from 6.5 and 6.7. □

**6.9. Remarks.** (1) It can happen that $\mathscr{M}(S) = \mathscr{P}^b(S)$ without $S$ being 2-divisible. Let $S = \{0, a, b\}$ be the commutative semigroup with neutral element 0 and $a + a = a + b = b + b = a$. Then $S$ is not 2-divisible, but $S^* = \hat{S}$ has two elements $\rho_0 \equiv 1$, $\rho_1(0) = 1$, $\rho_1(a) = \rho_1(b) = 0$ so $\mathscr{M}(S) = \mathscr{P}^b(S) = \mathscr{P}(S)$.

(2) Suppose $G$ is an abelian group considered as a semigroup with the identical involution. Then $\hat{G}$ is the group of homomorphisms of $G$ into the multiplicative group $\{-1, 1\}$. In this case $\mathscr{M}(G) = \mathscr{P}^b(G)$ if and only if $G$ is 2-divisible. In fact, if $G$ is not 2-divisible then $G/G_2$ has at least two elements, where $G_2 = \{2g \mid g \in G\}$, and every element $s \in G/G_2$ satisfies $2s = 0$. If $s \in G/G_2$, $s \neq 0$, then $\varphi\colon \{0, s\} \to \{1, -1\}$ defined by $\varphi(0) = 1$, $\varphi(s) = -1$ is an isomorphism, and it is easily seen (using Zorn's lemma if $G/G_2$ is infinite) that there exists a homomorphism $\tilde{\varphi}\colon G/G_2 \to \{1, -1\}$ extending $\varphi$. If $\pi\colon G \to G/G_2$ is the canonical map, the composition $\rho = \tilde{\varphi} \circ \pi$ belongs to $\hat{G} \setminus \hat{G}_+$ so $\mathscr{P}^b(G) \neq \mathscr{M}(G)$.

By Corollary 6.8 and Theorem 3.2.2 it follows that on a 2-divisible semigroup $S$ a function $\psi\colon S \to \mathbb{R}$ is completely alternating if and only if $\exp(-t\psi)$ is completely monotone for all $t > 0$. The next result shows that we can waive the assumption of 2-divisibility.

**6.10. Proposition.** *Let $\psi\colon S \to \mathbb{R}$. Then $\psi \in \mathscr{A}(S)$ if and only if $\exp(-t\psi) \in \mathscr{M}(S)$ for all $t > 0$.*

Proof. If $\exp(-t\psi) \in \mathscr{M}(S)$ for all $t > 0$ then $1 - \exp(-t\psi) \in \mathscr{A}(S)$ so $\psi = \lim_{t \to 0} t^{-1}(1 - \exp(-t\psi)) \in \mathscr{A}(S)$. For the converse it suffices to prove that $\exp(-\psi) \in \mathscr{M}(S)$ for $\psi \in \mathscr{A}(S)$ with the representation

$$\psi(s) = \psi(0) + q(s) + \int_{\hat{S}_+ \setminus \{1\}} (1 - \rho(s))\, d\mu(\rho).$$

Since $\mathscr{M}(S)$ is closed under multiplication and $\exp(-q) \in \hat{S}_+$, it suffices to prove that

$$s \mapsto \exp\left[-\int_{\hat{S}_+\setminus\{1\}} (1 - \rho(s))\, d\mu(\rho)\right]$$

belongs to $\mathscr{M}(S)$. Approximating $\mu$ by measures with compact support (2.3.9) and approximating these by molecular measures (2.3.5) the problem is reduced to the implication

$$\rho \in \hat{S}_+,\ c \geqq 0 \Rightarrow \exp(c\rho) \in \mathscr{M}(S).$$

This is in fact true because $\rho \in \mathscr{M}(S)$ and

$$\exp(c\rho) = \sum_{n=0}^{\infty} \frac{1}{n!} c^n \rho^n. \qquad \square$$

For the case $S = \mathbb{N}_0^k$ we have that $\hat{S}_+$ is homeomorphic with $[0, 1]^k$ via the mapping $x \mapsto \rho_x$, $\rho_x(n) = x^n = x_1^{n_1} \ldots x_k^{n_k}$, where $x = (x_1, \ldots, x_k) \in [0, 1]^k$, $n = (n_1, \ldots, n_k) \in \mathbb{N}_0^k$, cf. 4.8. The set $G = \{e_1, \ldots, e_k\}$, where $e_j = (0, \ldots, 0, 1, 0, \ldots, 0)$ with 1 on the $j$th place, is a generator set for $\mathbb{N}_0^k$. We put $E_j = E_{e_j}, j = 1, \ldots, k$, and

$$\binom{n}{p} = \binom{n_1}{p_1} \cdots \binom{n_k}{p_k}, \qquad |p| = p_1 + \cdots + p_k \qquad \text{for} \quad n, p \in \mathbb{N}_0^k.$$

Specializing 6.4, 6.6, 6.7 and 4.10 we get the following two results, the first of which is due to Hildebrandt and Schoenberg (1933); in the one-dimensional case it is the solution of the so-called Hausdorff moment problem.

**6.11. Proposition.** *For a function $\varphi\colon \mathbb{N}_0^k \to \mathbb{R}$ the following conditions are equivalent:*

(i) *$\varphi$ is completely monotone.*
(ii) $(I - E_1)^{n_1} \ldots (I - E_k)^{n_k} \varphi(m_1, \ldots, m_k)$

$$= \sum_{0 \leqq p \leqq n} (-1)^{|p|} \binom{n}{p} \varphi(m + p) \geqq 0 \qquad \textit{for all} \quad n, m \in \mathbb{N}_0^k.$$

(iii) *There exists $\mu \in M_+([0, 1]^k)$ such that*

$$\varphi(n) = \int_{[0,1]^k} x^n \, d\mu(x), \qquad n \in \mathbb{N}_0^k.$$

**6.12. Proposition.** *For a function $\psi\colon \mathbb{N}_0^k \to \mathbb{R}$ the following conditions are equivalent:*

(i) *$\psi$ is completely alternating.*
(ii) $(I - E_1)^{n_1} \ldots (I - E_k)^{n_k} \psi(m_1, \ldots, m_k)$

$$= \sum_{0 \leqq p \leqq n} (-1)^{|p|} \binom{n}{p} \psi(m + p) \leqq 0 \qquad \textit{for all} \quad m \in \mathbb{N}_0^k, \quad n \in \mathbb{N}_0^k \setminus \{0\}.$$

(iii) *There exist* $a \in \mathbb{R}$, $b \in \mathbb{R}_+^k$ *and* $\mu \in M_+([0,1]^k \setminus \{\mathbf{1}\})$ *such that*

$$\psi(n) = a + \langle n, b\rangle + \int_{[0,1]^k\setminus\{\mathbf{1}\}} (1 - x^n)\, d\mu(x).$$

In the case of the 2-divisible semigroup $S = \mathbb{R}_+$ the continuous functions in $\mathscr{M}(\mathbb{R}_+) = \mathscr{P}^b(\mathbb{R}_+)$ can be characterized in yet another way, cf. 4.5:

**6.13. Theorem.** *For a continuous function* $\varphi\colon \mathbb{R}_+ \to \mathbb{R}$ *the following conditions are equivalent*:

(i) $\varphi \in \mathscr{M}(\mathbb{R}_+)$.
(ii) $\varphi \in \mathscr{P}^b(\mathbb{R}_+)$.
(iii) *There exists* $\mu \in M_+^b(\mathbb{R}_+)$ *such that* $\varphi = \mathscr{L}\mu$.
(iv) $\varphi$ *is* $C^\infty$ *on* $]0, \infty[$ *and* $(-1)^n\varphi^{(n)}(s) \geqq 0$ *for* $n \geqq 0$, $s > 0$.

Proof. We have already established the equivalence of (i), (ii) and (iii), and "(iii) ⇒ (iv)" follows from 4.4. Suppose (iv) holds. For $a \geqq 0$ the function $\nabla_a\varphi$ is again continuous on $[0, \infty[$ and satisfies (iv). In fact, for $n \geqq 0$, $s > 0$ the mean value theorem gives

$$\begin{aligned}(-1)^n(\nabla_a\varphi)^{(n)}(s) &= (-1)^n[\varphi^{(n)}(s) - \varphi^{(n)}(s+a)]\\ &= (-1)^{n+1}a\varphi^{(n+1)}(\xi) \geqq 0\end{aligned}$$

for $\xi$ between $s$ and $a + s$. By iteration we find that $f := \nabla_{a_1}\ldots\nabla_{a_n}\varphi$ satisfies (iv) whenever $a_1, \ldots, a_n \geqq 0$, in particular, $f(s) \geqq 0$ for $s > 0$, and by continuity $f(s) \geqq 0$ for $s \geqq 0$, hence $\varphi \in \mathscr{M}(\mathbb{R}_+)$. □

The equivalence of (iii) and (iv) is a famous result of Bernstein, cf. Widder (1941), and can be given a slightly more general formulation:

**6.14. Corollary.** *For a function* $\varphi\colon ]0, \infty[ \to \mathbb{R}$ *the following conditions are equivalent*:

(i) *There exists* $\mu \in M_+(\mathbb{R}_+)$ *such that* $\varphi = \mathscr{L}\mu$.
(ii) $\varphi$ *is* $C^\infty$ *and* $(-1)^n\varphi^{(n)} \geqq 0$ *for* $n \geqq 0$.

*If* (i) *and* (ii) *are satisfied* $\lim_{s\to 0}\varphi(s) = \mu(\mathbb{R}_+) \leqq \infty$.

Proof. It is easy to see that (i) ⇒ (ii). Suppose that (ii) holds and let $h > 0$. The function $E_h\varphi$ is continuous on $\mathbb{R}_+$ and satisfies (iv) of 6.13, hence of the form $\mathscr{L}(\mu_h)$ for a uniquely determined $\mu_h \in M_+^b(\mathbb{R}_+)$. For $h, k > 0$ we then have

$$\varphi(s + h + k) = \int_0^\infty e^{-(s+k)x}\, d\mu_h(x) = \int_0^\infty e^{-sx}\, d\mu_{h+k}(x), \qquad s \geqq 0,$$

so by uniqueness of the representing measure

$$e^{-kx}\mu_h = e^{-hx}\mu_k = \mu_{h+k}.$$

Defining the measure $\mu$ on $[0, \infty[$ by $\mu = e^{hx}\mu_h$, which is independent of $h > 0$, we find

$$\varphi(s) = \int_0^\infty e^{-sx}\, d\mu(x), \qquad s > 0,$$

hence $\lim_{s\to 0} \varphi(s) = \mu(\mathbb{R}_+)$. □

The functions of the corollary are, in the literature, called completely monotone functions. In Chapter 8 we will see that they are the completely monotone functions on the semigroup without zero element $(]0, \infty[, +)$.

**6.15.** The abstract theory of capacities introduced by Choquet (1954) fits into the theory of completely monotone and alternating functions.

Let $X$ be a locally compact space and let $\mathscr{K} = \mathscr{K}(X)$ denote the set of compact subsets of $X$. Then $(\mathscr{K}, \cup)$ is an idempotent semigroup with $\varnothing$ as zero element and the induced ordering is inclusion. The set $\mathscr{S}$ of subsemigroups $I \subseteq \mathscr{K}$ which are hereditary on the left, i.e. families $I$ of compact sets verifying

$$K, L \in I \Rightarrow K \cup L \in I \qquad \text{and} \qquad K \subseteq L,\ \ L \in I \Rightarrow K \in I,$$

is a semigroup under intersection, and $I \mapsto 1_I$ is an isomorphism of $\mathscr{S}$ onto $\mathscr{K}^*$, cf. 4.16, 4.17. Defining

$$\tilde{K} = \{I \in \mathscr{S} \mid K \in I\} \qquad \text{for} \quad K \in \mathscr{K}$$

and equipping $\mathscr{S}$ with the coarsest topology in which the sets $\tilde{K}$ and $\mathscr{S}\backslash\tilde{K}$ are open (and hence clopen) for all $K \in \mathscr{K}$, the mapping $I \mapsto 1_I$ is a homeomorphism.

A function $\varphi\colon \mathscr{K} \to \mathbb{R}$ will be called *continuous on the right* if for every $K \in \mathscr{K}$ and $\varepsilon > 0$ there is an open neighbourhood $G$ of $K$ such that $|\varphi(K) - \varphi(L)| \leq \varepsilon$ for all $L \in \mathscr{V}(K, G) = \{L \in \mathscr{K} \mid K \subseteq L \subseteq G\}$. Note that the functions continuous on the right can be considered as the functions on $\mathscr{K}$ continuous in the coarsest topology for which the sets $\mathscr{V}(K, G)$ are open. This topology is not Hausdorff and $\varnothing$ is an isolated point. We see below that the completely monotone and alternating functions on $\mathscr{K}$ which are continuous on the right admit an integral representation over the set of semicharacters on $\mathscr{K}$ which are continuous on the right.

We first show that the set of semicharacters on $\mathscr{K}$ which are continuous on the right can be identified with the set $\mathscr{F} = \mathscr{F}(X)$ of closed subsets of $X$.

Let $\mathscr{S}_r$ denote the set of $I \in \mathscr{S}$ for which $1_I$ is continuous on the right.

**6.16. Lemma.** *For $F \in \mathscr{F}$ let $I_F = \{K \in \mathscr{K} \mid K \cap F = \varnothing\}$. Then $F \mapsto I_F$ is a bijection of $\mathscr{F}$ onto $\mathscr{S}_r$ and the inverse mapping is $I \mapsto X\backslash \bigcup_{K\in I} \mathring{K}$.*

Proof. It is obvious that $I_F \in \mathscr{S}$ and easy to see that $1_{I_F}$ is continuous on the right because a compact set $K$ not intersecting $F$ has an open neighbourhood which does not intersect $F$. Conversely, if $I \in \mathscr{S}_r$ we have

$$\bigcup_{K \in I} \mathring{K} = \bigcup_{K \in I} K,$$

for if $K \in I$ the continuity on the right implies that $I$ contains a compact neighbourhood of $K$. Therefore $F := X \setminus \bigcup_{K \in I} \mathring{K}$ is closed and $I \subseteq I_F$. For $L \in I_F$ we have $L \subseteq \bigcup_{K \in I} \mathring{K}$ so there exist finitely many $K_1, \ldots, K_n \in I$ with $L \subseteq \mathring{K}_1 \cup \cdots \cup \mathring{K}_n \subseteq K_1 \cup \cdots \cup K_n$, and $I$ being stable under finite unions and hereditary on the left we see that $L \in I$, hence $I = I_F$. Finally, if $F_1, F_2$ are different closed sets and $x \in F_1 \setminus F_2$ then $\{x\} \in I_{F_2} \setminus I_{F_1}$. □

We will now introduce a topology on $\mathscr{F}$. Let $\mathscr{G} = \mathscr{G}(X)$ denote the set of open subsets of $X$ and define for a subset $B \subseteq X$

$$\mathscr{F}^B = \{F \in \mathscr{F} \mid F \cap B = \varnothing\},$$
$$\mathscr{F}_B = \{F \in \mathscr{F} \mid F \cap B \neq \varnothing\} = \mathscr{F} \setminus \mathscr{F}^B.$$

We equip $\mathscr{F}$ with the coarsest topology in which the sets $\mathscr{F}^K$, $K \in \mathscr{K}$ and $\mathscr{F}_G$, $G \in \mathscr{G}$ are open. Since $\mathscr{F}^{K_1} \cap \mathscr{F}^{K_2} = \mathscr{F}^{K_1 \cup K_2}$ we see that sets $\mathscr{F}^K \cap \mathscr{F}_{G_1} \cap \cdots \cap \mathscr{F}_{G_n}$ with $K \in \mathscr{K}$, $G_1, \ldots, G_n \in \mathscr{G}$ form a base for the topology on $\mathscr{F}$. It is easy to see that $\mathscr{F}$ is a Hausdorff space, and using Alexander's subbase theorem it can be proved that $\mathscr{F}$ is compact, cf. Matheron (1975). The compactness is, however, also a consequence of the following result.

**6.17. Proposition.** *The mapping $c\colon \mathscr{S} \to \mathscr{F}$ defined by*

$$c(I) = X \setminus \bigcup_{K \in I} \mathring{K}, \qquad I \in \mathscr{S}$$

*is continuous and maps $\mathscr{S}_r$ bijectively onto $\mathscr{F}$.*

PROOF. By Lemma 6.16 we already know the last statement. For the continuity of $c$ it suffices to prove that $c^{-1}(\mathscr{F}^K)$ and $c^{-1}(\mathscr{F}_G)$ are open in $\mathscr{S}$ for $K \in \mathscr{K}$ and $G \in \mathscr{G}$. We first remark the following biimplication for $I \in \mathscr{S}$ and $K \in \mathscr{K}$:

$$c(I) \cap K = \varnothing \Leftrightarrow \exists L \in I\colon K \subseteq \mathring{L}. \tag{2}$$

Here "$\Leftarrow$" is clear, and if $c(I) \cap K = \varnothing$ there exist by compactness $K_1, \ldots, K_n \in I$ such that $K \subseteq \mathring{K}_1 \cup \cdots \cup \mathring{K}_n$ and then $K \subseteq \mathring{L}$ with $L = K_1 \cup \cdots \cup K_n \in I$. By (2) it follows immediately that

$$c^{-1}(\mathscr{F}^K) = \bigcup_{\substack{L \in \mathscr{K} \\ K \subseteq \mathring{L}}} \tilde{L} \tag{3}$$

which shows that $c^{-1}(\mathscr{F}^K)$ is open in $\mathscr{S}$.

Similarly, we have the following biimplication for $I \in \mathscr{S}$ and $G \in \mathscr{G}$:

$$c(I) \cap G \neq \varnothing \Leftrightarrow \exists K \in \mathscr{K} \setminus I\colon K \subseteq G,$$

hence

$$c^{-1}(\mathscr{F}_G) = \bigcup_{\substack{K \in \mathscr{K} \\ K \subseteq G}} (\mathscr{S} \setminus \tilde{K}), \tag{4}$$

showing that $c^{-1}(\mathscr{F}_G)$ is open in $\mathscr{S}$. □

**6.18. Theorem.** *For every $\varphi \in \mathscr{M}(\mathscr{K}, \cup)$ which is continuous on the right there exists a unique $\mu \in M_+(\mathscr{F})$ such that*

$$\varphi(K) = \mu(\mathscr{F}^K) \qquad \text{for} \quad K \in \mathscr{K},$$

*and for every $\mu \in M_+(\mathscr{F})$ this formula defines a function $\varphi \in \mathscr{M}(\mathscr{K}, \cup)$ which is continuous on the right.*

PROOF. A decreasing function $\varphi\colon \mathscr{K} \to \mathbb{R}$ is continuous on the right if and only if

$$\varphi(K) = \sup\{\varphi(L) \mid L \in \mathscr{K}, K \subseteq \mathring{L}\} \tag{5}$$

for all $K \in \mathscr{K}$.

Suppose first that $\varphi\colon \mathscr{K} \to \mathbb{R}$ is completely monotone and continuous on the right, in particular, decreasing. By Proposition 4.17, there exists $\nu \in M_+(\mathscr{S})$ such that

$$\varphi(K) = \nu(\{I \in \mathscr{S} \mid K \in I\}) = \nu(\tilde{K})$$

for $K \in \mathscr{K}$. Using that the family $\tilde{L}$, $L \in \mathscr{K}$, $K \subseteq \mathring{L}$ is upwards filtering we get by (3) and Theorem 2.1.5 that

$$\sup\{\nu(\tilde{L}) \mid L \in \mathscr{K}, K \subseteq \mathring{L}\} = \nu(c^{-1}(\mathscr{F}^K)),$$

hence $\varphi(K) = \mu(\mathscr{F}^K)$ where $\mu = \nu^c$ is the image measure of $\nu$ under the continuous mapping $c\colon \mathscr{S} \to \mathscr{F}$.

Conversely, if $\mu \in M_+(\mathscr{F})$ then $\varphi(K) = \mu(\mathscr{F}^K)$ is a positive decreasing function satisfying (5) because the family $\mathscr{F}^L$, $L \in \mathscr{K}$, $K \subseteq \mathring{L}$ is upwards filtering with union $\mathscr{F}^K$. To see that $\varphi$ is completely monotone we use the following formula which is easily established by induction:

$$\nabla_{K_1}\nabla_{K_2}\cdots\nabla_{K_n}\varphi(K) = \mu(\mathscr{F}^K \cap \mathscr{F}_{K_1} \cap \cdots \cap \mathscr{F}_{K_n})$$

for $K, K_1, \ldots, K_n \in \mathscr{K}$. By Theorem 2.1.5 we have for $K \in \mathscr{K}$, $G_1, \ldots, G_n \in \mathscr{G}$

$$\begin{aligned}
&\mu(\mathscr{F}^K \cap \mathscr{F}_{G_1} \cap \cdots \cap \mathscr{F}_{G_n}) \\
&\quad = \sup\{\mu(\mathscr{F}^K \cap \mathscr{F}_{\mathring{K}_1} \cap \cdots \cap \mathscr{F}_{\mathring{K}_n}) \mid K_i \in \mathscr{K}, K_i \subseteq G_i, i = 1, \ldots, n\} \\
&\quad = \sup\{\mu(\mathscr{F}^K \cap \mathscr{F}_{K_1} \cap \cdots \cap \mathscr{F}_{K_n}) \mid K_i \in \mathscr{K}, K_i \subseteq G_i, i = 1, \ldots, n\}.
\end{aligned}$$

If $\mu_1, \mu_2 \in M_+(\mathscr{F})$ agree on the sets $\mathscr{F}^K$, $K \in \mathscr{K}$, the above formulas show that they agree on all finite intersections of the sets $\mathscr{F}^K$, $K \in \mathscr{K}$ and $\mathscr{F}_G$, $G \in \mathscr{G}$, hence on the algebra $\mathscr{A}$ of sets they generate. Since any open set in $\mathscr{F}$ is union of an upwards filtering family of open sets from $\mathscr{A}$, another application of 2.1.5 shows that $\mu_1$ and $\mu_2$ agree on the open sets in $\mathscr{F}$, hence $\mu_1 = \mu_2$. □

In order to derive the corresponding integral representation for the set of completely alternating functions which are continuous on the right, we first notice that $c(I) = \varnothing$ for $I \in \mathscr{S}$ if and only if $I = \mathscr{K}$. Therefore, $c$ induces a continuous mapping of the locally compact space $\mathscr{S}\setminus\{\mathscr{K}\}$ onto the locally compact space $\mathscr{F}\setminus\{\emptyset\}$ and this mapping, still denoted $c$, is proper. It follows by Proposition 2.1.15 that every $\nu \in M_+(\mathscr{S}\setminus\{\mathscr{K}\})$ has a Radon image measure $\mu = \nu^c$ and every $\mu \in M_+(\mathscr{F}\setminus\{\emptyset\})$ is of the form $\nu^c$ for some $\nu \in M_+(\mathscr{S}\setminus\{\mathscr{K}\})$, cf. 2.2.13.

A completely alternating function $\psi: \mathscr{K} \to \mathbb{R}$ which is continuous on the right and satisfies $\psi(\emptyset) = 0$ is called a (completely alternating) *capacity* on $X$. By exactly the same method as in the preceding theorem it is possible to get the following result which contains Choquet's integral representation of the completely alternating capacities on $X$ (Choquet 1954, Chap. VII). The details are left to the reader.

**6.19. Theorem.** *For every $\psi \in \mathscr{A}(\mathscr{K}, \cup)$ which is continuous on the right there is a unique $\mu \in M_+(\mathscr{F}\setminus\{\emptyset\})$ such that*

$$\psi(K) = \psi(\emptyset) + \mu(\mathscr{F}_K), \qquad K \in \mathscr{K},$$

*and for every $\mu \in M_+(\mathscr{F}\setminus\{\emptyset\})$ this formula defines a function $\psi \in \mathscr{A}(\mathscr{K}, \cup)$ which is continuous on the right.*

**6.20. Exercise.** Show that $\nabla_a(fg) = \nabla_a f \cdot E_a g + f \cdot \nabla_a g$ for $f, g \in \mathbb{R}^S$. Use this for another proof that $\mathscr{M}(S)$ is stable under multiplication.

**6.21. Exercise.** Let $\varphi_1, \varphi_2 \in \mathscr{P}^b_+(S)$ and assume $\varphi_2(s) > 0$ for all $s$. Show that if $\varphi_1\varphi_2 \in \mathscr{M}(S)$ then $\varphi_1 \in \mathscr{M}(S)$. Show that if $\varphi \in \mathscr{P}^b_+(S)$ is such that $\varphi^k \in \mathscr{M}(S)$ for some $k \in \mathbb{N}$ then $\varphi \in \mathscr{M}(S)$.

**6.22. Exercise.** Assume $\varphi \in \mathscr{P}^b(S)$, $\psi \in \mathscr{N}^l(S)$. Show that $\varphi$ is completely monotone if and only if $n \mapsto \varphi(na)$ is completely monotone on $\mathbb{N}_0$ for each $a \in S$, and that $\psi$ is completely alternating if and only if $n \mapsto \psi(na)$ is completely alternating on $\mathbb{N}_0$ for each $a \in S$.

**6.23. Exercise.** Given $\varphi \in \mathscr{P}^b(\mathbb{N}_0)$ define $T\varphi: \mathbb{N}_0 \to \mathbb{R}$ by $T\varphi(0) = 0$, $T\varphi(n) := \sum_{j=0}^{n-1} \varphi(j)$ for $n \geqq 1$. Then $T$ is an affine isomorphism from $\mathscr{P}^b(\mathbb{N}_0)$ onto $\{\psi \in \mathscr{N}_+(\mathbb{N}_0) \mid \psi(0) = 0\}$, mapping $\mathscr{M}(\mathbb{N}_0)$ onto $\{\psi \in \mathscr{A}(\mathbb{N}_0) \mid \psi(0) = 0\}$.

**6.24. Exercise.** Show that

$$\varphi(n) = \frac{1}{n+1} - \log\frac{n+2}{n+1} = \int_0^1 x^n\left(1 + \frac{1-x}{\log x}\right) dx, \qquad n \geqq 0$$

is completely monotone on $(\mathbb{N}_0, +)$, and that

$$\psi(n) = 1 + \frac{1}{2} + \cdots + \frac{1}{n} - \log(n+1), \qquad n \geqq 0$$

$(\psi(0) := 0)$ is completely alternating with the Lévy measure

$$\mu = \left(\frac{1}{1-t} + \frac{1}{\log t}\right) 1_{]0,1[}(t)\, dt,$$

which is finite. It foilows that $\gamma := \lim_{n\to\infty} \psi(n)$ exists and that

$$\gamma = \int_0^1 \left(\frac{1}{1-t} + \frac{1}{\log t}\right) dt < \infty.$$

The number $\gamma$ is called Euler's constant.

**6.25. Exercise.** Let $\varphi \in \mathscr{P}^b(S)$. Show that the function $\varphi(2s)$ belongs to $\mathscr{M}(S)$.

**6.26. Exercise.** Try to find a decreasing nonnegative positive definite function on $\mathbb{N}_0$ which is not completely monotone.

**6.27. Exercise.** Show that Theorem 6.13 can be extended to $k$ dimensions: For a continuous function $\varphi\colon \mathbb{R}_+^k \to \mathbb{R}$ the following conditions are equivalent:

(i) $\varphi \in \mathscr{M}(\mathbb{R}_+^k)$;
(ii) $\varphi \in \mathscr{P}^b(\mathbb{R}_+^k)$;
(iii) $\varphi = \mathscr{L}\mu$ for some $\mu \in M_+^b(\mathbb{R}_+^k)$;
(iv) $\varphi$ is $C^\infty$ on $]0, \infty[^k$ and $(-1)^{|\alpha|}D^\alpha\varphi(s) \geqq 0$ for $\alpha \in \mathbb{N}_0^k$, $s \in ]0, \infty[^k$.

Extend Corollary 6.14 to $k$ dimensions.

**6.28. Exercise.** Let $\varphi\colon ]0, \infty[ \to \mathbb{R}$ be the Laplace transform of $\mu \in M_+(\mathbb{R}_+)$. Then for each $x \in ]0, \infty[$ the sequence $n \mapsto \varphi^{(n)}(x)$ belongs to $\mathscr{P}(\mathbb{N}_0)$.

**6.29. Exercise.** A function $\varphi\colon S \to \mathbb{R}$ is called *absolutely monotone* if $\varphi \geqq 0$ and $\Delta_{a_1} \cdots \Delta_{a_n}\varphi \geqq 0$ for all $a_1, \ldots, a_n \in S$; and $\psi\colon S \to \mathbb{R}$ is by definition *absolutely decreasing* if $\Delta_{a_1} \cdots \Delta_{a_n}\psi \leqq 0$ for all $a_1, \ldots, a_n \in S$.

(a) Show that both classes of functions are closed convex cones in $\mathbb{R}^S$ and that the absolutely monotone functions are, furthermore, stable under multiplication.
(b) A function $\psi$ is absolutely decreasing if and only if $\nabla_a\psi$ is absolutely monotone for each $a \in S$.
(c) Show that $\psi$ is absolutely decreasing if and only if $\exp(-t\psi)$ is absolutely monotone for each $t > 0$.
(d) A semicharacter $\rho \in S^*$ is absolutely monotone if and only if $\rho \geqq 1$.
(e) If $\alpha\colon S \to \mathbb{R}_+$ is additive and $k \in \mathbb{N}_0$ then $\alpha^k$ is absolutely monotone.

**6.30. Exercise.** Let $\psi\colon S \to \mathbb{R}$ be a lower bounded negative definite function and let $(\mu_t)_{t\geqq 0}$ be the corresponding convolution semigroup on $\hat{S}$, cf. Theorem 3.7. Show that $\psi$ is completely alternating if and only if $\operatorname{supp}(\mu_t) \subseteq \hat{S}_+$ for all $t \geqq 0$.

This result should be compared with Theorem 6.7. For a generalization of these results see Berg (1984).

## Notes and Remarks

The algebraic theory of semigroups is treated in Clifford and Preston (1961) and Redéi (1965). Concerning topological semigroups see Paalman-de Miranda (1964) and Hofmann and Mostert (1966). The concept of a semigroup with involution is general enough to include abelian semigroups and groups. It can be found in the appendix by Sz.-Nagy (1960) to the famous functional analysis monograph by Riesz and Sz.-Nagy. The concept is however so simple, that it might have appeared elsewhere long ago.

Positive definite functions on groups form a very important subject within harmonic analysis, and it is discussed in every monograph on the subject; see Stewart (1976) for an excellent survey of this topic. Operator valued functions of positive type occur in Sz.-Nagy (1960), and have been frequently studied since, see, e.g. Mlak (1978). It should be noticed that the negative definite functions $\psi$ in the group sense form a slightly more general class than those usually considered, cf., e.g. Berg and Forst (1975), where in addition it is required that $\psi(0) \geqq 0$.

The notion of exponentially bounded functions on a semigroup was introduced in Berg and Maserick (1984), but somewhat similar conditions are implicit in Szafraniec (1977). Gelfand's theory of commutative Banach algebras was applied in connection with commutative semigroups by Hewitt and Zuckermann (1956).

Theorem 2.8 was proved by Lindahl and Maserick (1971) and rediscovered by Berg *et al.* (1976) for semigroups with the identical involution. A special case was treated by Ressel (1974). The generalization of 2.8 given in 2.5–2.7 is due to Berg and Maserick (1984). Theorem 2.5 can also be proved by operator theory since the operators $\bar{\pi}(s)$, $s \in S$ from Theorem 1.14 generate a commutative $C^*$-algebra for which the spectral theorem can be applied, cf. Rudin (1973, Theorem 12.22). In this connection see also Schempp (1977).

The continuous negative definite functions on a locally compact abelian group $G$ are in one-to-one correspondence with the convolution semigroups on the dual group $\hat{G}$, cf. Berg and Forst (1975), which in turn are in one-to-one correspondence with the $\hat{G}$-valued stochastic processes with stationary and independent increments. The integral representation of the cone of negative definite functions is known in the literature as the Lévy–Khinchin formula, which was established for $G = \mathbb{R}$ in the late 1930's by Lévy and Khinchin. The theory of convolution semigroups has been developed for arbitrary locally compact groups in the monograph by Heyer (1977). Our most general Lévy–Khinchin formula is 3.19. The special case in 3.20, where the involution is identical, is due to Berg *et al.* (1976), and the special case of 3.19 (where the negative definite function is real-valued) is due to Maserick (1978).

Bernstein functions as defined after 4.3 occur at many places of analysis but often without the label Bernstein function. In analogy with Theorem 6.13 a continuous function $f: \mathbb{R}_+ \to \mathbb{R}_+$ is a Bernstein function if and only if $f \in C^\infty(]0, \infty[)$ and $(-1)^n f^{(n+1)} \geqq 0$ for $n = 0, 1, \ldots$. Bochner (1955) proved that Bernstein functions $f$ are characterized by the property that $\varphi \circ f$ is completely monotone for all (continuous) completely monotone functions $\varphi$ on $\mathbb{R}_+$ and used the name completely monotone mappings instead of Bernstein functions. Bochner's result is generalized in Exercise 4.28. Let us mention some places where Bernstein functions appear: In probability theory because a probability measure $\mu$ on $\mathbb{R}_+$ is infinitely divisible if and only if $-\log \mathscr{L}\mu$ is a Bernstein function. In potential theory because a positive measure $\kappa$ on $\mathbb{R}_+$ is a potential kernel if and only if $1/\mathscr{L}\kappa$ is a Bernstein function, cf. Berg and Forst (1975, 1979). Similarly, a bounded continuous function $p: \mathbb{R}_+ \to \mathbb{R}_+$ is proportional to a standard $p$-function in the theory of regenerative phenomena (cf. Kingman (1972)) if and only if $1/\mathscr{L}p$ is a Bernstein function. Micchelli and Willoughby (1979) have characterized the Bernstein functions as those functions which operate on the class of Stieltjes matrices. This has, in turn, interesting applications in the theory of orthogonal polynomials, cf. Micchelli (1978). An important class of Bernstein functions is the reciprocals of nonzero Stieltjes functions $\varphi$. A function $\varphi: ]0, \infty[ \to [0, \infty[$ is a Stieltjes function if it has the form

$$\varphi(s) = a + \int_0^\infty \frac{d\mu(x)}{s + x}, \qquad s > 0,$$

where $a \geqq 0$ and $\mu \in M_+(\mathbb{R}_+)$. The functions $\varphi(s) = s^{-\alpha}$, $0 < \alpha \leqq 1$ and $1/\log(1 + s)$ are Stieltjes functions, cf. 3.2.10. A survey of the proporties of Stieltjes functions can be found, e.g. in Berg (1980).

The results of Exercises 4.30 and 4.31 have some "infinite dimensional" analogues, see Hoffmann-Jørgensen and Ressel (1977).

For further applications of the theory of $\tau$-positive functions we refer to Maserick (1977), see also Tonev (1979) and Berg and Maserick (1984).

The theory of completely monotone functions on commutative semigroups goes back to Choquet (1954), who found the integral representation in 6.4 by proving $\mathrm{ex}(\mathscr{M}^1(S)) \subseteq \hat{S}_+$ and then using the Krein–Milman theorem. An elementary direct proof for the converse inclusion $\hat{S}_+ \subseteq \mathrm{ex}(\mathscr{M}^1(S))$ can be found in Fine and Maserick (1970) and it also follows from Corollary 2.5.12. Choquet also studied completely alternating functions—capacities are of this type—and proved an integral representation for bounded completely alternating functions. The general representation of functions $\psi \in \mathscr{A}(S)$ as a special case of the Lévy–Khinchin formula is due to Berg *et al.* (1976).

Choquet (1954, §49) obtained Theorem 6.19 from the Krein–Milman theorem by considering a suitable vague topology on the cone of increasing functions which are continuous on the right. See also Talagrand (1976).

Another approach to the theorem can be found in Matheron (1975), who considers a completely alternating capacity as "distribution function" for a random closed set. The present approach seems to be new.

The result of Exercise 6.27 can be generalized by replacing $\mathbb{R}_+^k$ by a closed convex cone in $\mathbb{R}^k$, cf. Hirsch (1972) and Bochner (1955).

Absolutely monotone functions on semigroups have hitherto not found particular attention. Widder (1941, p. 146) shows that any absolutely monotone function on $\mathbb{R}_+$ is analytic with nonnegative coefficients. The name "absolutely decreasing" has been introduced by us.

Functions $f: S \to \mathbb{R}$, which can be represented as $\varphi_1 - \varphi_2$ with $\varphi_i \in \mathscr{P}^b(S)$, can be characterized intrinsically and are called BV-functions by Maserick (1975, 1981 and references therein).

The beautiful duality theory for locally compact abelian groups does not seem to have been extended to topological semigroups, and it is probably difficult, if at all possible. The compact semigroup $S = [0, 1]$ with maximum as semigroup operation has only one continuous semicharacter, namely, the constant semicharacter. The right dual object to look at might be the set of semicharacters which are continuous at 0. In the group case, these are precisely the continuous group characters. Berg *et al.* (1976, §8) studied a semitopological semigroup having a continuous positive definite function, such that all semicharacters in the support of the representing measure are discontinuous at the neutral element.

For a study of representations of semitopological semigroups see Dunkl and Ramirez (1975).

See also the historical surveys by Williamson (1967) and Hofmann (1976) which both contain a rich bibliography.

CHAPTER 5

# Schoenberg-Type Results for Positive and Negative Definite Functions

## §1. Schoenberg Triples

We have already mentioned the three fundamental papers of Schoenberg (1938a, b, 1942), all of which are very closely related to positive and negative definite kernels. The main purpose in (1938b) was to show the close connection between (real-valued) negative definite kernels and Hilbert metrics, see Chapter 3, §3. In his first mentioned paper (1938a), entitled "Metric spaces and completely monotone functions", Schoenberg raises the question about the connection between the class of Fourier transforms of (finite, nonnegative) measures in euclidean spaces and the class of Laplace transforms of (finite, nonnegative) measures on the half-line $\mathbb{R}_+$. He states:

> In spite of the entirely different analytical character of these two classes, a certain kinship was to be expected for the following two reasons:
>
> 1. In both classes the defining kernel is the exponential function.
> 2. The less formal reason of the similarity of the closure properties of both classes, for both classes are convex, i.e. $a_1 f_1 + a_2 f_2$ ($a_1 \geqq 0, a_2 \geqq 0$) belongs to the class if $f_1$ and $f_2$ belong to it, multiplicative, i.e. also $f_1 \cdot f_2$ belongs to the class, and finally closed with respect to ordinary convergence to a continuous limit function.

The answer Schoenberg gave to the above question was the remarkable result that to each continuous function $\varphi: \mathbb{R} \to \mathbb{C}$, with the property that $\varphi \circ \|\cdot\|_n$ is a positive definite function on the group $\mathbb{R}^n$ for all $n$ ($\|\cdot\|_n$ denoting the euclidean norm), there exists a finite nonnegative measure on $\mathbb{R}_+$ with Laplace transform $\varphi(\sqrt{t})$; see (1938a, Theorem 2). The proof of this theorem (including that given in the more recent book of Donoghue (1969, pp. 201–206)), is, however, rather complicated and technical in nature. But there exists a further common feature of Fourier and Laplace transforms as shown in the preceding chapter: Laplace transforms, too, are characterized essentially by positive definiteness. Using this fact, we will prove a quite abstract form of Schoenberg's theorem, where we replace both the real line and the half-line $\mathbb{R}_+$ by arbitrary abelian $*$-semigroups with neutral element.

The key to most of the results in this chapter is given by the following, rather elementary, approximation lemma.

**1.1. Lemma.** *Let $(a_{jk})$ be a hermitian $p \times p$ matrix, let $X$ be a nonempty set and let $\varphi: X \times X \to \mathbb{C}$ be a given kernel. Suppose that for every $n \in \mathbb{N}$ there exists a finite subset $\{x_{j\alpha}^n \mid j = 1, \ldots, p; \alpha = 1, \ldots, n\} \subseteq X$ such that*

$$\varphi(x_{j\alpha}^n, x_{k\beta}^n) = \begin{cases} a_{jk} & \text{if } j \neq k, \\ a_{jj} & \text{if } j = k \text{ but } \alpha \neq \beta. \end{cases}$$

*If $\varphi$ is positive definite and bounded, then the matrix $(a_{jk})$ is positive definite; and if $\varphi$ is negative definite such that $\mathrm{Re}(\varphi)$ is bounded below, then $(a_{jk})$ is negative definite.*

Proof. Let $c_1, \ldots, c_p \in \mathbb{C}$ be given. For $n \in \mathbb{N}$ choose $\{x_{j\alpha}^n\}$ as indicated and put $d_{j\alpha} := c_j/n$, $1 \leqq j \leqq p$, $1 \leqq \alpha \leqq n$. If $\varphi$ is positive definite, then

$$\begin{aligned} 0 &\leqq \sum_{j,k=1}^{p} \sum_{\alpha,\beta=1}^{n} d_{j\alpha}\overline{d_{k\beta}}\varphi(x_{j\alpha}^n, x_{k\beta}^n) \\ &= \sum_{\substack{j,k=1 \\ j\neq k}}^{p} c_j\overline{c_k}a_{jk} + \frac{n^2-n}{n^2}\sum_{j=1}^{p}|c_j|^2 a_{jj} + \frac{1}{n^2}\sum_{j=1}^{p}\sum_{\alpha=1}^{n}|c_j|^2\varphi(x_{j\alpha}^n, x_{j\alpha}^n) \\ &= \sum_{j,k=1}^{p} c_j\overline{c_k}a_{jk} + R_n, \end{aligned}$$

where

$$R_n := -\frac{1}{n}\sum_{j=1}^{p}|c_j|^2 a_{jj} + \frac{1}{n^2}\sum_{j=1}^{p}\sum_{\alpha=1}^{n}|c_j|^2\varphi(x_{j\alpha}^n, x_{j\alpha}^n).$$

The kernel $\varphi$ being bounded, $R_n$ tends to zero for $n \to \infty$. The second statement follows immediately from Theorem 3.2.2. □

Suppose now we are given a triple $(T, S, v)$, where $T$ and $S$ are $*$-semigroups with neutral element, both abelian, and where $v: T \to S$ is a mapping satisfying:

($S_1$) $v(0) = 0$;
($S_2$) $v(t^*) = (v(t))^*$ for all $t \in T$;
($S_3$) $v(T)$ generates $S$.

Besides the finite powers $T^n$, $n \in \mathbb{N}$, we will also consider the infinite direct sum

$$T^{(\infty)} := \{(t_1, t_2, \ldots) \in T^{\mathbb{N}} \mid |\{i \in \mathbb{N} \mid t_i \neq 0\}| < \infty\},$$

which is again a $*$-semigroup with the canonical operations, and we associate with $v$ the mappings $v_n\colon T^n \to S$ defined by

$$v_n((t_1, t_2, \ldots)) := v(t_1) + v(t_2) + \cdots, \qquad n \in \mathbb{N} \cup \{\infty\}.$$

Condition ($S_3$) then means the same as $v_\infty(T^{(\infty)}) = S$.

For a function $\varphi\colon S \to \mathbb{C}$ it is equivalent to state that $\varphi \circ v_\infty$ is positive definite on $T^{(\infty)}$ and that $\varphi \circ v_n$ is positive definite on $T^n$ for all $n \in \mathbb{N}$.

Although the three properties ($S_1$)–($S_3$) seem to be rather weak, they nevertheless lead to the following surprisingly strong conclusion.

**1.2. Theorem.** *Let $(T, S, v)$ be a triple as described above and let $\varphi\colon S \to \mathbb{C}$ be a bounded function. Then if $\varphi \circ v_\infty$ is positive definite on $T^{(\infty)}$, it follows that $\varphi$ is positive definite on $S$.*

Proof. We fix a finite subset $\{s_1, \ldots, s_p\} \subseteq S$ and have to show that the $p \times p$ matrix

$$(a_{jk}) = (\varphi(s_j^* + s_k))$$

is positive definite. By condition ($S_3$) there exist $n_j \in \mathbb{N}$, $t_{jm} \in T$, $m = 1, \ldots, n_j$, $j = 1, \ldots, p$ such that

$$s_j = \sum_{m=1}^{n_j} v(t_{jm}), \qquad j = 1, \ldots, p.$$

For a fixed $n \in \mathbb{N}$ we define $x_{j\alpha}^n = (x_{j\alpha}^n(1), x_{j\alpha}^n(2), \ldots) \in T^{(\infty)}$ as the columns in the matrix on the next page, where empty places should be read as zeros. Formally, the coefficients $x_{j\alpha}^n(\sigma)$ are defined by

$$x_{j\alpha}^n(\sigma) := \begin{cases} t_{j\tau} & \text{if } \sigma = n(n_1 + \cdots + n_{j-1}) + (\alpha - 1)n_j + \tau,\ 1 \leqq \tau \leqq n_j; \\ 0 & \text{if } \sigma \text{ is not of this form}; \end{cases}$$

where $j = 1, \ldots, p$ and $\alpha = 1, \ldots, n$. On $X := T^{(\infty)}$ we consider the kernel $\Phi(x, y) := \varphi(v_\infty(x^* + y))$, being positive definite by assumption. We have

$$\Phi(x_{j\alpha}^n, x_{k\beta}^n) = \begin{cases} \varphi(s_j^* + s_k) & \text{if } j \neq k; \\ \varphi(s_j^* + s_j) & \text{if } j = k \text{ but } \alpha \neq \beta; \\ \varphi\left(\sum_{m=1}^{n_j} v(t_{jm} + t_{jm}^*)\right) & \text{if } j = k \text{ and } \alpha = \beta. \end{cases}$$

Consequently, $\Phi$ being bounded, the matrix $(a_{jk})$ is positive definite by Lemma 1.1. □

**1.3. Example.** Consider the triple $(T, S, v) = (\mathbb{R}, \mathbb{R}_+, x^2)$ with the involutions $t^* = -t$ on $\mathbb{R}$ and $s^* = s$ on $\mathbb{R}_+$. If $\varphi\colon \mathbb{R}_+ \to \mathbb{C}$ is a function such that $\varphi(\|\cdot\|_n)$ is a positive definite function on $\mathbb{R}^n$ for every $n \in \mathbb{N}$ (where $\|\cdot\|_n$ denotes the euclidean norm) then $\varphi$ is, of course, real-valued and bounded and the above theorem can be applied to $\varphi \circ \sqrt{\cdot}$, showing by 4.4.2 that

| $x_{11}^n$ | $x_{12}^n$ | $\cdots$ | $x_{1n}^n$ | $x_{21}^n$ | $x_{22}^n$ | $\cdots$ | $x_{2n}^n$ | $\cdots$ | $x_{p1}^n$ | $x_{p2}^n$ | $\cdots$ | $x_{pn}^n$ | |
|---|---|---|---|---|---|---|---|---|---|---|---|---|---|
| $t_{11}$ | 0 | $\cdots$ | 0 | | | | | | | | | | |
| $t_{12}$ | 0 | $\cdots$ | 0 | | | | | | | | | | |
| $\vdots$ | $\vdots$ | | $\vdots$ | | | | | | | | | | 1 |
| $t_{1n_1}$ | 0 | $\cdots$ | 0 | | | | | | | | | | |
| 0 | $t_{11}$ | $\cdots$ | 0 | | | | | | | | | | |
| 0 | $t_{12}$ | $\cdots$ | 0 | | | | | | | | | | |
| $\vdots$ | $\vdots$ | | $\vdots$ | | | | | | | | | | 2 |
| 0 | $t_{1n_1}$ | $\cdots$ | 0 | | | | | | | | | | |
| | | $\vdots$ | | | | | | | | | | | $\vdots$ |
| 0 | 0 | $\cdots$ | $t_{11}$ | | | | | | | | | | |
| 0 | 0 | $\cdots$ | $t_{12}$ | | | | | | | | | | |
| $\vdots$ | $\vdots$ | | $\vdots$ | | | | | | | | | | $n$ |
| 0 | 0 | $\cdots$ | $t_{1n_1}$ | | | | | | | | | | |
| | | | | $t_{21}$ | | | | | | | | | |
| | | | | $t_{22}$ | | | | | | | | | |
| | | | | $\vdots$ | | | | | | | | | $n + 1$ |
| | | | | $t_{2n_2}$ | | | | | | | | | |
| | | | | | $t_{21}$ | | | | | | | | |
| | | | | | $t_{22}$ | | | | | | | | |
| | | | | | $\vdots$ | | | | | | | | $n + 2$ |
| | | | | | $t_{2n_2}$ | | | | | | | | |
| | | $\vdots$ | | | | | | | | | | | $\vdots$ |
| | | | | | | | $t_{21}$ | | | | | | |
| | | | | | | | $t_{22}$ | | | | | | |
| | | | | | | | $\vdots$ | | | | | | $2n$ |
| | | | | | | | $t_{2n_2}$ | | | | | | |
| | | $\vdots$ | | | | | | | | | | | $\vdots$ |
| | | | | | | | | | $t_{p1}$ | | | | |
| | | | | | | | | | $t_{p2}$ | | | | |
| | | | | | | | | | $\vdots$ | | | | $(p - 1)n + 1$ |
| | | | | | | | | | $t_{pn_p}$ | | | | |
| | | | | | | | | | | $t_{p1}$ | | | |
| | | | | | | | | | | $t_{p2}$ | | | |
| | | | | | | | | | | $\vdots$ | | | $(p - 1)n + 2$ |
| | | | | | | | | | | $t_{pn_p}$ | | | |
| | | $\vdots$ | | | | | | | | | | | $\vdots$ |
| | | | | | | | | | | | | $t_{p1}$ | |
| | | | | | | | | | | | | $t_{p2}$ | |
| | | | | | | | | | | | | $\vdots$ | $pn$ |
| | | | | | | | | | | | | $t_{pn_p}$ | |

$\varphi(\sqrt{t})$ is the Laplace transform of some Radon measure on $[0, \infty]$. This measure is supported by $[0, \infty[$ if and only if $\varphi$ is continuous.

The above example also allows a statement in the other direction. Let $\varphi: \mathbb{R}_+ \to \mathbb{R}$ have the form

$$\varphi(t) = \int_0^\infty e^{-\lambda t^2}\, d\mu(\lambda)$$

for some nonnegative finite measure $\mu$ on $\mathbb{R}_+$. Then $\varphi(\|\cdot\|_n)$ is positive definite in the group sense on $\mathbb{R}^n$ because the square of the euclidean norm is negative definite, cf. 3.1.10 and 3.2.2. It will not surprise now, that the following condition

$$(S_4) \quad \rho \circ v \in \mathscr{P}(T) \text{ for all } \rho \in \hat{S}$$

for a triple $(T, S, v)$ as described above will play some role for a converse of Theorem 1.2.

**1.4. Definition.** A triple $(T, S, v)$, where $T$ and $S$ are abelian $*$-semigroups, and $v: T \to S$ is a mapping fulfilling the conditions $(S_1)$–$(S_4)$, will be called a *Schoenberg triple.*

**1.5. Theorem.** *Let $(T, S, v)$ be a Schoenberg triple. If the function $\varphi: S \to \mathbb{C}$ is bounded and positive definite, then $\varphi \circ v_n$ is positive definite on $T^n$ for every $n \in \mathbb{N}$.*

Proof. In view of Theorem 4.2.8 it is enough to show that $\rho \circ v_n \in \mathscr{P}(T^n)$ for all $\rho \in \hat{S}$. This, in fact, is true because

$$\begin{aligned}\rho \circ v_n(t_1, \ldots, t_n) &= \rho\left(\sum_{j=1}^n v(t_j)\right) = \prod_{j=1}^n \rho(v(t_j)) \\ &= \prod_{j=1}^n \rho \circ v \circ \pi_j(t_1, \ldots, t_n)\end{aligned}$$

(where $\pi_j$ denotes the canonical projection) is a product of positive definite functions. □

The restricted dual of the semigroup $(\mathbb{R}_+, +)$ is given by the exponentials $s \mapsto e^{-\lambda s}$, $0 \leqq \lambda \leqq \infty$, as we have seen earlier. This shows that a triple $(T, S, v)$ where $S = \mathbb{R}_+$, is a Schoenberg triple if and only if $v$ is a nonnegative negative definite function on $T$ for which $(S_1)$ and $(S_3)$ hold. In particular, if $T$ is a connected topological $*$-semigroup and $v: T \to [0, \infty[$ is a continuous negative definite function satisfying $v(0) = 0$ and $v \not\equiv 0$, then $(T, \mathbb{R}_+, v)$ is a Schoenberg triple.

**1.6. Example.** The space $l^p$ of all real sequences whose $p$th power is summable is a Banach space only for $p \geqq 1$, i.e. the function $\|x\|_p := [\sum |x_j|^p]^{1/p}$ is a norm only for $p \geqq 1$. Nevertheless $l^p$ as well as $\|\cdot\|_p$ is well defined also for

$0 < p < 1$, and $l^p$ is a linear space. We consider the $l^p$-spaces for $0 < p \leqq 2$. Let $\Phi: l^p \to \mathbb{R}$ be a given function depending only on the "norm" $\|\cdot\|_p$, i.e. $\Phi = \varphi \circ \|\cdot\|_p$, where $\varphi: \mathbb{R}_+ \to \mathbb{R}$ is an arbitrary function. We claim that $\Phi$ is positive definite in the group sense on $l^p$ if and only if $\varphi(t^{1/p})$ is positive definite and bounded on $\mathbb{R}_+$, i.e. if and only if

$$\varphi(s) = \int_0^{\infty} e^{-\lambda s^p}\, d\mu(\lambda)$$

for some $\mu \in M_+([0, \infty])$.

In fact, from 3.2.10 it follows that $x \mapsto |x|^p$ is negative definite in the group sense on $\mathbb{R}$, so $(\mathbb{R}, \mathbb{R}_+, |x|^p)$ is a Schoenberg triple, and

$$x \mapsto \|x\|_p^p = \lim_{n\to\infty} \sum_{j=1}^{n} |x_j|^p$$

is negative definite in the group sense on $l^p$. If $\Phi$ is positive definite then so is $\Phi|\mathbb{R}^{(\infty)}$, and Theorem 1.2 implies that $\varphi(t^{1/p})$ is positive definite and bounded on $\mathbb{R}_+$. Conversely, if this is the case, then the integral representation above and the negative definiteness of $\|\cdot\|_p^p$ on $l^p$ imply that $\Phi$ is positive definite.

Finally, let us remark that $\Phi$ is continuous if and only if $\varphi$ is continuous, if and only if $\mu(\{\infty\}) = 0$. The above result (assuming continuity of $\varphi$) has been obtained by Bretagnolle *et al.* (1966).

**1.7. Example.** Consider the triple $(\mathbb{Z}, \mathbb{N}_0^2, v)$ where $n^* = -n$ for $n \in \mathbb{Z}$, $(n, m)^* = (m, n)$ for $(n, m) \in \mathbb{N}_0^2$ and $v: \mathbb{Z} \to \mathbb{N}_0^2$ is given by $v(n) = (n, 0)$ for $n \geqq 0$, $v(n) = (0, -n)$ for $n \leqq 0$. We claim that $(\mathbb{Z}, \mathbb{N}_0^2, v)$ is a Schoenberg triple. In fact, we know from 4.4.11 that $(\mathbb{N}_0^2)\hat{}$ is isomorphic (and homeomorphic) with the unit disc $D$ in the complex plane. Let $z = re^{i\theta} \in D$ be given; then, writing $v(n) = (u(n), w(n))$,

$$z^{u(n)} \cdot \bar{z}^{w(n)} = r^{|n|} e^{in\theta}, \tag{1}$$

and since $n \mapsto |n|$ is negative definite in the group sense on $\mathbb{Z}$, the function $r^{|n|} = e^{(\log r)|n|}$ is positive definite in the group sense (for $r = 0$ this is the function $1_{\{0\}}$), and so is therefore the function (1). Given $\varphi: \mathbb{N}_0^2 \to \mathbb{C}$ we see that the functions

$$k = (k_1, \ldots, k_p) \mapsto \varphi(v(k_1) + \cdots + v(k_p))$$

are positive definite in the group sense on $\mathbb{Z}^p$ for all $p \in \mathbb{N}$ if and only if

$$\varphi(n, m) = \int_D z^n \bar{z}^m\, d\mu(z)$$

for some $\mu \in M_+(D)$.

A related example is given by the triple $(\mathbb{Z}, \mathbb{N}_0, |\cdot|)$ which is also easily seen to be a Schoenberg triple. That is, given an arbitrary sequence $\varphi$: $\mathbb{N}_0 \to \mathbb{R}$, the functions

$$k = (k_1, \ldots, k_p) \mapsto \varphi(|k_1| + \cdots + |k_p|)$$

are positive definite in the group sense on $\mathbb{Z}^p$ for all $p \in \mathbb{N}$ if and only if

$$\varphi(n) = \int_{-1}^{1} t^n \, d\mu(t)$$

for some $\mu \in M_+([-1, 1])$.

The reader will not be surprised that statements similar to 1.2 and 1.5 are also possible for negative definite functions.

**1.8. Corollary.** *Let $(T, S, v)$ be a Schoenberg triple and let $\psi: S \to \mathbb{C}$ be a function whose real part is bounded below. Then the following conditions are equivalent:*

(i) *$\psi$ is negative definite on $S$.*
(ii) *$\psi \circ v_n$ is negative definite on $T^n$ for all $n \in \mathbb{N}$.*
(iii) *$\psi \circ v_\infty$ is negative definite on $T^{(\infty)}$.*

PROOF. Immediate consequence of 1.2, 1.5 and 3.2.2. □

**1.9. Example.** Let $\psi: \mathbb{N}_0 \to \mathbb{R}$ be a sequence of real numbers. Then the functions

$$(k_1, \ldots, k_p) \mapsto \psi(|k_1| + \cdots + |k_p|)$$

are negative definite in the group sense on $\mathbb{Z}^p$ for all $p \in \mathbb{N}$ if and only if

$$\psi(m) = a + bm + \int_{[-1,1[} (1 - t^m) \, d\mu(t)$$

for certain $a \in \mathbb{R}$, $b \in \mathbb{R}_+$ and a Radon measure $\mu$ on $[-1, 1[$. This follows from Proposition 4.4.10.

Likewise for a doubly indexed sequence $\psi: \mathbb{N}_0^2 \to \mathbb{C}$ the functions

$$(k_1, \ldots, k_p) \mapsto \psi(v(k_1) + \cdots + v(k_p))$$

(with $v$ as in Example 1.7) are negative definite in the group sense on $\mathbb{Z}^p$ for all $p \geqq 1$ if and only if $\psi \in \mathcal{N}^l(\mathbb{N}_0^2)$. In this case $\psi$ has the Lévy–Khinchin decomposition given in Proposition 4.4.15.

**1.10. Exercise.** Show that the following triples $(T, S, v)$ are Schoenberg triples (where on the groups $\mathbb{Z}$, $\mathbb{Z}_2$, $\mathbb{R}$, $\mathbb{R}^k$ the involution $t^* = -t$ is used, whereas the other semigroups carry the identical involution apart from (i) where $S$ is given the "product-involution" $(s_1, s_2)^* := (-s_1, s_2)$).

(a) $(\mathbb{Z}, \mathbb{N}_0, 1_{2\mathbb{Z}+1})$.
(b) $(\mathbb{R}, \mathbb{R}_+, 1 - \cos x)$.
(c) $(\mathbb{Z}_2, \mathbb{N}_0, v)$ where $v(0) = 0, v(1) = 1$.
(d) $(\mathbb{R}^k, \mathbb{R}_+, \|x\|^\alpha)$ where $0 < \alpha \leqq 2$ and $\|\cdot\|$ is the euclidean norm.
(e) $(\mathbb{R}^k, \mathbb{R}_+^k, (|x_1|^{\alpha_1}, \ldots, |x_k|^{\alpha_k}))$ where $\{\alpha_1, \ldots, \alpha_k\} \subseteq ]0, 2]$.
(f) $(\mathbb{Z}, \mathbb{N}_0, n^2)$.
(g) $(\mathbb{R}_+, ]0, 1], (1 + t)^{-1})$.
(h) $([-1, 1], \mathbb{R}_+, 1 - t)$.
(i) $(\mathbb{R}, S, (t, t^2))$ where $S = (\mathbb{R} \times ]0, \infty[) \cup \{(0, 0)\}$.
(j) $((\mathbb{R}_+, \max), (\mathbb{R}_+, +), \mathrm{id})$.

**1.11. Exercise.** Let $h: T \to S$ be a $*$-homomorphism between the two semigroups with involution $T$ and $S$. Then $(T, S, h)$ is a Schoenberg triple if and only if $h$ is onto.

**1.12. Exercise.** If $(T_1, S_1, v_1)$ and $(T_2, S_2, v_2)$ both are Schoenberg triples, then so is $(T_1 \times T_2, S_1 \times S_2, v_1 \times v_2)$.

**1.13. Exercise.** Let $(T, S, v)$ and $(S, U, w)$ be two Schoenberg triples. Then $(T, U, w \circ v)$ is again a Schoenberg triple if $w(v(T))$ generates $U$. The special case $T = \mathbb{Z}$, $S = \mathbb{N}_0^2$, $v$ as in Example 1.7, $U = \mathbb{N}_0$ and $w(n, m) = n + m$ relates the two parts of 1.7.

**1.14. Exercise.** Let $\varphi$ be the Laplace transform of some probability measure $\mu$ on $\mathbb{R}_+$. Then $\varphi(\sum_{i=1}^n x_i^2)$ is the Fourier transform (or characteristic function) of some probability measure $\nu_n$ on $\mathbb{R}^n$. Show that $\mu$ is infinitely divisible if and only if all the $\nu_n$ are infinitely divisible.

## §2. Norm Dependent Positive Definite Functions on Banach Spaces

The classical result of Schoenberg mentioned at the beginning of §1 can also be reformulated in this way: If $H$ is an infinite dimensional Hilbert space and $\Phi: H \to \mathbb{R}$ is a positive definite continuous function on the group $H$ depending only on the norm, i.e. $\Phi(x) = \varphi(\|x\|)$ for some continuous $\varphi$: $\mathbb{R}_+ \to \mathbb{R}$, then $\varphi(\sqrt{t})$ is the Laplace transform of some finite measure $\mu$ on $\mathbb{R}_+$. Of course, the question arises as to how essential the assumption about $H$ actually is, i.e. can something similar be said if we only assume $H$ to be a Banach space? On finite dimensional spaces, a general result like this cannot be expected, since, for example, on the real line the positive definite function $\cos(t)$ only depends on $|t|$. More generally, the characteristic function of the uniform distribution on $\{x \in \mathbb{R}^n \mid \|x\| = 1\}$ depends only on the (euclidean) norm $\|x\|$, but not monotonically. Using a deep approximation theorem of Dvoretzky we can, however, show the following rather satisfactory result:

**2.1. Theorem.** *Let $B$ denote an infinite dimensional Banach space and let $\varphi\colon \mathbb{R}_+ \to \mathbb{C}$ be a continuous function such that $\varphi(\|x\|)$ is positive definite on the group $B$. Then $\varphi(\sqrt{t})$ is the Laplace transform of some finite measure on $\mathbb{R}_+$.*

PROOF. Of course, $\varphi$ is real-valued and bounded so that in view of Corollary 4.4.5 we are left with the problem of showing that $\varphi(\sqrt{t})$ is positive definite on the semigroup $\mathbb{R}_+$. Our approximation lemma (1.1) cannot directly be applied in this case, but a certain modification will suffice.

Let $\{t_1, \ldots, t_k\} \subseteq \mathbb{R}_+$ and $\{c_1, \ldots, c_k\} \subseteq \mathbb{R}$ be given and fixed, and let $\varepsilon > 0$. We put $t^* := \max_{1 \leq j \leq k} \sqrt{2t_j}$ and choose $\delta > 0$ such that

$$s, t \in [0, t^*], \qquad |s - t| \leqq \delta t^* \qquad \text{implies} \quad |\varphi(s) - \varphi(t)| \leqq \varepsilon.$$

Now by a famous theorem of Dvoretzky (1961) there are linear injections $T_N\colon \mathbb{R}^N \to B$, $N = 1, 2, \ldots$, such that

$$1 \leqq \|T_N\| \cdot \|T_N^{-1}\| \leqq 1 + \delta,$$

i.e. assuming without restriction $\|T_N\| = 1$, we have

$$\|T_N(x)\| \leqq \|x\| \leqq (1 + \delta)\|T_N(x)\| \tag{1}$$

for all $x \in \mathbb{R}^N$, where on $\mathbb{R}^N$ we use the euclidean norm.

Given $n \in \mathbb{N}$ we define vectors $x_{im} \in \mathbb{R}^{kn}$ and real numbers $a_{im}$ for $1 \leqq i \leqq k$ and $1 \leqq m \leqq n$ by

$$x_{im} := \sqrt{t_i}\, e_{(i-1)n+m} \qquad \text{and} \qquad a_{im} := c_i/n,$$

where $e_1, e_2, \ldots$ are the standard basis vectors in $\mathbb{R}^{kn}$. Then $\|x_{im} - x_{i'm'}\| = \sqrt{t_i + t_{i'}}$ for $(i, m) \neq (i', m')$ and from (1)

$$0 \leqq \|x_{im} - x_{i'm'}\| - \|T_{kn}(x_{im} - x_{i'm'})\| \leqq \delta\|x_{im} - x_{i'm'}\| \leqq \delta t^*,$$

whence

$$\left| \sum_{i,i'} \sum_{m,m'} a_{im} a_{i'm'} \varphi(\|T_{kn}(x_{im}) - T_{kn}(x_{i'm'})\|) - \sum_{i,i'} \sum_{m,m'} a_{im} a_{i'm'} \varphi(\|x_{im} - x_{i'm'}\|) \right| \leqq \varepsilon \left( \sum_{i,m} |a_{im}| \right)^2 = \varepsilon \left( \sum_{i=1}^{k} |c_i| \right)^2,$$

where on the left-hand side the first sum is nonnegative by assumption, whereas the second sum is equal to

$$\sum_{i,j=1}^{k} c_i c_j \varphi(\sqrt{t_i + t_j}) + \frac{1}{n} \left[ \varphi(0) \sum_{i=1}^{k} c_i^2 - \sum_{i=1}^{k} c_i^2 \varphi(\sqrt{2t_i}) \right].$$

Now letting first $n$ tend to infinity and then $\varepsilon$ to zero we get

$$0 \leqq \sum_{i,j=1}^{k} c_i c_j \varphi(\sqrt{t_i + t_j}),$$

thus proving the theorem. □

Being of a general nature, the above result may not in some special cases give the sharpest possible results. Consider, e.g. the Banach space $B = l^1$. If $\varphi(\|x\|)$ is positive definite on $l^1$, then not only is $\varphi(\sqrt{t})$ positive definite, but so also is $\varphi$ itself, as we have seen in Example 1.6.

Another question arises naturally: For which continuous bounded functions $\varphi: \mathbb{R}_+ \to \mathbb{R}$ is it true that $\varphi(\|x\|)$ is positive definite in the group sense on $B$? As we have seen, a necessary condition is that $\varphi(\sqrt{t})$ be a Laplace transform, and on a Hilbert space this condition is sufficient, too, because $\|\cdot\|^2$ is negative definite. Moreover, if this condition is (necessary and) sufficient for some Banach space, then by choosing $\varphi(t) = e^{-\lambda t^2}$, $\lambda > 0$, we see that $\|x\|^2$ is negative definite on $B$, and then $B$ is already a Hilbert space, as shown in the next lemma, which is taken from Wells and Williams (1975, Theorem 2.5).

**2.2. Lemma.** *Let $B$ be a Banach space such that the square of the norm is negative definite on $B$ (in the group sense). Then $B$ is already a Hilbert space.*

PROOF. Let $x, y \in B$ and $a \in \mathbb{R}$ and consider the four-point set $\{x_1, \ldots, x_4\}$ with $x_1 = x$, $x_2 = y$, $x_3 = -y$, $x_4 = 0$ as well as the coefficients $c_1 = 1 - 2a$, $c_2 = c_3 = a$, $c_4 = -1$. Then, by assumption,

$$0 \geqq \tfrac{1}{2} \sum_{j,k=1}^{4} c_j c_k \|x_j - x_k\|^2 = (1 - 2a)a\|x - y\|^2$$

$$+ (1 - 2a)a\|x + y\|^2 - (1 - 2a)\|x\|^2 + a^2\|2y\|^2 - a\|y\|^2 - a\|y\|^2,$$

or

$$(1 - 2a)a(\|x - y\|^2 + \|x + y\|^2) \leqq (1 - 2a)(\|x\|^2 + 2a\|y\|^2).$$

Assuming $a < \frac{1}{2}$, dividing out $1 - 2a$ and then letting $a$ tend to $\frac{1}{2}$, we get

$$\|x - y\|^2 + \|x + y\|^2 \leqq 2\|x\|^2 + 2\|y\|^2,$$

and the same procedure for $a > \frac{1}{2}$ gives the reverse inequality. Hence, the norm obeys the parallelogram law and is therefore induced by a scalar product. □

For any Banach space $B$ we introduce the convex cone

$$\mathscr{P}_r(B) = \{\varphi: \mathbb{R}_+ \to \mathbb{R} \mid \varphi \text{ continuous}, \varphi(\|\cdot\|) \in \mathscr{P}(B)\},$$

where positive definiteness refers to the group sense. The determination in general of $\mathscr{P}_r(B)$ seems to be open. If $\dim B = \infty$ then it is a subset of

$$\{\mathscr{L}\mu(t^2) \mid \mu \in M_+^b(\mathbb{R}_+)\},$$

and the sets are precisely equal when $B$ is a Hilbert space. The following result of Einhorn (1969) shows that $\mathscr{P}_r(B)$ may be degenerate.

**2.3. Theorem.** *Let $\varphi\colon \mathbb{R}_+ \to \mathbb{R}$ have the property that $\varphi(\|x\|)$ is positive definite on the Banach space $C[0, 1]$. For some $\alpha \in \mathbb{R}_+$ we then have $\varphi|]0, \infty[ \equiv \alpha$ and $\varphi(0) \geqq \alpha$. In particular, $\mathscr{P}_r(C[0, 1])$ consists only of the nonnegative constants.*

PROOF. For $n \geqq 2$ let $\{(i, j) | 1 \leqq i < j \leqq n\}$ be the disjoint union of $M$ and $N$. If $0 < s \leqq t \leqq 2s$ we define

$$d(i, j) := \begin{cases} s, & (i, j) \in M, \\ t, & (i, j) \in N, \end{cases}$$

and $d(j, i) := d(i, j)$ for $i < j$ as well as $d(i, i) := 0$. The finite set $\{1, \ldots, n\}$ together with $d$ is a metric space and can hence be embedded in $C[0, 1]$ by a famous theorem of Banach and Mazur (cf. Banach 1932, p. 169). This means that we can find $\{x_1, \ldots, x_n\} \subseteq C[0, 1]$ such that

$$\|x_i - x_j\| = \begin{cases} s, & (i, j) \in M, \\ t, & (i, j) \in N. \end{cases}$$

By assumption we have

$$\begin{aligned} 0 \leqq \sum_{i,j=1}^{n} c_i c_j \varphi(\|x_i - x_j\|) &= [\varphi(0) - \varphi(t)] \cdot \sum c_i^2 + \varphi(t)(\sum c_i)^2 \\ &\quad + 2[\varphi(s) - \varphi(t)] \cdot \sum_M c_i c_j \end{aligned} \tag{2}$$

for all $(c_1, \ldots, c_n) \in \mathbb{R}^n$. If $M = \varnothing$ and $\sum c_i = 0$ this shows $\varphi(0) \geqq \varphi(t)$. Choosing $n = 2m$ to be even, $c_1 = \cdots = c_m = 1, c_{m+1} = \cdots = c_n = -1$ and $M := \{(i, j) | i < j, c_i c_j = -1\}$, inequality (2) reduces to

$$[\varphi(0) - \varphi(t)] \cdot 2m \geqq 2[\varphi(s) - \varphi(t)] \cdot m^2,$$

and since this holds for all $m \in \mathbb{N}$, we see $\varphi(s) \leqq \varphi(t)$; analogously, the complementary choice of $M$ gives $\varphi(s) \geqq \varphi(t)$. Finally, if all $c_i = 1$, (2) implies $0 \leqq [\varphi(0) - \varphi(t)] \cdot n + \varphi(t) \cdot n^2$ for all $n$, hence $\varphi(t) \geqq 0$. □

It seems to be open again, if $\mathscr{P}_r(B)$ can be degenerate also for some finite dimensional Banach space $B$.

A closely related question was raised by Schoenberg (1938b): When is it true that $\exp(-\|x\|^\gamma)$ is positive definite on a Banach space for some $\gamma > 0$; if this is the case, which value of $\gamma$ then is the maximal possible? By homogeneity of the norm $\exp(-\|x\|^\gamma)$ is positive definite if and only if $\|x\|^\gamma$ is negative definite, so if we put

$$\gamma^*(B) := \sup\{\gamma \geqq 0 | \|x\|^\gamma \text{ is negative definite on } B\},$$

Schoenberg's problem is equivalent to the determination of $\gamma^*(B)$. By 3.3.3 we always have $\gamma^*(B) \leqq 2$ and Lemma 2.2 shows that $\gamma^*(B) = 2$ is characteristic for Hilbert spaces. Schoenberg was particularly interested in the maximum norm on $\mathbb{R}^n$ and stated the following explicit formula for $x, y \in \mathbb{R}$:

$$\exp[-(|x| \vee |y|)] = \frac{2}{\pi^2} \int_{-\infty}^{\infty} \int_{-\infty}^{\infty} \exp[i(xu + yv)]$$

$$\times \frac{du\, dv}{[1 + (u + v)^2][1 + (u - v)^2]},$$

showing the negative definiteness of the maximum norm on $\mathbb{R}^2$. In fact, as Herz (1963b) has shown, each norm on a two-dimensional real vector space is negative definite, but the maximum norm on $\mathbb{R}^3$ is not negative definite. Hence, $\gamma^*(\mathbb{R}^3, \|\cdot\|_\infty) < 1$; already Schoenberg found that

$$\gamma^*(\mathbb{R}^4, \|\cdot\|_\infty) < \tfrac{9}{10}.$$

**2.4. Exercise.** Let $B$ be a Banach space and let $\varphi \in \mathscr{P}_r(B)$ be nonconstant and have the form $\varphi(t) = \int_0^\infty \exp(-\lambda t^2)\, d\mu(\lambda)$ for some $\mu \in M_+^b(\mathbb{R}_+)$. (Note that this representation necessarily holds if $\dim(B) = \infty$.) Show that if $\int \lambda\, d\mu(\lambda) < \infty$ then $B$ is already a Hilbert space.

**2.5. Exercise.** If $B = L^p(\mu)$ for some measure space $(\Omega, \mathscr{A}, \mu)$ and $1 \leqq p \leqq 2$, then $\|x\|^p$ is a negative definite function in the group sense on $B$.

**2.6. Exercise.** A famous theorem of Mazur and Ulam (1932) says that an isometric mapping between normed spaces which maps 0 to 0, is automatically linear. Use this result and Proposition 3.3.2 to give another proof of Lemma 2.2.

# §3. Functions Operating on Positive Definite Matrices

In his paper "Positive definite functions on spheres" Schoenberg (1942) proved the following result: Every continuous function $\varphi: [-1, 1] \to \mathbb{R}$ having the property that $\varphi(\langle x, y \rangle)$ is a positive definite kernel on the unit sphere of some infinite dimensional real Hilbert space has a power series representation with nonnegative coefficients. He obtained this result by first proving finite dimensional analogues involving ultraspherical polynomials and then going to the limit. As a by-product he obtained this representation for all continuous functions operating on positive definite matrices.

We are now going to derive a slightly more general result (without continuity assumptions) in a completely different way. Our first main result will be that the set of all functions on $[-1, 1]$, operating on positive definite matrices and normalized in a suitable way, forms a Bauer simplex and that the monomials $t \mapsto t^n$, together with their two limit points, are precisely the extreme points of this simplex. The approximation lemma (1.1) will then give us Schoenberg's theorem for the real Hilbert sphere. A by-product of our considerations will be a new characterization of the so-called *generating functions*—well known and widely used in probability theory—which may be compared with Bochner's theorem for characteristic functions and with the characterization of the classical Laplace transforms given in Corollary 4.4.5. The functions $\psi$ on $[-1, 1]$ such that $\psi(\langle x, y\rangle)$ is negative definite on the Hilbert sphere turn out to have a very close connection with infinitely divisible probability measures on $\mathbb{N}_0$.

In order to prove the main results we need several lemmas which we prove first.

**3.1. Lemma.** *Let* $\frac{1}{2} < \delta < 1$, $\alpha \in \mathbb{R}_+$ *and put* $\varphi(x) := [\delta + (1 - \delta)e^{-x}]^\alpha$, $x \in \mathbb{R}_+$. *Then* $\varphi$ *is positive definite on the semigroup* $\mathbb{R}_+$ *if and only if* $\alpha \in \mathbb{N}_0$.

PROOF. The Laplace transform of $\delta\varepsilon_0 + (1 - \delta)\varepsilon_1$ is given by the function $\delta + (1 - \delta)e^{-x}$, therefore $\varphi$ is positive definite for all $\alpha = 0, 1, 2, \ldots$, cf. 4.4.5 and 3.1.12.

Suppose now that $\alpha$ is not an integer. We have

$$\varphi(x) = [1 + (1 - \delta)(e^{-x} - 1)]^\alpha = \sum_{n=0}^{\infty} \binom{\alpha}{n} (1 - \delta)^n (e^{-x} - 1)^n,$$

so that $\varphi$ is the Laplace transform of the signed measure

$$\mu := \sum_{n=0}^{\infty} \binom{\alpha}{n} (1 - \delta)^n (\varepsilon_1 - \varepsilon_0)^{*n}$$

(note that this series converges in total variation). In view of 4.2.10 we only have to show that $\mu$ is not a positive measure. Now

$$\mu = \sum_{n=0}^{\infty} \sum_{m=0}^{n} \binom{\alpha}{n}\binom{n}{m} (1 - \delta)^n (-1)^{n-m} \varepsilon_m$$

$$= \sum_{m=0}^{\infty} \sum_{n=0}^{\infty} \binom{\alpha}{n}\binom{n}{m} (1 - \delta)^n (-1)^{n-m} \varepsilon_m,$$

and therefore

$$\mu(\{m\}) = \sum_{n=0}^{\infty} \binom{\alpha}{n}\binom{n}{m} (1 - \delta)^n (-1)^{n-m} = \frac{(1 - \delta)^m}{m!} \sum_{n=0}^{\infty} (\delta - 1)^{n-m} \binom{\alpha}{n} n^{(m)}$$

$$= \frac{(1 - \delta)^m}{m!} f^{(m)}(\delta - 1),$$

where $f(t) := (1 + t)^\alpha$. We choose $m := [\alpha] + 2$; then

$$f^{(m)}(\delta - 1) = \alpha(\alpha - 1) \cdot \cdots \cdot (\alpha - [\alpha] - 1)\delta^{\alpha - [\alpha] - 2} < 0$$

implying $\mu(\{m\}) < 0$. □

**3.2. Lemma.** *If $n \in \mathbb{N}$ is odd then $|\cos(t)|^n$ is not positive definite in the group sense on $\mathbb{R}$. Likewise, if $n \in \mathbb{N}$ is even then $|\cos(t)|^n \operatorname{sgn}(\cos(t))$ is not positive definite.*

PROOF. Assume that $|\cos(t)|^n$ is positive definite; then by Bochner's theorem there exists a symmetric probability measure $\mu$ on $\mathbb{R}$ such that

$$|\cos(t)|^n = \int_{-\infty}^{\infty} e^{it\lambda}\, d\mu(\lambda) = \int_{-\infty}^{\infty} \cos(t\lambda)\, d\mu(\lambda), \qquad t \in \mathbb{R}.$$

From $|\cos(\pi)|^n = 1$ we see that $\mu(2\mathbb{Z}) = 1$. Now $(\cos(t))^{2n}$ is the characteristic function of $\kappa := \nu^{*2n}$ where $\nu := \frac{1}{2}(\varepsilon_1 + \varepsilon_{-1})$. The measure $\kappa$ is concentrated on $[-2n, 2n] \cap \mathbb{Z}$ and $\mu * \mu = \kappa$, hence $\mu([-n, n] \cap \mathbb{Z}) = 1$. If $n$ is odd then $\mu$ is even concentrated on $[-(n-1), n-1] \cap \mathbb{Z}$, contradicting the fact that $\mu * \mu(\{2n\}) = \kappa(\{2n\}) > 0$.

The second statement may be proved similarly. □

We now introduce $C_1$ to be the set of all functions $\varphi: [-1, 1] \to \mathbb{R}$ which operate on positive definite matrices in the sense that whenever $(a_{jk})$ is a positive definite matrix with entries $a_{jk} \in [-1, 1]$, then $(\varphi(a_{jk}))$ is again positive definite. By Schur's theorem (3.1.12) $C_1$ is a multiplicative closed convex cone in $\mathbb{R}^{[-1,1]}$. It is immediate that each $\varphi \in C_1$ is nonnegative on $[0, 1]$. For $t \in [-1, 1]$ the matrix $\begin{pmatrix} 1 & t \\ t & 1 \end{pmatrix}$ is positive definite, hence so is

$$\begin{pmatrix} \varphi(1) & \varphi(t) \\ \varphi(t) & \varphi(1) \end{pmatrix}$$

if $\varphi \in C_1$, and so $|\varphi(t)| \leqq \varphi(1)$ for all $t \in [-1, 1]$. Therefore

$$K_1 := \{\varphi \in C_1 \mid \varphi(1) = 1\}$$

is a compact convex base for the cone $C_1$.

**3.3. Lemma.** *Let $\varphi \in C_1$ and $t \in [-1, 1]$. Put $\varphi_1(s) := \varphi(s) + \varphi(st)$, $\varphi_2(s) := \varphi(s) - \varphi(st)$ for $s \in [-1, 1]$. Then $\varphi_1$ and $\varphi_2$ also belong to $C_1$.*

PROOF. Let $(a_{jk})$ be a fixed positive definite $n \times n$ matrix, where $a_{jk} \in [-1, 1]$ for all $j, k = 1, \ldots, n$, and let $(c_1, \ldots, c_n) \in \mathbb{R}^n$ also be fixed. Define $\Phi: [-1, 1] \to \mathbb{R}$ by

$$\Phi(s) := \sum_{j,k=1}^{n} c_j c_k \varphi(s a_{jk}).$$

We claim that $\Phi \in C_1$. In fact, for any positive definite $m \times m$ matrix $(b_{pq})$, $|b_{pq}| \leqq 1$ for all $p, q = 1, \ldots, m$, and for all $(d_1, \ldots, d_m) \in \mathbb{R}^m$ we get

$$\sum_{p,q=1}^{m} d_p d_q \Phi(b_{pq}) = \sum_{p,q=1}^{m} \sum_{j,k=1}^{n} d_p d_q c_j c_k \varphi(b_{pq} a_{jk}),$$

which is nonnegative because the tensor product $(a_{jk}) \otimes (b_{pq})$ is positive definite, too, see 3.1.13. In particular,

$$|\Phi(t)| \leqq \Phi(1) = \sum_{j,k=1}^{n} c_j c_k \varphi(a_{jk}) \qquad \text{for all} \quad t \in [-1, 1]$$

and this shows that $\varphi_1$, $\varphi_2$ both belong to $C_1$. □

**3.4. Theorem.** *The set* $K_1$ *is a Bauer simplex whose extreme points consist of the monomials* $t \mapsto t^n$, $n \in \mathbb{N}_0$, *and the two discontinuous functions* $1_{\{1\}} - 1_{\{-1\}}$ *and* $1_{\{-1,1\}}$.

PROOF. Let $\varphi \in \mathrm{ex}(K_1)$ and $t \in [-1, 1]$. By Lemma 3.3 there exists a constant $\lambda = \lambda(t) \geqq 0$ such that

$$\varphi(s) + \varphi(st) = \lambda\varphi(s) \qquad \text{for all} \quad s \in [-1, 1],$$

hence $\lambda = 1 + \varphi(t)$ and therefore

$$\varphi(st) = \varphi(s)\varphi(t) \qquad \text{for all} \quad s, t \in [-1, 1],$$

i.e. $\varphi$ is multiplicative. It is well known that there exists $\alpha \in [0, \infty]$ such that for all $s \in ]0, 1]$, $\varphi(s) = s^\alpha$ where $1^\infty := 1$, $s^\infty := 0$ for $s \in ]0, 1[$. If $\alpha \in \mathbb{R}_+ \setminus \mathbb{N}_0$ then by Lemma 3.1 there is a positive definite matrix $(a_{jk})$ with $a_{jk} \in ]0, 1]$ such that $(a_{jk}^\alpha)$ is not positive definite. We conclude that $\alpha \in \mathbb{N}_0 \cup \{\infty\}$. Of course, $\varphi(0) \in \{0, 1\}$ and $\varphi(0) = 1$ only if $\varphi \equiv 1$, hence for $\alpha = \infty$ we get $\varphi = 1_{\{-1,1\}}$ or $\varphi = 1_{\{1\}} - 1_{\{-1\}}$. Let $\alpha = n \in \mathbb{N}_0$ be finite. If $\varphi(-1) = 1$ then $\varphi(t) = |t|^n$; by Lemma 3.2 $n$ has to be even, therefore $\varphi(t) = t^n$ for all $t \in [-1, 1]$, except for the special case that $n = 0$ and $\varphi = 1_{[-1,1]\setminus\{0\}}$. If $\varphi(-1) = -1$ then $\varphi(t) = |t|^n \cdot \mathrm{sgn}(t)$ and for $n > 0$ we get from Lemma 3.2 that $n$ must be odd, i.e. $\varphi(t) = t^n$ for all $t \in [-1, 1]$ holds again.

Two cases are left:

$$\varphi(t) = \mathrm{sgn}(t) \qquad \text{and} \qquad \varphi(t) = \mathrm{sgn}^2(t) = 1_{[-1,1]\setminus\{0\}}(t).$$

We shall show that these two functions do not belong to $K_1$. Let $t_1 = 0$, $t_2 = \pi/2$, $t_3 = \pi$ and put $a_{jk} := \frac{1}{2} + \frac{1}{2}\cos(t_j - t_k)$, $j, k = 1, 2, 3$. The matrix

$$(a_{jk}) = \begin{pmatrix} 1 & \frac{1}{2} & 0 \\ \frac{1}{2} & 1 & \frac{1}{2} \\ 0 & \frac{1}{2} & 1 \end{pmatrix}$$

is positive definite, but

$$A := (1_{]0,1]}(a_{jk})) = \begin{pmatrix} 1 & 1 & 0 \\ 1 & 1 & 1 \\ 0 & 1 & 1 \end{pmatrix}$$

has not this property, because $(1, -1, 1)A(1, -1, 1)' = -1 < 0$. This finishes the first half of the proof.

Let $E := \{t \mapsto t^n | n \in \mathbb{N}_0\} \cup \{1_{\{1\}} - 1_{\{-1\}}, 1_{\{1,-1\}}\}$. By Theorem 3.1.12 the monomials belong to $K_1$ and so do their two limit points, hence $E \subseteq K_1$. Let $S$ denote the multiplicative semigroup $[-1, 1]$, then $K_1 \subseteq \mathscr{P}_1^b(S)$ and $E \subseteq \hat{S}$. In Theorem 4.2.8 we have seen $\hat{S} = \mathrm{ex}(\mathscr{P}_1^b(S))$, so that

$$\mathrm{ex}(K_1) \subseteq E \subseteq K_1 \cap \mathrm{ex}(\mathscr{P}_1^b(S))$$

and finally $\mathrm{ex}(K_1) = E$.

It is very easy to see that every function in $K_1$ has a unique integral (= series) representation over $E$. Therefore $K_1$ is a Bauer simplex. □

**3.5. Corollary.** *Every $\varphi \in C_1$ admits a unique series representation of the form*

$$\varphi(t) = \sum_{n=0}^{\infty} a_n t^n + a_{-1}[1_{\{1\}}(t) - 1_{\{-1\}}(t)] + a_{-2} 1_{\{-1,1\}}(t),$$

*where all $a_n \geqq 0$ and $\sum a_n < \infty$. The function $\varphi$ is continuous if and only if $a_{-1} = a_{-2} = 0$, and $\varphi$ is then even an analytic function.*

Now let $H$ denote an infinite dimensional real Hilbert space with scalar product $\langle \cdot, \cdot \rangle$ and denote by $X := \{x \in H \mid \|x\| = 1\}$ the unit sphere of $H$. Restricted to $X \times X$, the scalar product is a mapping onto the interval $[-1, 1]$, and plainly for every $\varphi \in C_1$ the composed map $\varphi(\langle x, y \rangle)$ is a positive definite kernel on $X \times X$.

The next theorem shows that the converse of this result also holds.

**3.6. Theorem.** *Let $\varphi: [-1, 1] \to \mathbb{R}$ be a function such that $\varphi(\langle x, y \rangle)$ is a positive definite kernel on $X \times X$, then $\varphi$ belongs to the cone $C_1$.*

PROOF. Let $(a_{jk})$ be a positive definite $p \times p$ matrix with entries $a_{jk} \in [-1, 1]$. We have to show that the matrix $(\varphi(a_{jk}))$ is positive definite, too, and shall again apply Lemma 1.1. First we choose vectors $x_1, \ldots, x_p \in \mathbb{R}^p$ such that $a_{jk} = \langle x_j, x_k \rangle$ for all $j, k = 1, \ldots, p$. Without loss of generality we assume $H = l^2(\mathbb{N})$ and define $x_{j\alpha}^n = (x_{j\alpha}^n(1), x_{j\alpha}^n(2), \ldots) \in H$, $1 \leqq j \leqq p$, $1 \leqq \alpha \leqq n$ as the columns in the matrix on the next page, where empty places should be read as zeros.

Formally, this means

$$x_{j\alpha}^n(\sigma) := \begin{cases} x_j(\sigma) & \text{if } 1 \leqq \sigma \leqq p, \\ \sqrt{1 - \|x_j\|^2} & \text{if } \sigma = p + (j-1)n + \alpha, \\ 0 & \text{otherwise.} \end{cases}$$

(Note that $\|x_j\|^2 = a_{jj} \leqq 1$.) Then

$$\langle x_{j\alpha}^n, x_{k\beta}^n \rangle = \begin{cases} \langle x_j, x_k \rangle = a_{jk} & \text{if } j \neq k, \\ \|x_j\|^2 = a_{jj} & \text{if } j = k \text{ but } \alpha \neq \beta, \\ 1 & \text{if } j = k \text{ and } \alpha = \beta, \end{cases}$$

$$
\begin{array}{cccc|cccc|c|cccc}
x^n_{11} & x^n_{12} & \cdots & x^n_{1n} & x^n_{21} & x^n_{22} & \cdots & x^n_{2n} & \cdots & x^n_{p1} & x^n_{p2} & \cdots & x^n_{pn} \\
\hline
x_1(1) & x_1(1) & \cdots & x_1(1) & x_2(1) & x_2(1) & \cdots & x_2(1) & \cdots & x_p(1) & x_p(1) & \cdots & x_p(1) \\
x_1(2) & x_1(2) & \cdots & x_1(2) & x_2(2) & x_2(2) & \cdots & x_2(2) & \cdots & x_p(2) & x_p(2) & \cdots & x_p(2) \\
\vdots & \vdots & & \vdots & \vdots & \vdots & & \vdots & \ddots & \vdots & \vdots & & \vdots \\
x_1(p) & x_1(p) & \cdots & x_1(p) & x_2(p) & x_2(p) & \cdots & x_2(p) & \cdots & x_p(p) & x_p(p) & \cdots & x_p(p) \\
\hline
\sqrt{1-\|x_1\|^2} & 0 & \cdots & 0 & & & & & & & & & \\
0 & \sqrt{1-\|x_1\|^2} & \cdots & 0 & & & & & & & & & \\
0 & 0 & \cdots & 0 & & 0 & & & 0 & & 0 & & \\
\vdots & \vdots & & \vdots & & & & & & & & & \\
0 & 0 & \cdots & \sqrt{1-\|x_1\|^2} & & & & & & & & & \\
\hline
 & & & & \sqrt{1-\|x_2\|^2} & & & & & & & & \\
 & 0 & & & & \sqrt{1-\|x_2\|^2} & & & 0 & & 0 & & \\
 & & & & & & \ddots & & & & & & \\
 & & & & & & & \sqrt{1-\|x_2\|^2} & & & & & \\
\hline
 & \vdots & & & \vdots & & & & \vdots & & \vdots & & \\
\hline
 & & & & & & & & & \sqrt{1-\|x_p\|^2} & & & \\
 & 0 & & & & 0 & & & 0 & & \sqrt{1-\|x_p\|^2} & & \\
 & & & & & & & & & & & \ddots & \\
 & & & & & & & & & & & & \sqrt{1-\|x_p\|^2}
\end{array}
$$

so that $\{x_{j\alpha}^n\} \subseteq X$ and $\varphi(\langle x_{j\alpha}^n, x_{k\beta}^n\rangle) = \varphi(a_{jk})$ for all $(j, \alpha) \neq (k, \beta)$. Moreover $\varphi$ is obviously bounded by $\varphi(1)$ and therefore Lemma 1.1 implies the positive definiteness of $(\varphi(a_{jk}))$. □

Let us insert here a short discussion of the so-called *generating functions*—an adequate and widely used transform method in probability theory. They are defined for probability measures on $\mathbb{N}_0$ (resp. on $\mathbb{N}_0^k$ in the multivariate case). If $\mu \in M_+^1(\mathbb{N}_0)$ then its generating function $\hat{\mu}: [-1, 1] \to \mathbb{R}$ is given by

$$\hat{\mu}(t) := \sum_{n=0}^{\infty} \mu(\{n\})t^n.$$

It is immediate that $\hat{\mu} = \hat{\nu}$ implies $\mu = \nu$ and that $\widehat{\mu * \nu} = \hat{\mu} \cdot \hat{\nu}$, so that this transform already shares some well-known properties of Fourier and Laplace transformation. We can add the following Bochner-type characterization theorem:

**3.7. Proposition.** *A function $\varphi: [-1, 1] \to \mathbb{R}$ is the generating function of some probability measure $\mu$ on $\mathbb{N}_0$ if and only if*

(i) $\varphi(1) = 1$;
(ii) $\varphi$ *is continuous*;
(iii) *for each positive definite matrix $(a_{jk})$ with entries in $[-1, 1]$ the matrix $(\varphi(a_{jk}))$ is also positive definite.*

PROOF. As a consequence of Corollary 3.5 the three conditions (i)–(iii) describe exactly the continuous functions in $K_1$, i.e. the set of functions representable as a power series with nonnegative coefficients summing up to 1. □

**3.8. Remark.** In view of Theorem 3.6 condition (iii) above may be replaced by

(iii)′ $\varphi(\langle x, y\rangle)$ is a positive definite kernel on $X \times X$, $X$ being the unit sphere of some infinite dimensional Hilbert space.

**3.9. Remark.** If we replace condition (iii) in Proposition 3.7 instead of (iii)′ by

(iii)″ for all finite subsets $\{s_1, \ldots, s_n\} \subseteq [-1, 1]$, $n \in \mathbb{N}$, the matrix $(\varphi(s_j s_k))$ is positive definite

and remove condition (i), then we obtain exactly the family of continuous positive definite functions on the multiplicative semigroup $[-1, 1]$. Such functions $\varphi$ have the unique integral representation

$$\varphi(t) = \int_{[0, \infty[} |t|^\alpha \, d\mu(\alpha) + \int_{]0, \infty[} |t|^\beta \cdot \operatorname{sgn}(t) \, d\nu(\beta)$$

with finite Radon measures $\mu$, $\nu$ (see Exercise 4.2.17). Sharpening (iii)″ to (iii) therefore corresponds to the requirement that $\mu$ is concentrated on $2\mathbb{N}_0$ and $\nu$ is concentrated on $2\mathbb{N}_0 + 1$.

As an application of our Bochner-type characterization, we give a short proof of the continuity theorem for generating functions:

**3.10. Proposition.** *Let $\mu_1, \mu_2, \ldots$ be a sequence of probability measures on $\mathbb{N}_0$ and suppose the corresponding generating functions $\hat{\mu}_1, \hat{\mu}_2, \ldots$ to be pointwise convergent to a limit function $\varphi$. Then, if $\varphi$ is continuous at $t = 1$, $\varphi$ is also the generating function of some probability measure $\mu$ on $\mathbb{N}_0$ and, moreover, $\mu_n \to \mu$ weakly, i.e.*

$$\mu_n(\{m\}) \to \mu(\{m\}) \qquad \textit{for all} \quad m \in \mathbb{N}_0.$$

Proof. Looking at Proposition 3.7 we see that the properties (i) and (iii) are preserved in the limit whereas the integral representation given in 3.5 shows that $\varphi$ is continuous as soon as it is continuous at $t = 1$. Hence $\varphi = \hat{\mu}$ for some $\mu \in M_+^1(\mathbb{N}_0)$.

Identifying a natural number with the monomial of the corresponding degree we may view $\mathbb{N}_0$ as a subset of $[-1, 1]^\wedge$. By Theorem 4.2.11 the measures $\mu_n$ converge to $\mu$ as elements of $M_+^1([-1, 1]^\wedge)$ and they are all concentrated on the Borel subset $\mathbb{N}_0$ of $[-1, 1]^\wedge$, therefore by Exercise 2.3.6 $\lim \mu_n = \mu$ weakly in the space $M_+^1(\mathbb{N}_0)$, and this convergence is equivalent with $\mu_n(B) \to \mu(B)$ for all $B \subseteq \mathbb{N}_0$, or with

$$\mu_n(\{m\}) \to \mu(\{m\}) \qquad \text{for all} \quad m \in \mathbb{N}$$

because of the portmanteau theorem (2.3.1), noting that all subsets of $\mathbb{N}_0$ are open and closed. □

Again let $(a_{jk})$ be a positive definite matrix with entries $a_{jk} \in [-1, 1]$, and let $\varphi \in K_1$. Then the new matrix $(1 - \varphi(a_{jk}))$ is negative definite, $\sum c_j = 0$ implying

$$\sum c_j c_k (1 - \varphi(a_{jk})) = -\sum c_j c_k \varphi(a_{jk}) \leqq 0.$$

This shows that $1 - \varphi$ belongs to the closed convex cone $C_2$ of all functions $\psi: [-1, 1] \to \mathbb{R}$ which satisfy the two conditions:

(i) $\psi(1) = 0$;
(ii) $\psi(a_{jk})$ is negative definite whenever $(a_{jk})$ is positive definite and $a_{jk} \in [-1, 1]$ for all $j, k$.

If $\psi \in C_2$ then in particular the $2 \times 2$ matrix $\begin{pmatrix} \psi(1) & \psi(t) \\ \psi(t) & \psi(1) \end{pmatrix}$ is negative definite, implying $\psi \geqq 0$ by 3.1.7. It is immediate that $C_2$ is a (proper) subcone of $\mathcal{N}_+([-1, 1])$, and since the semigroup $[-1, 1]$ contains an absorbing element, $\psi$ is bounded by $\sqrt{5}\psi(0)$, see Proposition 4.3.4, so that $K_2 := \{\psi \in C_2 \mid \psi(0) \leqq 1\}$ is a compact convex base for the cone $C_2$.

**3.11. Theorem.** *The transformation $T: K_1 \to K_2$ defined by $T(\varphi) := 1 - \varphi$ is an affine homeomorphism.*

PROOF. From Corollary 4.3.16 we know the two Bauer simplices $\mathscr{P}_1^b([-1, 1])$ and $\{\psi \in \mathscr{N}^l([-1, 1]) \mid \psi(1) = 0,\ \psi(0) \leqq 1\}$ to be isomorphic under the mapping $\varphi \mapsto 1 - \varphi$. Therefore, if $\psi \in K_2$ then $\varphi := 1 - \psi \in \mathscr{P}_1^b([-1, 1])$ and it remains to be shown that $\varphi$ even belongs to the smaller set $K_1$, i.e. denoting by $\nu \in M_+^1([-1, 1]\hat{\ })$ the measure representing $\varphi$, we have to make sure that $\nu$ is concentrated on $\text{ex}(K_1)$, the closure of the monomials in $[-1, 1]\hat{\ }$.

Since $\psi \in K_2$ it follows $e^{-t\psi} \in K_1$ for all $t > 0$, hence $e^{-t\psi} = \hat{\mu}_t$ where $\mu_t \in M_+^1(\text{ex}(K_1))$ by Theorem 3.4. The negative definite function $\psi$ is bounded and has the integral representation

$$\psi(s) = \int_{[-1,1]\hat{\ }\setminus\{1\}} (1 - \rho(s))\, d\mu(\rho),$$

where the finite Radon measure $\mu$ is the vague limit

$$\mu = \lim_{0 < t \to 0} \frac{1}{t} (\mu_t | [-1, 1]\hat{\ } \setminus \{1\})$$

on the locally compact space $[-1, 1]\hat{\ } \setminus \{1\}$, see Proposition 4.3.15. Since all the measures $\mu_t | [-1, 1]\hat{\ } \setminus \{1\}$ are concentrated on the closed subset $\text{ex}(K_1) \setminus \{1\}$ of $[-1, 1]\hat{\ } \setminus \{1\}$, the limit measure $\mu$ is concentrated there, too, by Exercise 2.4.13. Now

$$\mu + [1 - \mu(\text{ex}(K_1) \setminus \{1\})] \cdot \varepsilon_1 = \nu$$

is the representing measure for $\varphi$, and it is concentrated on $\text{ex}(K_1)$ as asserted. □

**3.12. Corollary.** *Every $\psi \in C_2$ admits a unique series representation of the form*

$$\psi(s) = \sum_{n=1}^{\infty} a_n(1 - s^n) + a_{-1}[2 \cdot 1_{\{-1\}}(s) + 1_{]-1,1[}(s)] + a_{-2} 1_{]-1,1[}(s),$$

*where all $a_n \geqq 0$ and $\sum a_n < \infty$. The function $\psi$ is continuous if and only if $a_{-1} = a_{-2} = 0$.*

**3.13. Corollary.** *A real-valued function $\psi$ on $[-1, 1]$ belongs to the cone $C_2$ if and only if $\psi(\langle x, y \rangle)$ is a negative definite kernel on the infinite dimensional Hilbert sphere and $\psi(1) = 0$.*

In the preceding chapters we saw already the close connection between negative definite and infinitely divisible positive definite kernels. We shall now give a further example of this kind. Let us agree that calling a probability measure $\mu$ on $\mathbb{N}_0$ *infinitely divisible* will always tacitly assume $\mu$ to be infinitely divisible inside $M_+^1(\mathbb{N}_0)$, i.e. $\mu = (\mu_n)^{*n}$ for suitable $\mu_n \in M_+^1(\mathbb{N}_0)$ and all $n \in \mathbb{N}$. The following result may be considered as a Lévy–Khinchin representation theorem for infinitely divisible generating functions.

**3.14. Proposition.** *A probability measure $\mu$ on $\mathbb{N}_0$ is infinitely divisible if and only if $\hat{\mu}(t) > 0$ for all $t \in [-1, 1]$ and $\psi := -\log \hat{\mu}$ fulfils the three conditions:*

(i) $\psi(1) = 0$;
(ii) *$\psi$ is continuous;*
(iii) *for each positive definite matrix $(a_{jk})$ with entries in $[-1, 1]$ the matrix $(\psi(a_{jk}))$ is negative definite.*

*If the conditions are fulfilled there exist uniquely determined numbers $b_1, b_2, \ldots \geqq 0$, $\sum b_n < \infty$, such that*

$$\psi(s) = \sum_{n=1}^{\infty} b_n(1 - s^n) \qquad \textit{for all} \quad s \in [-1, 1].$$

Proof. If $\hat{\mu} > 0$ and $\psi := -\log \hat{\mu}$ fulfils (i)–(iii) then $\psi$ is a continuous function in $C_2$, hence by 3.2.2 $\exp(-t\psi)$ is a continuous function in $K_1$, i.e. a generating function for all $t > 0$. In particular

$$\hat{\mu} = [\exp(-\psi/n)]^n$$

for all $n \in \mathbb{N}$, showing that $\mu$ is infinitely divisible.

Now let us assume $\mu = (\mu_n)^{*n}$ for suitable probability measures $\mu_1, \mu_2, \ldots$ on $\mathbb{N}_0$; equivalently, $\hat{\mu} = (\hat{\mu}_n)^n$ for the corresponding generating functions. If $\hat{\mu}(t_0) = 0$ for some $t_0 \in ]-1, 1[$, then all derivatives of $\hat{\mu}$ in $t_0$ would vanish and then $\hat{\mu} \equiv 0$ in contradiction to $\hat{\mu}(1) = 1$ and to $\hat{\mu}$'s continuity on $[-1, 1]$. Hence $\hat{\mu}(t) > 0$ at least for all $t \in ]-1, 1]$ and consequently all the functions $\hat{\mu}_n$ are strictly positive on $]-1, 1]$. Now $\psi := -\log \hat{\mu}$ is well defined, continuous on $[-1, 1]$ and finite on $]-1, 1]$, but still we have to exclude that $\psi(-1) = \infty$. Let $(a_{jk})$ be a positive definite matrix, $a_{jk} \in ]-1, 1]$ for all $j, k$. Then $(\psi(a_{jk}))$ is negative definite, because for all $n \in \mathbb{N}$

$$(e^{-(1/n)\psi(a_{jk})}) = (\hat{\mu}(a_{jk})^{1/n}) = (\hat{\mu}_n(a_{jk}))$$

is positive definite by Proposition 3.7. In particular, $\psi | ]-1, 1]$ is a non-negative negative definite function on the semigroup $]-1, 1]$, therefore bounded by Proposition 4.3.4, and since $\psi$ is continuous on $[-1, 1]$ we get $\psi(-1) < \infty$, i.e. $\hat{\mu}(-1) > 0$, so that $\psi$ is indeed a (real-valued) continuous function in the cone $C_2$.

The asserted representation of $\psi$ follows immediately from Corollary 3.12. □

**3.15. Corollary.** *Let $\mu \in M_+^1(\mathbb{N}_0)$ be infinitely divisible. Then $\mu$ is the distribution of a random variable $X = \sum_{n=1}^{\infty} nX_n$, where the $X_1, X_2, \ldots$ are independent, and where each $X_n$ is Poisson distributed with parameter $b_n$, and $\sum_{n=1}^{\infty} b_n < \infty$.*

Proof. By 3.14 $\hat{\mu} = \exp(-\psi)$ with $\psi(s) = \sum_{n=1}^{\infty} b_n(1 - s^n)$, $b_n \geqq 0$ and $\sum b_n < \infty$. The Poisson distribution $\pi_b = \sum_{k=0}^{\infty} e^{-b}(b^k/k!)\varepsilon_k$ with parameter

$b \geqq 0$ has the generating function

$$\hat{\pi}_b(s) = \sum_{k=0}^{\infty} e^{-b} \frac{(bs)^k}{k!} = e^{-b(1-s)},$$

and if $Y$ is a random variable with distribution $\pi_b$, then $nY$ has the distribution $\sum_{k=0}^{\infty} e^{-b}(b^k/k!)\varepsilon_{nk}$ with generating function $\exp[-b(1 - s^n)]$. Choosing now a sequence of independent random variables $X_1, X_2, \ldots$, where $X_n$ has the distribution $\pi_{b_n}$, the (well-defined) infinite series $X := \sum nX_n$ has the generating function

$$\prod_{n=1}^{\infty} \exp[-b_n(1 - s^n)] = \exp\left[-\sum_{n=1}^{\infty} b_n(1 - s^n)\right] = \exp(-\psi(s))$$

implying that $X$ has the prescribed distribution $\mu$. □

**3.16. Exercise.** Let $\mu_n$ be the binomial distribution $B(n, p_n)$, i.e.

$$\mu_n = \sum_{k=0}^{n} \binom{n}{k} p_n^k (1 - p_n)^{n-k} \varepsilon_k,$$

where $n \in \mathbb{N}$ and $p_n \in [0, 1]$. Show by using 3.10 that if $\lim_{n\to\infty} np_n = b \in \mathbb{R}_+$, then $\mu_n$ tends weakly to the Poisson distribution $\pi_b$.

**3.17. Exercise.** Let $\mu_1, \mu_2, \ldots$ be a sequence of infinitely divisible probability measures on $\mathbb{N}_0$ tending weakly to $\mu \in M_+^1(\mathbb{N}_0)$. Show that $\mu$ is infinitely divisible inside $M_+^1(\mathbb{N}_0)$.

**3.18. Exercise.** If $\mu \in M_+^1(\mathbb{N}_0)$ is infinitely divisible then

$$\mu(2\mathbb{N}_0) > \mu(2\mathbb{N}_0 + 1).$$

**3.19. Exercise.** Let $K$ be the set of all real-valued functions $\varphi$ on the unit interval $[0, 1]$ operating on positive definite matrices with entries from $[0, 1]$ and normalized by $\varphi(1) = 1$. Show that $K$ is a Bauer simplex and determine the extreme points of $K$.

**3.20. Exercise.** For a probability measure $\mu \in M_+^1(\mathbb{N}_0^2)$ the (bivariate) generating function $\hat{\mu}: [-1, 1]^2 \to \mathbb{R}$ is defined by

$$\hat{\mu}(s, t) := \sum_{n,m=0}^{\infty} \mu(\{(n, m)\}) s^n t^m.$$

Show that a given function $\varphi: [-1, 1]^2 \to \mathbb{R}$ is a bivariate generating function if and only if

(i) $\varphi(1, 1) = 1$;
(ii) $\varphi$ is continuous;
(iii) for each pair of positive definite matrices $(a_{jk})$, $(b_{jk})$ of equal size and with entries in $[-1, 1]$ the new matrix $(\varphi(a_{jk}, b_{jk}))$ is also positive definite.

**3.21. Exercise.** If $\mu \in M_+^1(\mathbb{N}_0^k)$ is infinitely divisible (inside $\mathbb{N}_0^k$!), then $\mu$ is the distribution of some random vector $X$ where

$$X = (X_1, \ldots, X_k) = \sum_{n_1, \ldots, n_k = 1}^{\infty} (n_1 X_{n_1, \ldots, n_k}, \ldots, n_k X_{n_1, \ldots, n_k}),$$

the $\{X_{n_1, \ldots, n_k} | (n_1, \ldots, n_k) \in \mathbb{N}^k\}$ being independent and Poisson distributed with parameters $b_{n_1, \ldots, n_k}$, and $\sum b_{n_1, \ldots, n_k} < \infty$.

**3.22. Exercise.** Let $\varphi: [-1, 1] \to \mathbb{R}$ be given such that for all $n \in \mathbb{N}$ the function

$$\mathbb{R}^n \to \mathbb{R}$$

$$(t_1, \ldots, t_n) \mapsto \varphi\left(\prod_{j=1}^{n} \cos t_j\right)$$

is positive definite in the group sense. Then $\varphi$ belongs to the cone $C_1$ (and therefore has the representation stated in Corollary 3.5).

Conversely, if $\varphi \in C_1$, then $\varphi(\prod_{j=1}^n \cos t_j)$ is positive definite on $\mathbb{R}^n$ for all $n$.

**3.23. Exercise.** Show that the extreme points of the Bauer simplex $K_1$ are characterized by the three conditions:

(i) $\varphi \in C_1$;
(ii) $\varphi(1) > 0$;
(iii) $\varphi^2(t) = \varphi(t^2)$ for all $t \in [-1, 1]$.

# §4. Schoenberg's Theorem for the Complex Hilbert Sphere

In the preceding section we have seen that a continuous function $\varphi: [-1, 1] \to \mathbb{R}$, such that $\varphi(\langle x, y \rangle)$ is positive definite on the unit sphere of some infinite dimensional real Hilbert space, has a power series representation $\varphi(s) = \sum a_n s^n$ with $a_n \geqq 0$ and $\sum a_n < \infty$. We are now going to extend this result to the case of a complex Hilbert space $H$ (always infinite dimensional), and of course we have to replace the interval $[-1, 1]$ by the closed unit disc $D := \{z \in \mathbb{C} \mid |z| \leqq 1\}$ in the complex plane. Let $\mathbb{T} := \{z \in \mathbb{C} \mid |z| = 1\}$ denote the torus group. When we consider $\mathbb{T}$ as a discrete group, we shall write $\mathbb{T}_d$ instead of $\mathbb{T}$.

In analogy with the real case we introduce $C$ to be the set of all functions $\varphi: D \to \mathbb{C}$ operating on positive definite matrices in the sense that $(\varphi(a_{jk}))$ is positive definite whenever $(a_{jk})$ is a positive definite matrix with entries in $D$. By the Schur Product Theorem (3.1.12) $C$ is a multiplicative closed convex cone in $\mathbb{C}^D$. It is immediately seen that each $\varphi \in C$ is nonnegative on $[0, 1]$.

For $z \in D$ the matrix $\begin{pmatrix} 1 & z \\ \bar{z} & 1 \end{pmatrix}$ is positive definite, hence so is

$$\begin{pmatrix} \varphi(1) & \varphi(z) \\ \varphi(\bar{z}) & \varphi(1) \end{pmatrix}$$

if $\varphi \in C$, and therefore $\varphi(\bar{z}) = \overline{\varphi(z)}$ as well as $|\varphi(z)| \leqq \varphi(1)$ for all $z \in D$, showing, in particular, that

$$K := \{\varphi \in C \,|\, \varphi(1) = 1\}$$

is a compact convex base for $C$.

For the proof of Theorem 4.5 below, in which the extreme points of $K$ are determined, we need several lemmas which we prove first.

**4.1. Lemma.** *The set of numbers $c \in \mathbb{C}$ for which the matrix*

$$M_c := \begin{pmatrix} 1 & c & c \\ \bar{c} & 1 & c \\ \bar{c} & \bar{c} & 1 \end{pmatrix}$$

*is positive definite is the "pear-like" subset $\Omega$ of the unit disc (Figure 2)*

$$\Omega = \{c \in D \,|\, [3 - 2\,\mathrm{Re}(c)]|c|^2 \leqq 1\},$$

*having the following properties:*

(i) $\{c \in \mathbb{C} \,|\, |c| \leqq \frac{1}{2}\} \subseteq \Omega$.

(ii) *$\Omega$ is symmetric about the x-axis and its boundary curve in the upper half-plane is*

$$y = y(x) = (1 - x)\sqrt{\frac{2x + 1}{3 - 2x}} \qquad \textit{for} \quad x \in [-\tfrac{1}{2}, 1]$$

*with $y'(1) = -\sqrt{3}$.*

(iii) *For $t \in [\frac{1}{2}, 1]$ and $\theta \in [-\pi, \pi]$ we have $c = te^{i\theta} \in \Omega$ if and only if*

$$|\theta| \leqq \theta(t) := \arccos(\tfrac{3}{2}t^{-1} - \tfrac{1}{2}t^{-3}).$$

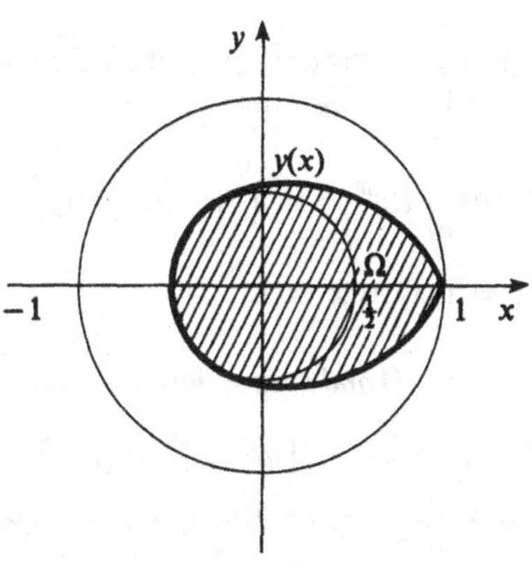


Figure 2

PROOF. We have $\det\begin{pmatrix} 1 & c \\ \bar{c} & 1 \end{pmatrix} = 1 - |c|^2$ and

$$\det M_c = 1 - [3 - 2\,\mathrm{Re}(c)]|c|^2$$

so the description of $\Omega$ follows from Theorem 3.1.17.

(i) is clear.

(ii) If $c = x + iy \in \Omega$ then $x \in [-1, 1]$ and $\det M_c = 1 - (3 - 2x)(x^2 + y^2) \geqq 0$, hence

$$y^2 \leqq \frac{(1 - x)^2(2x + 1)}{3 - 2x},$$

which shows that $x \in [-\frac{1}{2}, 1]$ and

$$|y| \leqq (1 - x)\sqrt{\frac{2x + 1}{3 - 2x}},$$

and the converse is also true. By differentiation of the function $y = y(x)$ we get $y'(1) = -\sqrt{3}$.

(iii) For $c = te^{i\theta}$ we have $\det M_c = 1 - [3 - 2t\cos\theta]t^2 \geqq 0$ if and only if $\cos\theta \geqq \frac{3}{2}t^{-1} - \frac{1}{2}t^{-3}$. Since $t \mapsto \frac{3}{2}t^{-1} - \frac{1}{2}t^{-3}$ maps $[\frac{1}{2}, 1]$ onto $[-1, 1]$ the assertion follows. □

**4.2. Lemma.** *Let $\varphi(te^{i\theta}) = t^k e^{ip\theta}$, where $k \in \mathbb{N}_0$, $p \in \mathbb{Z}$. If $\varphi \in K$ then $|p| \leqq k$.*

PROOF. If $\varphi \in K$ we get by Lemma 4.1 that $\varphi(\Omega) \subseteq \Omega$, in particular

$$\varphi(te^{i\theta(t)}) = t^k e^{ip\theta(t)} \in \Omega \qquad \text{for } \tfrac{1}{2} < t < 1,$$

where $\theta(t) = \arccos(\frac{3}{2}t^{-1} - \frac{1}{2}t^{-3})$. If $t^k > \frac{1}{2}$ and $|p\theta(t)| < \pi$, which is true for $t$ sufficiently close to 1, we have $|p\theta(t)| \leqq \theta(t^k)$, hence

$$|p| \leqq \lim_{t\to 1} \frac{\theta(t^k)}{\theta(t)} = \lim_{t\to 1} \frac{kt^{k-1}\theta'(t^k)}{\theta'(t)} = k$$

by l'Hôpital's rule because

$$\lim_{t\to 1} \theta'(t) = -\sqrt{3}.$$

The direct evaluation of this limit is tedious but elementary. It is, however, no coincidence that this limit is $y'(1)$ because

$$\lim_{t\to 1} \frac{d}{dt}(te^{i\theta(t)}) = 1 + i\lim_{t\to 1}\theta'(t)$$

is a tangent vector to $y = y(x)$ at $x = 1$. □

**4.3. Lemma.** *Let $\varphi \in C$ and $z \in D$ and define the functions $\varphi_1, \varphi_2, \varphi_3, \varphi_4\colon D \to \mathbb{C}$ by*

$$\varphi_{1,2}(w) = \varphi(w) \pm \tfrac{1}{2}[\varphi(zw) + \varphi(\bar{z}w)]$$

$$\varphi_{3,4}(w) = \varphi(w) \pm (1/2i)[\varphi(zw) - \varphi(\bar{z}w)].$$

*Then $\varphi_1, \ldots, \varphi_4$ all belong to $C$, too.*

PROOF. Let $(a_{jk})$ be a fixed positive definite $n \times n$ matrix with entries in $D$, and let $(c_1, \ldots, c_n) \in \mathbb{C}^n$ also be fixed. Define the function $\Phi: D \to \mathbb{C}$ by

$$\Phi(z) := \sum_{j,k=1}^{n} c_j \bar{c}_k \varphi(z a_{jk}).$$

We claim that $\Phi \in C$. In fact, for any positive definite $m \times m$ matrix $(b_{pq})$ with entries in $D$ and for all $(d_1, \ldots, d_m) \in \mathbb{C}^m$ we have

$$\sum_{p,q=1}^{m} d_p \bar{d}_q \Phi(b_{pq}) = \sum_{p,q=1}^{m} \sum_{j,k=1}^{n} d_p c_j \overline{d_q c_k} \varphi(b_{pq} a_{jk}) \geqq 0$$

because the tensor product $(b_{pq}) \otimes (a_{jk})$ is positive definite by Corollary 3.1.13.

Therefore

$$|\Phi(z)| \leqq \Phi(1) = \sum_{j,k=1}^{n} c_j \overline{c_k} \varphi(a_{jk}),$$

hence, denoting $G = \mathrm{Re}(\Phi)$ and $H = \mathrm{Im}(\Phi)$,

$$|G(z)| \leqq \Phi(1) \qquad \text{and} \qquad |H(z)| \leqq \Phi(1).$$

But $G(z) = \frac{1}{2} \sum c_j \overline{c_k} [\varphi(z a_{jk}) + \varphi(\bar{z} a_{jk})]$ and

$$H(z) = (1/2i) \sum c_j \overline{c_k} [\varphi(z a_{jk}) - \varphi(\bar{z} a_{jk})]$$

and thus

$$\sum c_j \overline{c_k} \varphi_\alpha(a_{jk}) \geqq 0$$

for $\alpha = 1, 2, 3, 4$, i.e., $\{\varphi_1, \varphi_2, \varphi_3, \varphi_4\} \subseteq C$. □

**4.4. Lemma.** *The set of continuous group characters on the torus $\mathbb{T}$ is dense in the set of all group characters on $\mathbb{T}$, i.e. in $\hat{\mathbb{T}}_d$.*

PROOF. The group $G$ of all continuous group characters on $\mathbb{T}$ is known to be canonically isomorphic to $\mathbb{Z}$. Assume that $\bar{G} \subsetneqq \hat{\mathbb{T}}_d$. Since the continuous characters on a locally compact abelian group separate points, there is a nontrivial continuous character $\xi$ on $\hat{\mathbb{T}}_d/\bar{G}$. Composing $\xi$ with the canonical projection we obtain a nontrivial continuous character $\eta$ on $\hat{\mathbb{T}}_d$, being 1 on $G$. By Pontryagin's duality theorem, cf. Rudin (1962), there is some $z \in \mathbb{T}$ such that $\eta(\rho) = \rho(z)$ for all $\rho \in \hat{\mathbb{T}}_d$. The special $\rho = \mathrm{id}_{\mathbb{T}}$ gives $1 = \eta(\rho) = \rho(z) = z$, so that $\eta \equiv 1$. This contradiction proves our lemma. □

**4.5. Theorem.** *The set $K$ of normalized operating functions is a Bauer simplex whose extreme points consist of the closure of $\{z^n \bar{z}^m \mid n, m \in \mathbb{N}_0\}$ in the pointwise topology of $\mathbb{C}^D$. More precisely, denoting by $\rho_0$ the zero extension of any group character $\rho$ on the discrete torus $\mathbb{T}_d$ (i.e. $\rho_0 = \rho$ on $\mathbb{T}$ and $\rho_0 | \mathring{D} = 0$), we have*

$$\mathrm{ex}(K) = \{z^n \bar{z}^m \mid n, m \in \mathbb{N}_0\} \cup \{\rho_0 \mid \rho \in \hat{\mathbb{T}}_d\}.$$

PROOF. Let $\varphi \in \text{ex}(K)$ and $z \in D$. By Lemma 4.3 there exist $\alpha = \alpha(z) \geqq 0$ and $\beta = \beta(z) \geqq 0$ such that

$$\varphi(w) + \tfrac{1}{2}[\varphi(zw) + \varphi(\bar{z}w)] = \alpha\varphi(w) \qquad \text{for} \quad w \in D,$$

$$\varphi(w) + \frac{1}{2i}[\varphi(zw) - \varphi(\bar{z}w)] = \beta\varphi(w) \qquad \text{for} \quad w \in D.$$

This implies $\alpha = 1 + \frac{1}{2}[\varphi(z) + \varphi(\bar{z})]$ and $\beta = 1 + (1/2i)[\varphi(z) - \varphi(\bar{z})]$ and further

$$\varphi(zw) + \varphi(\bar{z}w) = \varphi(z)\varphi(w) + \varphi(\bar{z})\varphi(w)$$

as well as

$$\varphi(zw) - \varphi(\bar{z}w) = \varphi(z)\varphi(w) - \varphi(\bar{z})\varphi(w),$$

so that finally $\varphi(zw) = \varphi(z)\varphi(w)$ for all $z, w \in D$, that is, $\varphi$ is multiplicative.

We conclude that $\rho = \varphi | \mathbb{T}$ is a character on the (discrete) torus, and from the proof of Theorem 3.4 we know that either $\varphi | ]-1, 1[ \equiv 0$ in which case the multiplicativity of $\varphi$ implies $\varphi | \mathring{D} \equiv 0$, hence $\varphi = \rho_0$, or there is a $k \in \mathbb{N}_0$ such that $\varphi(t) = t^k$ for all $t \in [-1, 1]$, and then $\varphi(te^{i\theta}) = t^k \rho(e^{i\theta})$ for $0 \leqq t \leqq 1$, $\theta \in \mathbb{R}$. We will now show that $\rho$ is continuous in the latter case. Since $\varphi(\Omega) \subseteq \Omega$, cf. Lemma 4.1, we have

$$t^k \rho(e^{i\theta}) \in \Omega \qquad \text{for} \quad \tfrac{1}{2} \leqq t < 1, \quad |\theta| \leqq \theta(t).$$

If $k = 0$ this implies that $\rho(e^{i\theta}) = 1$ for $|\theta| \leqq \theta(t)$, i.e. $\rho \equiv 1$. Suppose $k > 0$ and let $\varepsilon > 0$ be given. If $t$ is chosen so close to 1 that $t^k > \frac{1}{2}$ and $\theta(t^k) < \varepsilon$ then

$$|\arg(\rho(e^{i\theta}))| \leqq \theta(t^k) < \varepsilon \qquad \text{for} \quad |\theta| \leqq \theta(t)$$

by Lemma 4.1, and this shows that $\rho$ is continuous at 1, hence continuous. Therefore there is $p \in \mathbb{Z}$ such that $\rho(e^{i\theta}) = e^{ip\theta}$, and from Lemma 4.2 we get $|p| \leqq k$. Now since $\varphi(-1) = (-1)^k = \rho(e^{i\pi}) = e^{ip\pi}$, we see that $k$ and $p$ are of equal parity. Consequently, $n := (k + p)/2$ and $m := (k - p)/2$ are non-negative integers and for each $z = te^{i\theta} \in D$ we get

$$\varphi(z) = t^k e^{ip\theta} = t^{n+m} e^{i\theta(n-m)} = z^n \bar{z}^m.$$

We have now proved

$$\text{ex}(K) \subseteq \{z^n \bar{z}^m \mid n, m \in \mathbb{N}_0\} \cup \{\rho_0 \mid \rho \in \hat{\mathbb{T}}_d\}.$$

To prove the reverse inclusion we first remark that Theorem 3.1.12 implies that $(z \mapsto z^n \bar{z}^m) \in K$ for $n, m \in \mathbb{N}_0$. Furthermore, we see that

$$1_{\mathbb{T}}(z) = \lim_{n \to \infty} z^n \bar{z}^n$$

belongs to $K$, so since $K$ is closed under multiplication, the zero extension of every continuous group character belongs to $K$. Lemma 4.4 shows that the zero extension of arbitrary group characters on $\mathbb{T}$ also belongs to $K$. This shows that

$$\{z^n\bar{z}^m \mid n, m \in \mathbb{N}_0\} \cup \{\rho_0 \mid \rho \in \hat{\mathbb{T}}_d\} \subseteq \Gamma,$$

where

$$\Gamma = \{\varphi \in K \mid |\varphi(z)|^2 = \varphi(|z|^2) \text{ for all } z \in D\}$$

is defined in accordance with Neumann's result for $\gamma(z) = |z|^2$, cf. 2.5.11. By Corollary 2.5.12 we conclude that $\mathrm{ex}(K) = \Gamma$. This shows that $\mathrm{ex}(K)$ is compact and that

$$\mathrm{ex}(K) = \overline{\{z^n\bar{z}^m \mid n, m \in \mathbb{N}_0\}} = \{z^n\bar{z}^m \mid n, m \in \mathbb{N}_0\} \cup \{\rho_0 \mid \rho \in \hat{\mathbb{T}}_d\}.$$

The fact that $\mathrm{ex}(K)$ is closed under multiplication together with the Stone–Weierstrass theorem shows that $K$ is a Bauer simplex. □

Let $H$ denote a fixed infinite dimensional complex Hilbert space with scalar product $\langle \cdot, \cdot \rangle$ and denote by $S := \{x \in H \mid \|x\| = 1\}$ the unit sphere of $H$. Restricted to $S$ the scalar product is a mapping into the closed unit disc $D$ and plainly for every $\varphi \in K$ the composed map $\varphi(\langle x, y \rangle)$ is a positive definite kernel. The next theorem shows that the converse also holds.

**4.6. Theorem.** *Let $\varphi: D \to \mathbb{C}$ be a function such that:*

(i) $\varphi(1) = 1$;
(ii) $\varphi(\langle x, y \rangle)$ *is a positive definite kernel on* $S \times S$.

*Then $\varphi$ belongs to $K$.*

The proof is very similar to that of Theorem 3.6 and therefore omitted, cf. Christensen and Ressel (1982).

**4.7. Corollary.** *A continuous complex-valued function $\varphi$ on $D$ such that $\varphi(\langle x, y \rangle)$ is positive definite on the unit sphere of some infinite dimensional complex Hilbert space, has the unique series representation*

$$\varphi(z) = \sum_{n,m=0}^{\infty} a_{n,m} z^n \bar{z}^m \qquad \text{for all} \quad z \in D,$$

*where all coefficients $a_{n,m}$ are nonnegative and $\sum_{n,m=0}^{\infty} a_{n,m} < \infty$.*

PROOF. By Theorem 4.5 there is a uniquely determined finite Radon measure $\mu$ on $\mathrm{ex}(K)$ such that

$$\varphi(z) = \int_{\mathrm{ex}(K)} \rho(z)\, d\mu(\rho) \qquad \text{for all} \quad z \in D.$$

Let $\tilde{K}$ be the countable set of functions $z \mapsto z^n\bar{z}^m$, $n, m \in \mathbb{N}_0$. Of course, $\tilde{K}$ is measurable and with $a_{n,m} := \mu(\{z^n\bar{z}^m\})$ we have

$$\varphi(z) - \sum_{n,m=0}^{\infty} a_{n,m} z^n \bar{z}^m = \int_{\mathrm{ex}(K)\setminus\tilde{K}} \rho(z)\, d\mu(\rho)$$

showing that the function on the right-hand side is also continuous on all of $D$. The description of $\mathrm{ex}(K)$ given in Theorem 4.5, however, shows that this function is identically zero in $\mathring{D}$; therefore $\mu(\mathrm{ex}(K)\backslash\tilde{K}) = 0$. □

**4.8. Corollary.** *Let $\varphi_1, \varphi_2, \ldots$ be a sequence of continuous functions of the type described in Corollary 4.7 and let them have the series representations*

$$\varphi_k(z) = \sum_{n,m=0}^{\infty} a_{n,m}^{(k)} z^n \bar{z}^m, \qquad z \in D, \quad k = 1, 2, \ldots .$$

*If the pointwise limit $\varphi(z) = \lim_{k\to\infty} \varphi_k(z)$ exists for all $z \in D$ and if $\varphi | [0, 1]$ is continuous in 1, then $\varphi$ is continuous everywhere and for the coefficients $a_{n,m}$ in*

$$\varphi(z) = \sum_{n,m=0}^{\infty} a_{n,m} z^n \bar{z}^m$$

*we have $a_{n,m} = \lim_{k\to\infty} a_{n,m}^{(k)}$ for all $n, m \in \mathbb{N}_0$.*

Proof. The cone $C$ being closed we have of course $\varphi \in C$. Let $\varphi$ have the integral representation

$$\varphi(z) = \int_{\mathrm{ex}(K)} \rho(z)\, d\mu(\rho).$$

Then

$$0 = \int_{\mathrm{ex}(K)\backslash\tilde{K}} \rho\left(1 - \frac{1}{j}\right) d\mu(\rho) \to \int_{\mathrm{ex}(K)\backslash\tilde{K}} \rho(1)\, d\mu(\rho) = \mu(\mathrm{ex}(K)\backslash\tilde{K}),$$

so that $\mu$ is in fact concentrated on $\tilde{K}$. The affine isomorphism between the two Bauer simplices $K$ and $M_+^1(\mathrm{ex}(K))$ and the fact that the functions $z \mapsto z^n \bar{z}^m$ are isolated points in $\mathrm{ex}(K)$, which is easily seen, imply the convergence of the corresponding coefficients. □

**4.9. Exercise.** Let $\varphi$ be a real-valued continuous function on $D$ such that $(\varphi(a_{jk}))$ is positive definite for each positive definite matrix $(a_{jk})$ with entries in $D$. Show that in the representation of $\varphi$, as stated in Corollary 4.7, we have $a_{n,m} = a_{m,n}$ for all $n, m$.

**4.10. Exercise.** Let $\psi: D \to \mathbb{C}$ have the property that for each positive definite matrix $(a_{jk})$ with $a_{jk} \in D$ the transformed matrix $(\psi(a_{jk}))$ is negative definite, and suppose $\psi$ is normalized by the two conditions $\psi(1) = 0$, $\psi(0) \leqq 1$. Then $1 - \psi \in K$. (Conversely, but this is trivial, if $\varphi \in K$, then $\psi := 1 - \varphi$ has the properties stated above.)

# §5. The Real Infinite Dimensional Hyperbolic Space

In 1974 Faraut and Harzallah proved some results on positive and negative definite kernels defined on the real infinite dimensional hyperbolic space $X$ and which depend only on a kind of natural distance on $X$. These results are very similar to Schoenberg's theorems discussed earlier in this chapter, and in fact their method of proof also uses the corresponding finite dimensional results to obtain the infinite dimensional version by a rather complicated limiting procedure. However, our approximation lemma (1.1) again gives easy and short direct access to these relations, slightly more general while not assuming continuity.

The space $X$ is formally defined by

$$X := \left\{ x = (x_k)_{k \geqq 0} \in l^2_{\mathbb{R}}(\mathbb{N}_0) \,|\, x_0 \geqq 1, x_0^2 - \sum_{k=1}^{\infty} x_k^2 = 1 \right\}.$$

The kernels $\sum_{k=1}^{\infty} x_k y_k$ and $(x_0 y_0)^{-1}$ are both positive definite on $X \times X$, hence so is

$$\varphi(x, y) := \frac{\sum_{k=1}^{\infty} x_k y_k}{x_0 y_0},$$

and the inequality 3.1.8 gives

$$\varphi^2(x, y) \leqq \varphi(x, x)\varphi(y, y) = \frac{x_0^2 - 1}{x_0^2} \cdot \frac{y_0^2 - 1}{y_0^2}$$

so that

$$\left( \sum_{k=1}^{\infty} x_k y_k \right)^2 \leqq (x_0^2 - 1)(y_0^2 - 1) \leqq (x_0 y_0 - 1)^2.$$

Hence

$$[x, y] := x_0 y_0 - \sum_{k=1}^{\infty} x_k y_k \geqq 1$$

for all $x, y \in X$. The distance $d(x, y)$ between $x$ and $y$ is defined as the positive solution of

$$\cosh d(x, y) = [x, y].$$

From $[x, y] = x_0 y_0(1 - \varphi(x, y))$ we conclude that for all $t > 0$

$$[x, y]^{-t} = e^{-t \log[x, y]} = (x_0 y_0)^{-t}(1 - \varphi(x, y))^{-t}.$$

The right-hand side is positive definite by Corollary 3.1.14 so that $\psi(x, y) := \log[x, y]$ is a negative definite nonnegative kernel on $X \times X$. The equation

$\cosh d = \exp(\psi)$ has for $d, \psi \geqq 0$ the unique solution

$$d = \psi + \log(1 + \sqrt{1 - e^{-2\psi}}),$$

and therefore $d$ is negative definite, too, by Corollary 3.2.10.

Now we can state the above-mentioned result of Faraut and Harzallah (1974).

**5.1. Theorem.** *Let $f\colon \mathbb{R}_+ \to \mathbb{R}$ be an arbitrary given function. Then*

$$f(\log \cosh d(x, y))$$

*is positive definite on $X \times X$ if and only if $f \in \mathscr{P}^b(\mathbb{R}_+)$. The kernel*

$$f(\log \cosh d(x, y))$$

*is negative definite if and only if $f \in \mathscr{N}^l(\mathbb{R}_+)$.*

Proof. If $f \in \mathscr{P}^b(\mathbb{R}_+)$, then $f(s) = \int_{[0,\infty]} e^{-\lambda s}\, d\mu(\lambda)$ for some $\mu \in M_+([0, \infty])$ and then

$$f(\log \cosh d(x, y)) = f(\log[x, y]) = \int e^{-\lambda \log[x, y]}\, d\mu(\lambda)$$

$$= \int [x, y]^{-\lambda}\, d\mu(\lambda)$$

is positive definite as a mixture of positive definite kernels.

Suppose now that $f(\log[x, y])$ is positive definite. For all $x, y \in X$ the $2 \times 2$ matrix

$$\begin{pmatrix} f(\log[x, x]) & f(\log[x, y]) \\ f(\log[x, y]) & f(\log[y, y]) \end{pmatrix}$$

is then positive definite, hence

$$\{f(\log[x, y])\}^2 \leqq (f(0))^2$$

and because of $\{[x, y] \mid x, y \in X\} = [1, \infty[$, $f$ is bounded.

Let $s_1, \ldots, s_p \geqq 0$ be given. We fix $n \in \mathbb{N}$ and define for $1 \leqq j \leqq p$ and $1 \leqq \alpha \leqq n$ the vectors $x^n_{j\alpha} \in X$ as the columns in the matrix on the next page where empty places should be read as zeros.

That is,

$$x^n_{j\alpha}(\sigma) := \begin{cases} e^{s_j} & \text{for } \sigma = 0, \\ \sqrt{e^{2s_j} - 1} & \text{for } \sigma = (j-1)n + \alpha, \\ 0 & \text{otherwise.} \end{cases}$$

Then

$$\log[x^n_{j\alpha}, x^n_{k\beta}] = \begin{cases} s_j + s_k & \text{if } j \neq k, \\ 2s_j & \text{if } j = k \text{ but } \alpha \neq \beta, \\ 0 & \text{if } j = k \text{ and } \alpha = \beta, \end{cases}$$

| $x_{11}^n$ | $x_{12}^n$ | $\cdots$ | $x_{1n}^n$ | $x_{21}^n$ | $x_{22}^n$ | $\cdots$ | $x_{2n}^n$ | $\cdots$ | $x_{p1}^n$ | $x_{p2}^n$ | $\cdots$ | $x_{pn}^n$ |
|---|---|---|---|---|---|---|---|---|---|---|---|---|
| $e^{s_1}$ | $e^{s_1}$ | $\cdots$ | $e^{s_1}$ | $e^{s_2}$ | $e^{s_2}$ | $\cdots$ | $e^{s_2}$ | $\cdots$ | $e^{s_p}$ | $e^{s_p}$ | $\cdots$ | $e^{s_p}$ |
| $\sqrt{e^{2s_1}-1}$ | 0 | $\cdots$ | 0 | | | | | | | | | |
| 0 | $\sqrt{e^{2s_1}-1}$ | $\cdots$ | 0 | | 0 | | | 0 | | 0 | | |
| $\vdots$ | $\vdots$ | | $\vdots$ | | | | | | | | | |
| 0 | 0 | $\cdots$ | $\sqrt{e^{2s_1}-1}$ | | | | | | | | | |
| | | | | $\sqrt{e^{2s_2}-1}$ | 0 | $\cdots$ | 0 | | | | | |
| | 0 | | | 0 | $\sqrt{e^{2s_2}-1}$ | $\cdots$ | 0 | 0 | | 0 | | |
| | | | | $\vdots$ | $\vdots$ | | $\vdots$ | | | | | |
| | | | | 0 | 0 | $\cdots$ | $\sqrt{e^{2s_2}-1}$ | | | | | |
| | $\vdots$ | | | | | | | $\ddots$ | | | | |
| | | | | | | | | | $\sqrt{e^{2s_p}-1}$ | 0 | $\cdots$ | 0 |
| | | | | | | | | | 0 | $\sqrt{e^{2s_p}-1}$ | $\cdots$ | 0 |
| | 0 | | | | 0 | | | 0 | 0 | 0 | $\cdots$ | 0 |
| | | | | | | | | | $\vdots$ | $\vdots$ | | $\vdots$ |
| | | | | | | | | | 0 | 0 | $\cdots$ | $\sqrt{e^{2s_p}-1}$ |

so that Lemma 1.1 can again be applied and shows the positive definiteness of $(f(s_j + s_k))_{j,k=1,\ldots,p}$. Hence $f \in \mathscr{P}^b(\mathbb{R}_+)$. The second statement follows in the usual way. □

## Notes and Remarks

The idea of using an approximation like that in Lemma 1.1 appears for the first time in Ressel (1976). The name *Schoenberg triple*—restricted, however, to the case where $T$ is an abelian group and $S$ is a semigroup with the identical involution—was introduced in Berg and Ressel (1978), where also the second part of Example 1.7 was given which may be considered as a discrete counterpart to Schoenberg's classical result (cf. Example 1.3). That $(\mathbb{Z}_2, \mathbb{N}_0, v)$ with $v(0) = 0$, $v(1) = 1$ is a Schoenberg triple was shown independently by Kahane (1979, 1981).

The results of §1 have a very close connection to de Finetti-type theorems in probability theory; see Ressel (1983) where also Theorem 1.2 is sharpened in the sense that the measure representing $\varphi$ is shown to be concentrated on the closed subsemigroup $\{\rho \in \hat{S} \mid \rho \circ v \in \mathscr{P}(T)\}$ of $\hat{S}$.

Theorem 2.1 was published in Christensen and Ressel (1983). Dvoretzky's theorem on almost spherical sections has later found some simpler proofs, see for ex. Szankowski (1974) and Figiel (1976). The fact that negative definiteness of the norm square characterizes Hilbert spaces has a counterpart for $L^p$-spaces, $1 \leqq p \leqq 2$. If $\|x\|^p$ is negative definite on a Banach space $B$, then $B$ can be linearly and isometrically imbedded into some $L^p$-space. This was proved by Bretagnolle *et al.* (1966, Théorème 2), the proof however is complicated and (ten pages) long, and a shorter one seems not to be available. Bretagnolle *et al.* also show in this paper that $\mathscr{P}_r(l^p)$ is degenerate if $p > 2$.

For some related results see Kuelbs (1973) where in particular Theorem 2.3 is extended from $C[0, 1]$ to Banach spaces $C^b(X)$ which contain a sequence of functions with norm one and disjoint supports, $X$ being an otherwise arbitrary topological space.

There are many papers dealing with functions operating on positive definite matrices or positive definite functions on groups. Rudin (1959) showed that a function defined on the open interval $]-1, 1[$ and leaving invariant only the positive definite Toeplitz matrices can be represented as a power series with nonnegative coefficients. Horn (1969a) proved that a function defined on $]0, \infty[$ which leaves invariant the cone of positive definite matrices with positive entries must be absolutely monotonic and therefore (Widder 1941, p. 146) analytic with nonnegative coefficients. Theorem 3.4 and some other results of §3 were published in Christensen and Ressel (1978).

Rudin (1959) also conjectured that a complex-valued function defined on the open unit disc operating on positive definite Toeplitz matrices must be of the form $\sum a_{n,m} z^n \bar{z}^m$ with nonnegative coefficients $a_{n,m}$. The conjecture was proved by Herz (1963a, Théorème 1) in even greater generality, replacing the group $\mathbb{Z}$ (on which the positive definite functions define the Toeplitz matrices) by other abelian groups, called *illimités*. Although these operating functions are real analytic in the open unit disc, their boundary behaviour may be quite complicated as Graham (1977) has shown; see also the recent book of Graham and McGehee (1979, Chap. 9.6), where more references in this connection may be found. Harzallah (1969) determined the functions which operate on negative definite functions on abelian groups. Roughly speaking it is the Bernstein functions, cf. the remark following 4.4.3.

Lukacs (1960) posed the question of determining the class of functions which operate on the set of characteristic functions, i.e. on the normalized continuous positive definite functions on the real line. Konheim and Weiss (1965) answered this question and proved (without assuming continuity of the operating function) that precisely functions of the form $\sum a_{n,m} z^n \bar{z}^m$, where $a_{n,m} \geqq 0$ and $\sum a_{n,m} = 1$, have this property.

The main part of §4 was recently published in Christensen and Ressel (1982).

Faraut and Harzallah (1974) proved Theorem 5.1 assuming continuity of $f$; they state that this result is originally due to Krein.

CHAPTER 6
# Positive Definite Functions and Moment Functions

## §1. Moment Functions

Throughout this section, $S = (S, +, *)$ will be an abelian semigroup with involution and zero element 0. In Chapter 4 we introduced the convex cones $\mathscr{P}^b(S) \subseteq \mathscr{P}^e(S) \subseteq \mathscr{P}(S) \subseteq \mathbb{C}^S$, and every function $\varphi \in \mathscr{P}^e(S)$ has a representation

$$\varphi(s) = \int_{S^*} \rho(s)\, d\mu(\rho)$$

with a uniquely determined measure $\mu \in M_+^c(S^*)$. Our aim is to examine for which $\varphi \in \mathscr{P}(S)$ there exists a representation as above with $\mu \in M_+(S^*)$ of not necessarily compact support.

We start by giving a simple necessary and sufficient condition for a function $\varphi \in \mathscr{P}(S)$ to belong to $\mathscr{P}^b(S)$ in the case of which the above representation is possible with $\mu$ supported by $\hat{S}$.

**1.1. Proposition.** *Let $G$ be a generator set for $S$ and $\varphi: S \to \mathbb{C}$. Then $\varphi \in \mathscr{P}^b(S)$ if and only if $\varphi \in \mathscr{P}(S)$ and $(I - E_{a+a^*})\varphi \in \mathscr{P}(S)$ for all $a \in G$.*

PROOF. If $\varphi \in \mathscr{P}^b(S)$ has the integral representation

$$\varphi(s) = \int_{\hat{S}} \rho(s)\, d\mu(\rho), \qquad s \in S,$$

we find

$$(I - E_{a+a^*})\varphi(s) = \int_{\hat{S}} \rho(s)(1 - |\rho(a)|^2)\, d\mu(\rho),$$

so $(I - E_{a+a^*})\varphi$ is positive definite (and bounded) for all $a \in S$. If, conversely, $\varphi \in \mathscr{P}(S)$ and $(I - E_{a+a^*})\varphi \in \mathscr{P}(S)$ for $a \in G$ we have

$$\varphi(s + s^*) \geqq 0 \qquad \text{and} \qquad (I - E_{a+a^*})\varphi(s + s^*) \geqq 0 \qquad \text{for} \quad s \in S,$$

hence

$$0 \leqq \varphi(s + s^* + a + a^*) \leqq \varphi(s + s^*) \qquad \text{for} \quad a \in G, s \in S.$$

For $a_1, \ldots, a_n \in G \cup \{a^* | a \in G\}$ we then have

$$\begin{aligned} 0 &\leqq \varphi(a_1 + a_1^* + \cdots + a_n + a_n^*) \\ &\leqq \varphi(a_2 + a_2^* + \cdots + a_n + a_n^*) \\ &\leqq \cdots \leqq \varphi(a_n + a_n^*) \leqq \varphi(0), \end{aligned}$$

and applying $|\varphi(s)|^2 \leqq \varphi(s + s^*)\varphi(0)$ we find

$$|\varphi(a_1 + \cdots + a_n)|^2 \leqq \varphi(0)^2,$$

hence $|\varphi(s)| \leqq \varphi(0)$ for all $s \in S$. □

**1.2. Remark.** For the case of $S = (\mathbb{N}_0^2, +, *)$ and $G = \{(1, 0)\}$, cf. 4.4.11, we get that $\varphi: \mathbb{N}_0^2 \to \mathbb{C}$ belongs to $\mathscr{P}^b(\mathbb{N}_0^2)$ if and only if $\varphi$ and $(I - E_{(1,1)})\varphi$ belong to $\mathscr{P}(\mathbb{N}_0^2)$, a result given by Atzmon (1975).

For the case of $S = (\mathbb{N}_0, +)$ and $G = \{1\}$ we get that $\varphi: \mathbb{N}_0 \to \mathbb{R}$ belongs to $\mathscr{P}^b(\mathbb{N}_0)$ if and only if $\varphi$ and $(I - E_2)\varphi$ belong to $\mathscr{P}(\mathbb{N}_0)$, cf. Haviland (1936). More generally, for the case of $S = (\mathbb{N}_0^k, +)$ and $G = \{e_1, \ldots, e_k\}$, where $e_j = (0, \ldots, 0, 1, 0, \ldots, 0)$ with 1 on the $j$th place, we get that $\varphi: \mathbb{N}_0^k \to \mathbb{R}$ belongs to $\mathscr{P}^b(\mathbb{N}_0^k)$ if and only if $\varphi$ and $(I - E_{2e_j})\varphi$ belong to $\mathscr{P}(\mathbb{N}_0^k)$ for $j = 1, \ldots, k$.

These results should also be compared with Propositions 4.4.12 and 4.4.9.

Let $E_+(S^*)$ denote the set of Radon measures $\mu \in M_+(S^*)$ such that

$$\int_{S^*} |\rho(s)| \, d\mu(\rho) < \infty \qquad \text{for all} \quad s \in S.$$

Clearly $M_+^c(S^*) \subseteq E_+(S^*) \subseteq M_+^b(S^*)$, and if $\mu, \nu \in E_+(S^*)$ then $\mu * \nu \in E_+(S^*)$.

**1.3. Definition.** A function $\varphi: S \to \mathbb{C}$ is called a *moment function* if there exists a measure $\mu \in E_+(S^*)$ such that

$$\varphi(s) = \int_{S^*} \rho(s) \, d\mu(\rho) \qquad \text{for} \quad s \in S. \tag{1}$$

The set of moment functions is a convex cone denoted $\mathscr{H}(S)$ and it is clear that $\mathscr{P}^e(S) \subseteq \mathscr{H}(S) \subseteq \mathscr{P}(S)$, cf. 4.2.5. Furthermore, $\mathscr{H}(S)$ is stable under products. In fact, if $\varphi_1$ and $\varphi_2$ in $\mathscr{H}(S)$ are represented via (1) by $\mu_1$ and $\mu_2$ then $\varphi_1 \cdot \varphi_2$ is represented by $\mu_1 * \mu_2$.

**1.4.** Two interesting problems turn up:

(a) For which semigroups $S$ is it true that $\mathscr{H}(S) = \mathscr{P}(S)$, and if $\mathscr{H}(S) \neq \mathscr{P}(S)$ find the necessary and sufficient conditions for $\varphi \in \mathscr{P}(S)$ to belong to $\mathscr{H}(S)$.
(b) For $\varphi \in \mathscr{H}(S)$ describe the set $E_+(S^*, \varphi)$ of *representing measures* for $\varphi$, i.e. the set of $\mu \in E_+(S^*)$ such that (1) holds.

These questions seem to be very difficult due to their generality. Theorem 1.11 below gives a necessary and sufficient condition for $\mathscr{H}(S) = \mathscr{P}(S)$ under fairly general assumptions, and this theorem is used for special semi-groups in the following sections.

Clearly $E_+(S^*, \varphi)$ is a convex set in $M_+(S^*)$. The moment function $\varphi \in \mathscr{H}(S)$ will be called *determinate* if $E_+(S^*, \varphi)$ is a one-point set, and *indeterminate* if it consists of more than one point. Similarly $\mu \in E_+(S^*)$ will be called determinate or indeterminate if the corresponding moment function $\varphi$ given by (1) is so.

By Theorem 4.2.5 we have

**1.5. Theorem.** *Every exponentially bounded positive definite function is a determinate moment function.*

In general $S^*$ is not locally compact, however we have the following sufficient condition for this to be true:

**1.6. Proposition.** *If $S$ is finitely generated then $S^*$ is homeomorphic and isomorphic with a closed subsemigroup of $(\mathbb{C}^n, \cdot)$ for a suitable $n \in \mathbb{N}$, in particular $S^*$ is locally compact with a countable base for the topology.*

PROOF. By definition there exist finitely many elements $e_1, \ldots, e_n \in S$ such that every $s \in S$ has a representation of the form

$$s = k_1 e_1 + \cdots + k_n e_n \qquad \text{with} \quad k_j \in \mathbb{N}_0, \quad j = 1, \ldots, n. \tag{2}$$

For $\rho \in S^*$ we define $I(\rho) \in \mathbb{C}^n$ by $I(\rho) = (\rho(e_1), \ldots, \rho(e_n))$ and it is clear that $I: S^* \to \mathbb{C}^n$ is a continuous homomorphism and one-to-one. To see that $I$ is a homeomorphism and that $I(S^*)$ is closed in $\mathbb{C}^n$ we consider a net $(\rho_\alpha)_{\alpha \in A}$ from $S^*$ such that $\lim_\alpha I(\rho_\alpha)$ exists in $\mathbb{C}^n$, i.e. $\lim_\alpha \rho_\alpha(e_j)$ exists for each $j = 1, \ldots, n$. If $s \in S$ is given by (2) we find for $\rho \in S^*$ that

$$\rho(s) = \prod_{j=1}^{n} \rho(e_j)^{k_j},$$

so $\lim_\alpha \rho_\alpha(s)$ exists for each $s \in S$. Defining $\rho(s) := \lim_\alpha \rho_\alpha(s)$, $s \in S$, it is clear that $\rho \in S^*$, $I(\rho) = \lim_\alpha I(\rho_\alpha)$ and $\rho_\alpha \to \rho$ in $S^*$. □

For $s \in S$ we denote by $\chi_s$ the function of $S^*$ into $\mathbb{C}$ given by

$$\chi_s(\rho) = \rho(s) \qquad \text{for} \quad \rho \in S^*, \tag{3}$$

i.e. the point evaluation at $s$, and we denote by $V(\mathbb{C})$ the complex vector space spanned by the functions $\chi_s$, $s \in S$, i.e.

$$V(\mathbb{C}) := \left\{ \sum_{j=1}^{n} a_j \chi_{s_j} \,|\, n \geqq 1, a_j \in \mathbb{C}, s_j \in S, j = 1, \ldots, n \right\}.$$

Notice that $\chi_0 \equiv 1$. Since $\chi_{s^*} = \bar{\chi}_s$ we have that $V(\mathbb{C})$ is stable by complex conjugation and $V(\mathbb{C}) = V(\mathbb{R}) + iV(\mathbb{R})$ where $V(\mathbb{R})$ is the real vector space of real-valued functions in $V(\mathbb{C})$. If the involution on $S$ is the identical it is, of course, sufficient to consider the real vector space $V(\mathbb{R})$ spanned by the real functions $\chi_s$, $s \in S$.

**1.7. Proposition.** *If $S$ is finitely generated then $V(\mathbb{R})$ is an adapted space and $\mathscr{H}(S)$ is closed in $\mathbb{C}^S$. The set $E_+(S^*, \varphi)$ of representing measures for $\varphi \in \mathscr{H}(S)$ is a metrizable compact convex set in $M_+(S^*)$ with the vague topology.*

Proof. Let $e_1, \ldots, e_n$ be elements of $S$ such that every $s \in S$ has a representation (2). To show that $V(\mathbb{R}) = V(\mathbb{R})_+ - V(\mathbb{R})_+$ it suffices to verify that $\operatorname{Re} \chi_s, \operatorname{Im} \chi_s \in V(\mathbb{R})_+ - V(\mathbb{R})_+$ for $s \in S$. But this follows from the equation

$$\chi_s = \chi_s + 1 + \chi_{s+s^*} - (1 + \chi_{s+s^*}),$$

since $|\chi_s| \leqq 1 + \chi_{s+s^*}$.

Next let $s \in S$ be given. We define

$$f = 1 + \chi_{s+s^*} + \sum_{j=1}^{n} \chi_{e_j+e_j^*} \tag{4}$$

and will show that $\chi_s/f$ tends to zero at infinity on the locally compact space $S^*$. Having proved this, it follows easily that $V(\mathbb{R})$ is an adapted space.

Given $\varepsilon > 0$ there exists $N \in \mathbb{N}$ such that $0 < x/(1 + x^2) \leqq \varepsilon$ for $x \geqq N$. The set

$$K = \{\rho \in S^* \,|\, |\rho(s)| \leqq N, |\rho(e_j)| \leqq N, j = 1, \ldots, n\}$$

is compact, cf. the proof of Proposition 1.6, and for $\rho \in S^* \backslash K$ we find

$$\frac{|\chi_s(\rho)|}{f(\rho)} = |\rho(s)| \left\{ 1 + |\rho(s)|^2 + \sum_{j=1}^{n} |\rho(e_j)|^2 \right\}^{-1}$$

$$\leqq \begin{cases} \dfrac{|\rho(s)|}{1 + |\rho(s)|^2} \leqq \varepsilon & \text{if } |\rho(s)| > N, \\[2ex] \dfrac{N}{1 + N^2} \leqq \varepsilon & \text{if } |\rho(s)| \leqq N. \end{cases}$$

In fact, if $|\rho(s)| \leqq N$ and $\rho \in S^* \backslash K$ there exists $j \in \{1, \ldots, n\}$ such that $|\rho(e_j)| > N$.

Since $S$ is finitely generated $S$ is countable, so $\mathbb{C}^S$ is metrizable. To prove that $\mathscr{H}(S)$ is closed we consider a sequence $(\varphi_n)_{n \geqq 1}$ from $\mathscr{H}(S)$ converging pointwise to $\varphi \in \mathscr{P}(S)$. If $\mu_n \in E_+(S^*)$ is a representing measure for $\varphi_n$ we have for $s \in S$

$$\lim_{n\to\infty} \int |\rho(s)|^2 \, d\mu_n(\rho) = \lim_{n\to\infty} \varphi_n(s + s^*) = \varphi(s + s^*),$$

so the measures in the sequence $(|\rho(s)|^2 \mu_n)$ have bounded total mass. Putting $s = 0$ we get by 2.4.6 and 2.4.10 that $(\mu_n)$ has a vaguely convergent subsequence $(\mu_{n_p})_{p \geqq 1}$ with limit $\mu \in M_+^b(S^*)$, and therefore

$$\int |\rho(s)|^2 \, d\mu(\rho) \leqq \liminf_{p\to\infty} \int |\rho(s)|^2 \, d\mu_{n_p}(\rho) = \varphi(s + s^*),$$

hence

$$\left(\int |\rho(s)| \, d\mu(\rho)\right)^2 \leqq \mu(S^*) \int |\rho(s)|^2 \, d\mu(\rho) < \infty,$$

so $\mu \in E_+(S^*)$. Let $s \in S$ be fixed. With $f$ as in (4) $(f\mu_{n_p})_{p \geqq 1}$ converges vaguely to $f\mu$, and since

$$\sup_n \int f \, d\mu_n < \infty$$

we even have

$$\lim_{p\to\infty} \int g f \, d\mu_{n_p} = \int g f \, d\mu$$

for $g \in C^0(S^*)$, cf. 2.4.4. If this is applied to the function $g = \chi_s/f$ we get

$$\int \rho(s) \, d\mu(\rho) = \lim_{p\to\infty} \int \rho(s) \, d\mu_{n_p}(\rho) = \lim_{p\to\infty} \varphi_{n_p}(s) = \varphi(s),$$

hence $\varphi \in \mathscr{H}(S)$.

For $\varphi \in \mathscr{H}(S)$ and $\mu \in E_+(S^*, \varphi)$ the linear functional $L: V(\mathbb{C}) \to \mathbb{C}$ defined by $L(f) = \int f \, d\mu$ is independent of $\mu \in E_+(S^*, \varphi)$ and we have

$$L(\chi_s) = \int \rho(s) \, d\mu(\rho) = \varphi(s), \qquad s \in S.$$

It follows that $E_+(S^*, \varphi)$ is the set of representing measures for the positive linear functional $L$ restricted to the adapted space $V(\mathbb{R})$, hence vaguely compact by Proposition 2.4.8. We have already remarked that $M_+(S^*)$ is metrizable in the vague topology. □

In general the functions $\chi_s$, $s \in S$ need not be linearly independent in $\mathbb{C}^{S^*}$, cf. Remark 1 in 4.6.9. The following result is inspired by an argument in Hewitt and Zuckermann (1956).

**1.8. Proposition.** *For an abelian semigroup with involution $S$ the following conditions are equivalent:*

(i) *$S^*$ separates the points of $S$.*
(ii) *The functions $\chi_s$, $s \in S$ are linearly independent in $\mathbb{C}^{S^*}$.*

PROOF. "(i) ⇒ (ii)" Suppose $s_1, \ldots, s_n \in S$ are distinct points and that $\sum a_j \chi_{s_j} = 0$ on $S^*$, i.e.

$$\sum_{j=1}^{n} a_j \rho(s_j) = 0 \qquad \text{for} \quad \rho \in S^*.$$

In order to prove that $a_1 = 0$ we choose $\rho_j \in S^*$ for $j = 2, \ldots, n$ such that $\rho_j(s_1) \neq \rho_j(s_j)$, which is possible by (i), and define

$$f(s) = \prod_{j=2}^{n} (\rho_j(s) - \rho_j(s_j))(\rho_j(s_1) - \rho_j(s_j))^{-1}, \qquad s \in S.$$

Then $f$ belongs to the linear span of $S^*$ in $\mathbb{C}^S$ and $f(s_1) = 1, f(s_j) = 0$ for $j = 2, \ldots, n$, hence

$$a_1 = \sum_{j=1}^{n} a_j f(s_j) = 0.$$

The implication "(ii) ⇒ (i)" is trivial. □

**1.9.** Suppose that the equivalent conditions of Proposition 1.8 are verified. Then $\mathbb{C}^S$ and $V(\mathbb{C})$ form a dual pair under the bilinear mapping

$$\left\langle \varphi, \sum_{j=1}^{n} a_j \chi_{s_j} \right\rangle = \sum_{j=1}^{n} a_j \varphi(s_j), \tag{5}$$

which is well-defined because of 1.8(ii). The topology of pointwise convergence on $\mathbb{C}^S$ is compatible with this duality. In fact, if $L: \mathbb{C}^S \to \mathbb{C}$ is a continuous linear functional there exist a finite set $D \subseteq S$ and $\varepsilon > 0$ such that $|L(\varphi)| \leqq 1$ for $\varphi \in \mathscr{U}(D, \varepsilon)$, where $\mathscr{U}(D, \varepsilon)$ is the neighbourhood of 0 given by

$$\mathscr{U}(D, \varepsilon) = \{\varphi \in \mathbb{C}^S \mid |\varphi(s)| \leqq \varepsilon \text{ for } s \in D\}.$$

Then clearly $n\varphi 1_{S \setminus D} \in \mathscr{U}(D, \varepsilon)$ for all $n \in \mathbb{N}$ and $\varphi \in \mathbb{C}^S$ so that $L(\varphi 1_{S\setminus D}) = 0$, and it follows that

$$L(\varphi) = L(\varphi 1_D) = \sum_{j=1}^{n} a_j \varphi(s_j),$$

where $D = \{s_1, \ldots, s_n\}$ and $a_j = L(1_{\{s_j\}})$.

Similarly the finest locally convex topology on $V(\mathbb{C})$, cf. 1.1.9, is compatible with the duality.

In $V(\mathbb{C})$ we consider the convex cone $\Sigma$ generated by the absolute squares of elements of $V(\mathbb{C})$, i.e.

$$\Sigma = \{|f_1|^2 + \cdots + |f_n|^2 \mid n \geqq 1, f_j \in V(\mathbb{C}), j = 1, \ldots, n\}. \tag{6}$$

For $f = f_1 + if_2 \in V(\mathbb{C})$ with $f_1, f_2 \in V(\mathbb{R})$ we have $|f|^2 = f_1^2 + f_2^2$, so $\Sigma$ is equal to the convex cone of finite sums of squares of elements in $V(\mathbb{R})$. It is clear that $\Sigma \subseteq V(\mathbb{R})_+$.

For a convex cone $C \subseteq V(\mathbb{C})$ we have introduced the closed convex cone $C^\perp \subseteq \mathbb{C}^S$ by

$$C^\perp = \{\varphi \in \mathbb{C}^S \mid \langle \varphi, f\rangle \geqq 0 \text{ for all } f \in C\},$$

cf. Remark 1.3.7. With this notation we can describe $\mathscr{H}(S)$ and $\mathscr{P}(S)$ as the following theorem shows.

**1.10. Theorem.** *Let $S$ be a finitely generated abelian semigroup with involution in which $S^*$ separates the points. Under the duality between $\mathbb{C}^S$ and $V(\mathbb{C})$ given by* (5) *we have:*

(i) $\mathscr{P}(S) = \Sigma^\perp$;
(ii) $\mathscr{H}(S) = (V(\mathbb{R})_+)^\perp$.

PROOF. For $\varphi \in \mathbb{C}^S$ and $f = \sum_{j=1}^n c_j \chi_{s_j} \in V(\mathbb{C})$ we have

$$\langle \varphi, |f|^2\rangle = \sum_{j,k=1}^n c_j \bar{c}_k \varphi(s_j + s_k^*),$$

which shows that $\varphi \in \mathscr{P}(S)$ if and only if $\langle \varphi, |f|^2\rangle \geqq 0$ for all $f \in V(\mathbb{C})$, and this establishes (i).

For $\varphi \in \mathscr{H}(S)$ there exists a measure $\mu \in E_+(S^*)$ such that

$$\varphi(s) = \int \rho(s)\, d\mu(\rho) \qquad \text{for} \quad s \in S,$$

so for $f \in V(\mathbb{R})_+$ we get

$$\langle \varphi, f\rangle = \int f(\rho)\, d\mu(\rho) \geqq 0,$$

hence $\varphi \in (V(\mathbb{R})_+)^\perp$. Conversely, if $\varphi \in (V(\mathbb{R})_+)^\perp$ then the mapping $f \mapsto \langle \varphi, f\rangle$ is a $\mathbb{C}$-linear functional on $V(\mathbb{C})$ and a positive $\mathbb{R}$-linear functional on $V(\mathbb{R})$ which is adapted by Proposition 1.7. By Theorem 2.2.7 there exists $\mu \in M_+(S^*)$ such that $V(\mathbb{R}) \subseteq \mathscr{L}^1(\mu)$ and

$$\langle \varphi, f\rangle = \int f\, d\mu \qquad \text{for} \quad f \in V(\mathbb{R}),$$

and it follows easily that $\mu \in E_+(S^*)$ and

$$\varphi(s) = \langle \varphi, \chi_s\rangle = \int \rho(s)\, d\mu(\rho) \qquad \text{for} \quad s \in S,$$

i.e. $\varphi \in \mathscr{H}(S)$. □

**1.11. Theorem.** *Let S be a finitely generated abelian semigroup with involution in which $S^*$ separates the points. The following conditions are equivalent:*

(i) $\mathscr{H}(S) = \mathscr{P}(S)$;
(ii) $\bar{\Sigma} = V(\mathbb{R})_+$.

*Here the closure refers to any locally convex topology on $V(\mathbb{C})$ compatible with the duality between $\mathbb{C}^S$ and $V(\mathbb{C})$ given by* (5).

PROOF. By Proposition 1.3.8 the closure $\bar{\Sigma}$ of the convex set $\Sigma$ is the same for all locally convex topologies on $V(\mathbb{C})$ compatible with the duality. We may therefore assume that $V(\mathbb{C})$ is equipped with the finest locally convex topology. Since this topology is finer than the topology of pointwise convergence on $S^*$, we have that $V(\mathbb{R})$ and $V(\mathbb{R})_+$ are closed in $V(\mathbb{C})$, hence $\bar{\Sigma} \subseteq V(\mathbb{R})_+$.

First assume that there exists $f_0 \in V(\mathbb{R})_+ \setminus \bar{\Sigma}$. By the theorem of separation (1.2.3) applied to $V(\mathbb{R})$ there exists a $\mathbb{R}$-linear functional $L: V(\mathbb{R}) \to \mathbb{R}$ such that $L(f_0) < 0 \leqq L(f)$ for all $f \in \bar{\Sigma}$. Defining $\tilde{L}: V(\mathbb{C}) \to \mathbb{C}$ by $\tilde{L}(f_1 + if_2) = L(f_1) + iL(f_2)$ for $f_1, f_2 \in V(\mathbb{R})$ and $\varphi(s) = \tilde{L}(\chi_s)$ for $s \in S$, we see that $\tilde{L}$ is a $\mathbb{C}$-linear functional on $V(\mathbb{C})$ such that $\tilde{L}(f) = \langle \varphi, f \rangle$ for $f \in V(\mathbb{C})$. It follows that $\langle \varphi, |f|^2 \rangle = L(|f|^2) \geqq 0$ for $f \in V(\mathbb{C})$, hence $\varphi \in \mathscr{P}(S)$. On the other hand, the inequality $\langle \varphi, f_0 \rangle < 0$ shows that $\varphi \notin \mathscr{H}(S)$, and we have proved that "(i) $\Rightarrow$ (ii)."

Finally, if $\bar{\Sigma} = V(\mathbb{R})_+$ then $\Sigma^\perp = (\bar{\Sigma})^\perp = (V(\mathbb{R})_+)^\perp$, and $\mathscr{P}(S) = \mathscr{H}(S)$ follows from Theorem 1.10. □

In the following sections, we shall see examples where $\mathscr{H}(S) = \mathscr{P}(S)$ and also where $\mathscr{H}(S) \neq \mathscr{P}(S)$.

**1.12. Exercise** (The Method of Moments). Let $S$ be a finitely generated abelian semigroup with involution and let $(\mu_n)$ be a sequence from $E_+(S^*)$ with corresponding sequence of moment functions $(\varphi_n)$. Prove that if $\varphi \in \mathscr{H}(S)$ is a determinate moment function with representing measure $\mu \in E_+(S^*)$ and if $\lim_{n\to\infty} \varphi_n(s) = \varphi(s)$ for each $s \in S$ then $\lim_{n\to\infty} \mu_n = \mu$ weakly.

## §2. The One-Dimensional Moment Problem

In this section, we will apply the results of §1 to the semigroup $S = (\mathbb{N}_0, +)$, the involution being the identical. For $x \in \mathbb{R}$ the mapping $\rho_x: \mathbb{N}_0 \to \mathbb{R}$ given by $\rho_x(n) = x^n, n \geqq 0$, is an element of $\mathbb{N}_0^*$ and it is easy to see that the mapping $x \mapsto \rho_x$ is a topological semigroup isomorphism of $(\mathbb{R}, \cdot)$ onto $\mathbb{N}_0^*$. The functions $\chi_n$, $n \in \mathbb{N}_0$, cf. (3) in §1, will be identified with the monomials $x \mapsto x^n$, which are linearly independent, so $V(\mathbb{R})$ is the vector space of polynomials in one variable with real coefficients. The following lemma shows that the cone $\Sigma$ (see (6) in §1) is equal to $V(\mathbb{R})_+$.

**2.1. Lemma.** *Let $p$ be a polynomial with real coefficients and suppose $p(x) \geqq 0$ for $x \in \mathbb{R}$. Then there exist two polynomials with real coefficients $p_1$, $p_2$ such that $p(x) = p_1(x)^2 + p_2(x)^2$ for $x \in \mathbb{R}$.*

PROOF. Complex roots of $p$ appear in conjugate pairs and real roots have an even multiplicity. Therefore we may write $p(x) = r(x)\overline{r(x)}$, $x \in \mathbb{R}$ for some polynomial $r$ with complex coefficients. Writing $r = p_1 + ip_2$, where $p_1$, $p_2$ are real polynomials, we get $p(x) = p_1(x)^2 + p_2(x)^2$ for $x \in \mathbb{R}$. □

By Theorem 1.11 it follows that $\mathscr{P}(\mathbb{N}_0) = \mathscr{H}(\mathbb{N}_0)$, so we have proved the following theorem of Hamburger, called the solution of Hamburger's moment problem, cf. Akhiezer (1965).

**2.2. Theorem.** *A sequence $s = (s_n)_{n \geqq 0}$ of real numbers is the sequence of moments of a measure $\mu \in E_+(\mathbb{R})$ where*

$$E_+(\mathbb{R}) = \left\{\mu \in M_+(\mathbb{R}) \,\middle|\, \int |x^n| \, d\mu(x) < \infty \text{ for all } n \geqq 0\right\},$$

*i.e. is of the form*

$$s_n = \int x^n \, d\mu(x), \qquad n = 0, 1, \ldots, \tag{1}$$

*if and only if $s \in \mathscr{P}(\mathbb{N}_0)$. The set $E_+(\mathbb{R}, s)$ of measures $\mu \in E_+(\mathbb{R})$ satisfying (1) is a nonempty compact convex set in $M_+(\mathbb{R})$ in the vague topology.*

**2.3. Remark.** A sequence $s = (s_n)_{n \geqq 0}$ belongs to $\mathscr{P}(\mathbb{N}_0)$ if and only if for every $n \geqq 0$ the matrix $((s_{j+k})_{0 \leqq j,k \leqq n})$ is positive definite. Using Theorem 3.1.16 it follows that $s$ is strictly positive definite if and only if

$$\det((s_{j+k})_{0 \leqq j,k \leqq n}) > 0 \qquad \text{for} \quad n = 0, 1, 2, \ldots .$$

Let $s \in \mathscr{P}(\mathbb{N}_0)$ and $\mu \in E_+(\mathbb{R}, s)$. For $\{c_0, c_1, \ldots, c_n\} \subseteq \mathbb{R}$ we define $p(x) = \sum_{k=0}^n c_k x^k$ and have

$$\sum_{j,k=0}^{n} c_j c_k s_{j+k} = \int p(x)^2 \, d\mu(x),$$

so $s$ is strictly positive definite precisely when $\operatorname{supp}(\mu)$ is an infinite set.

**2.4.** Let $F$ be a closed subset of $\mathbb{R}$. The *$F$-moment problem* consists in characterizing the *$F$-moment sequences*, i.e. the real sequences $s = (s_n)_{n \geqq 0}$ of the form

$$s_n = \int_F x^n \, d\mu(x), \qquad n \geqq 0,$$

where $\mu \in E_+(\mathbb{R})$ is supported by the closed set $F$.

Hamburger's moment problem corresponds to $F = \mathbb{R}$. The case of $F = [-a, a]$, $a \geqq 0$ has been solved in Proposition 4.4.9, and in the special case $a = 1$ which we will call Haviland's moment problem, two different characterizations of $[-1, 1]$-moment sequences $s$ have been found, (a) $s \in \mathscr{P}^b(\mathbb{N}_0)$, (b) $s \in \mathscr{P}(\mathbb{N}_0)$ and $s - E_2 s \in \mathscr{P}(\mathbb{N}_0)$, the latter going back to Haviland (1936), cf. 1.2. The case of $F = [0, 1]$ is Hausdorff's moment problem and the $[0, 1]$-moment sequences are precisely the completely monotone functions on the semigroup $(\mathbb{N}_0, +)$, see 4.6.11.

We shall now consider the case $F = [0, \infty[$ which is called Stieltjes' moment problem, solved by Stieltjes in 1894. The $[0, \infty[$-moment sequences are also called Stieltjes moment sequences.

**2.5. Theorem.** *For a sequence $s = (s_n)_{n \geqq 0}$ of real numbers the following conditions are equivalent:*

(i) $s, E_1 s \in \mathscr{P}(\mathbb{N}_0)$.
(ii) $t = (s_0, 0, s_1, 0, s_2, \ldots) \in \mathscr{P}(\mathbb{N}_0)$.
(iii) *There exists $\mu \in M_+([0, \infty[)$ such that*

$$s_n = \int_0^\infty x^n \, d\mu(x), \qquad n \geqq 0.$$

Proof. "(i) $\Rightarrow$ (ii)" Let $t_n = s_{n/2}$ for $n$ even and $t_n = 0$ for $n$ odd. For

$$\{c_0, c_1, \ldots, c_n\} \subseteq \mathbb{R}$$

we then have

$$\sum_{j,k=0}^{n} c_j c_k t_{j+k} = \sum_{\substack{j,k=0\\ j,k \text{ even}}}^{n} c_j c_k t_{j+k} + \sum_{\substack{j,k=0\\ j,k \text{ odd}}}^{n} c_j c_k t_{j+k}$$
$$= \sum_{p,q=0}^{[n/2]} c_{2p} c_{2q} s_{p+q} + \sum_{p,q=0}^{[(n-1)/2]} c_{2p+1} c_{2q+1} s_{p+q+1} \geqq 0$$

showing that $t \in \mathscr{P}(\mathbb{N}_0)$.

"(ii) $\Rightarrow$ (iii)" By Hamburger's theorem (2.2) there is $\sigma \in E_+(\mathbb{R})$ such that

$$t_{2n} = s_n = \int x^{2n} \, d\sigma(x), \qquad t_{2n+1} = \int x^{2n+1} \, d\sigma(x) = 0, \qquad n \geqq 0,$$

hence

$$s_n = \int_0^\infty x^n \, d\mu(x), \qquad n \geqq 0,$$

where $\mu$ is the image measure of $\sigma$ under the continuous mapping $x \mapsto x^2$ of $\mathbb{R}$ into $[0, \infty[$.

"(iii) $\Rightarrow$ (i)" is clear since $E_1 s$ is represented by the positive Radon measure $x\mu$. □

In analogy with Theorem 2.2 we can give an integral representation of all functions $\psi \in \mathcal{N}(\mathbb{N}_0)$ with a representing measure which is not necessarily unique.

**2.6. Theorem.** *A function $\psi\colon \mathbb{N}_0 \to \mathbb{R}$ is negative definite if and only if it has a representation of the form*

$$\psi(n) = a + bn - cn^2 + \int_{\mathbb{R}\setminus\{1\}} (1 - x^n - n(1 - x))\, d\mu(x), \qquad n \in \mathbb{N}_0,$$

*where $a, b \in \mathbb{R}$, $c \geqq 0$ and $\mu \in M_+(\mathbb{R}\setminus\{1\})$ satisfies*

$$\int_{0<|x-1|<1} (1 - x)^2\, d\mu(x) < \infty,$$

$$\int_{|x-1|\geqq 1} |x|^n\, d\mu(x) < \infty \qquad \textit{for all} \quad n \in \mathbb{N}_0.$$

PROOF. We note that

$$\lim_{x\to 1} \frac{1 - x^n - n(1 - x)}{(1 - x)^2} = -\binom{n}{2} \qquad \text{for} \quad n \geqq 2,$$

and it follows easily that any function $\psi$ of the above form is negative definite.

If $\psi \in \mathcal{N}(\mathbb{N}_0)$ we put $a = \psi(0)$ and $\psi - a$ is again negative definite. Without restriction we may therefore assume that $\psi(0) = 0$. By Theorem 3.2.2 there exist (possibly many) $\mu_t \in E_+(\mathbb{R})$ such that

$$e^{-t\psi(n)} = \int_{-\infty}^{\infty} x^n\, d\mu_t(x) \qquad \text{for} \quad n \in \mathbb{N}_0, \quad t > 0,$$

and $\mu_t$ is a probability measure. We find

$$\frac{1}{t}\int_{-\infty}^{\infty} (1 - x^n - n(1 - x))\, d\mu_t(x) = \frac{1}{t}(1 - e^{-t\psi(n)}) - \frac{n}{t}(1 - e^{-t\psi(1)})$$

and

$$0 \leqq \frac{1}{t}\int_{-\infty}^{\infty} (1 - x^n)^2\, d\mu_t(x) = \frac{1}{t}(1 - 2e^{-t\psi(n)} + e^{-t\psi(2n)}),$$

which shows that

$$\lim_{t\to 0} \frac{1}{t}\int_{-\infty}^{\infty} (1 - x^n - n(1 - x))\, d\mu_t(x) = \psi(n) - n\psi(1)$$

and

$$0 \leqq \lim_{t\to 0} \frac{1}{t}\int_{-\infty}^{\infty} (1 - x^n)^2\, d\mu_t(x) = 2\psi(n) - \psi(2n).$$

The last equation shows that there is a constant $A_n$ depending on $n \in \mathbb{N}_0$ such that

$$\frac{1}{t}\int_{-\infty}^{\infty}(1 - x^n)^2\, d\mu_t(x) \leqq A_n \qquad \text{for} \quad 0 < t \leqq 1,$$

so by Proposition 2.4.6 there is a measure $\sigma \in M_+(\mathbb{R})$ and a sequence $(t_j)$ tending to zero such that

$$\lim_{j\to\infty} \frac{1}{t_j}(1 - x)^2 \mu_{t_j} = \sigma \quad \text{vaguely.}$$

Since $(1 - x^n)^2(1 - x)^{-2}$ is continuous on $\mathbb{R}$ we get

$$\lim_{j\to\infty} \frac{1}{t_j}\int_{-\infty}^{\infty}(1 - x^n)^2 f(x)\, d\mu_{t_j}(x) = \int_{-\infty}^{\infty}\left(\frac{1 - x^n}{1 - x}\right)^2 f(x)\, d\sigma(x)$$

for all $f \in C^c(\mathbb{R})$, and by Proposition 2.4.4 this holds even for $f \in C^0(\mathbb{R})$.

Choosing a function $\varphi \in C^c_+(\mathbb{R})$ with $\varphi(x) = 1$ for $x \in [0, 2]$ we now find for $n \geqq 1$

$$\frac{1}{t_j}\int_{-\infty}^{\infty}(1 - x^n - n(1 - x))\, d\mu_{t_j}(x)$$

$$= \frac{1}{t_j}\int_{-\infty}^{\infty}\frac{1 - x^n - n(1 - x)}{(1 - x)^2}\varphi(x)(1 - x)^2\, d\mu_{t_j}(x)$$

$$+ \frac{1}{t_j}\int_{-\infty}^{\infty} f(x)(1 - x^n)^2\, d\mu_{t_j}(x),$$

where

$$f(x) := \frac{1 - x^n - n(1 - x)}{(1 - x^n)^2}(1 - \varphi(x)) \in C^0(\mathbb{R}).$$

Letting $j \to \infty$ we therefore get

$$\psi(n) - n\psi(1) = \int_{-\infty}^{\infty}\frac{1 - x^n - n(1 - x)}{(1 - x)^2}\varphi(x)\, d\sigma(x)$$

$$+ \int_{-\infty}^{\infty}\left(\frac{1 - x^n}{1 - x}\right)^2 f(x)\, d\sigma(x)$$

$$= \int_{-\infty}^{\infty}\frac{(1 - x^n - n(1 - x))}{(1 - x)^2}\, d\sigma(x)$$

$$= -\binom{n}{2}\sigma(\{1\}) + \int_{\mathbb{R}\setminus\{1\}}(1 - x^n - n(1 - x))\, d\mu(x),$$

with $\mu = (1 - x)^{-2}(\sigma | \mathbb{R}\backslash\{1\})$, hence,

$$\psi(n) = n(\psi(1) + \tfrac{1}{2}\sigma(\{1\})) - \tfrac{1}{2}n^2\sigma(\{1\}) + \int_{\mathbb{R}\backslash\{1\}} (1 - x^n - n(1 - x))\, d\mu(x).$$

By Proposition 2.4.6 we have

$$\int \left(\frac{1 - x^n}{1 - x}\right)^2 d\sigma(x) \leqq A_n \qquad \text{for all} \quad n \in \mathbb{N},$$

which shows that $\mu$ has the asserted properties. □

**2.7. Exercise.** Show that if $\mu \in M_+(\mathbb{R})$ satisfies $\int \exp(ax^2)\, d\mu(x) < \infty$ for some $a > 0$ then $\mu \in E_+(\mathbb{R})$ and $\mu$ is determinate. *Hint*: Show that if $\nu \in E_+(\mathbb{R})$ has the same moments as $\mu$ then $\int \exp(ax^2)\, d\nu(x) < \infty$ and $\int \exp(ixy)\, d\mu(x) = \int \exp(ixy)\, d\nu(x)$ for all $y \in \mathbb{R}$.

**2.8. Exercise.** Show that the normal distribution $\mu = (2\pi)^{-1/2} \exp(-\frac{1}{2}x^2)\, dx$ belongs to $E_+(\mathbb{R})$ and is determinate. The corresponding moment sequence is $s_0 = 1$, $s_{2n} = 1 \cdot 3 \cdot 5 \cdots (2n - 1)$, $s_{2n-1} = 0$ for $n \geqq 1$.

## §3. The Multi-Dimensional Moment Problem

In this section, we will apply the results of §1 to the semigroup $(\mathbb{N}_0^k, +)$, the involution being the identical. We use the notation from 4.4.8 so $(\mathbb{R}^k, \cdot) \approx (\mathbb{N}_0^k)^*$ via the mapping $x \mapsto \rho_x$, where $\rho_x(n) = x^n$. The functions $\chi_n$, $n \in \mathbb{N}_0^k$, cf. (3) in §1, will be identified with the monomials $x \mapsto x^n$, which are linearly independent, so $V(\mathbb{R})$ is the vector space of polynomials in $k$ variables with real coefficients, which from now on will be denoted $A^{(k)}$. The convex cone $\Sigma$ of finite sums of squares of polynomials, cf. (6) in §1, will be denoted $\Sigma^{(k)}$.

In contrast to the one-dimensional case we have $\Sigma^{(k)} \neq A_+^{(k)}$ for $k \geqq 2$ as the following shows:

**3.1. Lemma.** *The polynomial*

$$p(x, y) = x^2y^2(x^2 + y^2 - 1) + 1$$

*belongs to* $A_+^{(2)}\backslash\Sigma^{(2)}$.

**Proof.** It is easy to see that $p \geqq 0$ and in fact its greatest lower bound is 26/27.

Suppose that $p = q_1^2 + \cdots + q_n^2$ for certain polynomials $q_1, \ldots, q_n$. Since $p(x, 0) = p(0, y) = 1$ we get that $q_i(x, 0)$ and $q_i(0, y)$ must be constant for $i = 1, \ldots, n$. Therefore each $q_i$ can be written

$$q_i(x, y) = a_i + xyh_i(x, y),$$

where $a_i$ is a constant and $h_i$ is of degree at most 1. By comparing terms of the same degree we find

$$\sum_{i=1}^{n} a_i^2 = 1, \qquad 2xy \sum_{i=1}^{n} a_i h_i(x, y) = 0$$

and

$$x^2y^2(x^2 + y^2 - 1) = x^2y^2 \sum_{i=1}^{n} h_i^2(x, y),$$

hence

$$x^2 + y^2 - 1 = \sum_{i=1}^{n} h_i^2(x, y)$$

which is a contradiction. □

The space $A^{(k)}$ is equipped with its finest locally convex topology. For $d = 0, 1, \ldots$ let $A_d^{(k)}$ denote the subspace of polynomials of degree $\leqq d$. Being a finite dimensional space $A_d^{(k)}$ has a canonical topology, cf. 1.3.9.

**3.2. Theorem.** *The convex cone $\Sigma^{(k)}$ is closed in $A^{(k)}$ for $k \geqq 1$.*

Proof. First we show that $\Sigma^{(k)} \cap A_d^{(k)}$ is closed in $A_d^{(k)}$ for any fixed $d \geqq 0$. Let us choose $d + 1$ different points $a_i$, $i = 1, \ldots, d + 1$ in $\mathbb{R}$. The points $a = (a_{i_1}, \ldots, a_{i_k}) \in \mathbb{R}^k$, where $i_1, \ldots, i_k \in \{1, \ldots, d + 1\}$ form a set $H$ of $(d + 1)^k$ points in $\mathbb{R}^k$ and they separate $A_d^{(k)}$: If two polynomials $p_1, p_2$ in $A_d^{(k)}$ agree in every point of $H$ they are identical. Therefore the topology on $A_d^{(k)}$ of pointwise convergence on the set $H$ is a Hausdorff topology compatible with the vector space structure, hence equal to the canonical topology. In particular, if a sequence $(p_n)$ of polynomials from $A_d^{(k)}$ converges at each point $a \in H$, then there is $p \in A_d^{(k)}$ such that $p_n$ converges to $p$ in the canonical topology of $A_d^{(k)}$, which is equivalent to convergence of the coefficients. Suppose that $p \in \Sigma^{(k)} \cap A_d^{(k)}$ is written as

$$p = p_1^2 + \cdots + p_r^2,$$

where $p_1, \ldots, p_r \in A^{(k)}$. Writing a polynomial as the sum of its homogeneous parts, it is easy to see that the degree of $p$ is an even number $2q \leqq d$ and that the degree of each $p_i$ is at most $q$. We can always assume that $p$ is written as a sum of $N$ squares where $N = \dim A_q^{(k)}$. For if $r < N$ then zeros can be added, and if $r > N$ there is a nonzero set of real numbers $c_1, \ldots, c_r$ such that $\sum_{i=1}^r c_i p_i^2 = 0$. By rearranging, if necessary, we may assume that $|c_1| \geqq |c_2| \geqq \cdots \geqq |c_r|$. Then

$$p = \sum_{i=2}^{r} \left(1 - \frac{c_i}{c_1}\right) p_i^2,$$

which is a sum of $r - 1$ squares. Suppose now that $(p_n)$ is a sequence from $A_d^{(k)} \cap \Sigma^{(k)}$ converging in $A_d^{(k)}$ to $p \in A_d^{(k)}$. We can find polynomials $p_{ni} \in A_{[d/2]}^{(k)}$, $n = 1, 2, \ldots, i = 1, \ldots, N$, such that

$$p_n = \sum_{i=1}^{N} p_{ni}^2.$$

Since $p_n(a)$ converges to $p(a)$ for every $a$ in the finite set $H$ there exists a constant $C > 0$ such that

$$p_{ni}^2(a) \leqq p_n(a) \leqq C \qquad \text{for} \quad a \in H, \quad n \in \mathbb{N}, \quad i = 1, \ldots, N.$$

Consequently, there is an increasing sequence $(n_k)$ in $\mathbb{N}$ such that

$$\lim_{k \to \infty} p_{n_k i}(a)$$

exists for all $a \in H$ and $i = 1, \ldots, N$. This implies the existence of polynomials $p_1, \ldots, p_N \in A_{[d/2]}^{(k)}$ such that $\lim_{k \to \infty} p_{n_k i} = p_i$, $i = 1, \ldots, N$, and hence

$$p = \sum_{i=1}^{N} p_i^2.$$

To finish the proof we may either appeal to the theorem of Krein–Šmulian as in Berg *et al.* (1979) or more simply make use of the following lemma:

**3.3. Lemma.** *Let $(E_n)_{n \geqq 1}$ be a strictly increasing sequence of finite dimensional subspaces of a vector space $E$ such that $E = \bigcup_{i=1}^{\infty} E_n$. A convex subset $C \subseteq E$ is closed in the finest locally convex topology on $E$ if and only if $C \cap E_n$ is closed in the canonical topology on $E_n$ for each $n$.*

Proof. The "only if" part is clear. Next suppose that $C$ is a convex set such that $C \cap E_n$ is closed in $E_n$ for each $n$ and let $a \notin C$. We shall construct a neighbourhood of $a$ disjoint with $C$. Without loss of generality we may assume $a = 0$. Since $C \cap E_1$ is a closed convex subset of $E_1$ not containing 0 there is an absolutely convex and compact neighbourhood $V_1$ of 0 in $E_1$ such that $V_1 \cap C = \emptyset$. Inductively, we now construct a sequence $(V_n)_{n \geqq 2}$ where $V_n$ is an absolutely convex and compact neighbourhood of 0 in $E_n$ such that

$$V_n \cap C = \emptyset, \qquad V_n \cap E_{n-1} = V_{n-1} \qquad \text{for} \quad n \geqq 2. \tag{1}$$

Suppose that $V_1, \ldots, V_{n-1}$ are constructed. There exists a closed and absolutely convex neighbourhood $V_n'$ of 0 in $E_n$ such that $V_n' \cap E_{n-1} = V_{n-1}$, and by the separation theorem (1.2.3) applied to the disjoint sets $V_{n-1}$ and $C \cap E_n$ in $E_n$ there exists a compact, absolutely convex neighbourhood $V_n''$ of $V_{n-1}$ in $E_n$ such that $V_n'' \cap (C \cap E_n) = \emptyset$. Then $V_n = V_n' \cap V_n''$ satisfies (1). Finally, $V = \bigcup_{n=1}^{\infty} V_n$ is absolutely convex and absorbing in $E$, hence a neighbourhood of 0 in the finest locally convex topology on $E$, and $V \cap C = \emptyset$. □

Since $\Sigma^{(k)} = \Sigma^{(k)} \mp A_+^{(k)}$ for $k \geqq 2$ we get by specialization of Theorem 1.11:

**3.4. Theorem.** *For $k \geqq 2$ there exist positive definite functions on $\mathbb{N}_0^k$ which are not moment functions.*

For the case of the complex moment problem, cf. 4.4.11, where $S = (\mathbb{N}_0^2, +, *)$ and $(n, m)^* = (m, n)$ the situation is analogous to the two-dimensional moment problem.

**3.5. Theorem.** *There exist positive definite functions on $(\mathbb{N}_0^2, +, *)$ which are not moment functions.*

Proof. If $S^*$ is identified with $\mathbb{C}$ as in 4.4.11 then $\chi_{(n,m)}$ is identified with the function $z \mapsto z^n \bar{z}^m$ and $V(\mathbb{C})$ is the complex vector space spanned by these functions. Putting $z = x + iy$, $(x, y) \in \mathbb{R}^2$ we can identify $V(\mathbb{C})$ (resp. $V(\mathbb{R})$) with the complex (resp. real) vector space of polynomials in $x, y$ with complex (resp. real) coefficients. Under this identification the convex cone $\Sigma$, cf. (6) in §1, is equal to the cone $\Sigma^{(2)}$ above. By 3.2 and 3.1 we know that $\Sigma^{(2)}$ is closed and different from $A_+^{(2)} = V(\mathbb{R})_+$, hence $\mathscr{H}(S) \neq \mathscr{P}(S)$ by 1.11. □

If $s = (s_n)_{n \geqq 0}$ is a Stieltjes moment sequence then $E_a s$ is obviously positive definite for all $a \in \mathbb{N}_0$, and this property implies on the other hand, that $s$ is a Stieltjes moment sequence, cf. Theorem 2.5. It is natural to examine if this result extends to $k$ dimensions.

**3.6. Definition.** A function $\varphi\colon \mathbb{N}_0^k \to \mathbb{R}$ is called *completely positive definite* if $E_a\varphi$ is positive definite for all $a \in \mathbb{N}_0^k$, and $\varphi$ is called a *Stieltjes moment function* if there exists $\mu \in M_+([0, \infty[^k)$ such that

$$\varphi(n) = \int x^n \, d\mu(x) \qquad \text{for} \quad n \in \mathbb{N}_0^k.$$

A Stieltjes moment function is completely positive definite. The converse is, however, false if $k \geqq 2$ because the following generalization of Theorem 3.4 holds.

**3.7. Theorem.** *For $k \geqq 2$ there exist completely positive definite functions on $\mathbb{N}_0^k$ which are not moment functions.*

The proof of Theorem 3.7 depends on several lemmas. We introduce the convex cone $C^{(k)} \subseteq A^{(k)}$ generated by the polynomials of the form $x^a p(x)^2$ where $a \in \mathbb{N}_0^k$, $p \in A^{(k)}$. Since we can put even powers of the coordinates into the square we have

$$C^{(k)} = \sum_{a \in \{0,1\}^k} x^a \Sigma^{(k)}.$$

**3.8. Lemma.** *A function $\varphi\colon \mathbb{N}_0^k \to \mathbb{R}$ is completely positive definite if and only if*

$$\langle \varphi, p \rangle \geqq 0 \qquad \textit{for all} \quad p \in C^{(k)},$$

*where $\langle \varphi, p \rangle = \sum \varphi(n) c_n$ if $p(x) = \sum c_n x^n$.*

PROOF. For $a \in \mathbb{N}_0^k$, $p \in A^{(k)}$ we have

$$\langle E_a \varphi, p^2 \rangle = \langle \varphi, x^a p^2 \rangle$$

and the assertion follows since $E_a \varphi \in \mathscr{P}(\mathbb{N}_0^k)$ if and only if $\langle E_a \varphi, p^2 \rangle \geqq 0$ for all $p \in A^{(k)}$, cf. 1.10. □

**3.9. Lemma.** *The convex cone $C^{(k)}$ is closed in $A^{(k)}$.*

PROOF. We use the same idea of proof as in Theorem 3.2, so it suffices to prove that $C^{(k)} \cap A_d^{(k)}$ is closed in $A_d^{(k)}$ for every $d \geqq 0$. The point set $H$ is chosen so that each $a_i \geqq 1$. Every polynomial $p \in C^{(k)} \cap A_d^{(k)}$ can be written

$$p(x) = \sum_{a \in \{0,1\}^k} x^a p(a, x),$$

where $p(a, \cdot) \in \Sigma^{(k)}$, and we see that the degree of $p(a, \cdot)$ is at most $d - (a_1 + \cdots + a_k)$. If $(p_n)$ is a sequence from $C^{(k)} \cap A_d^{(k)}$ converging in $A_d^{(k)}$ to $p \in A_d^{(k)}$, the choice of $H$ implies that each sequence $(p_n(a, \cdot))$ is bounded on $H$, and we conclude as in 3.2 that a subsequence $p_{n_k}(a, \cdot)$ converges to $p(a, \cdot) \in \Sigma^{(k)}$ for each $a \in \{0, 1\}^k$ and finally that $p = \sum x^a p(a, \cdot) \in C^{(k)}$. □

We will next show that $A_+^{(k)} \setminus C^{(k)} \neq \emptyset$ for $k \geqq 2$, and it suffices to do this for $k = 2$ since a polynomial of two variables may be considered as a polynomial in more variables. Once we have constructed $p_0 \in A_+^{(k)} \setminus C^{(k)}$, we can separate $p_0$ from $C^{(k)}$ by a continuous linear functional $L: A^{(k)} \to \mathbb{R}$ using the Hahn–Banach theorem (1.2.3). Therefore, there is a function $\varphi: \mathbb{N}_0^k \to \mathbb{R}$ such that

$$\langle \varphi, p \rangle \geqq 0 \quad \text{for} \quad p \in C^{(k)} \quad \text{and} \quad \langle \varphi, p_0 \rangle < 0,$$

showing by Lemma 3.8 that $\varphi$ is completely positive definite. The inequality $\langle \varphi, p_0 \rangle < 0$ is inconsistent with $\varphi$ being a moment sequence.

It is plausible that the polynomial $p$ of Lemma 3.1 does not belong to $C^{(2)}$, but we have not been able to decide it. We will instead modify an example of Robinson (1969) in order to construct a polynomial $p_0 \in A_+^{(2)} \setminus C^{(2)}$.

**3.10. Lemma.** *If a polynomial $p \in A_3^{(2)}$ is zero at the eight points of*

$$\{-1, 0, 1\}^2 \setminus \{(0, 0)\},$$

*then automatically $p(0, 0) = 0$.*

PROOF. We may write

$$p(x, y) = \sum_{i=0}^{3} p_i(y) x^i,$$

where $p_i \in A_{3-i}^{(1)}$. By assumption the polynomials

$$p(1, y) = p_0(y) + p_1(y) + p_2(y) + p_3(y),$$
$$p(-1, y) = p_0(y) - p_1(y) + p_2(y) - p_3(y),$$

have zeros at $y = -1, 0, 1$, and so have half of their sum $p_0(y) + p_2(y)$ and half of their difference $p_1(y) + p_3(y)$. Since $p_1 + p_3 \in A_2^{(1)}$ we must have $p_1 + p_3 = 0$, and we may write $p_0(y) + p_2(y) = ay(y^2 - 1)$. However, $p_0(y) = p(0, y)$ has zeros at $y = \pm 1$, hence $p_2(\pm 1) = 0$. Since $p_2 \in A_1^{(1)}$ we get $p_2 = 0$. The polynomial $p_3$ is a constant $b \in \mathbb{R}$, hence $p_1(y) = -b$ and finally

$$p(x, y) = ay(y^2 - 1) + bx(x^2 - 1),$$

which shows $p(0, 0) = 0$. □

**3.11. Lemma.** *Let $p(x, y)$ be the polynomial of Robinson*

$$p(x, y) = x^2(x^2 - 1)^2 + y^2(y^2 - 1)^2 - (x^2 - 1)(y^2 - 1)(x^2 + y^2 - 1).$$

*Then $q(x, y) := p(x - 2, y - 2) \in A_+^{(2)} \setminus C^{(2)}$.*

Proof. To see that $p(x, y) \geqq 0$ it suffices to consider $x \geqq 0$, $y \geqq 0$. This domain is divided by the unit circle and the lines $x = 1$ and $y = 1$ in five domains denoted $D_1, \ldots, D_5$ in Figure 3. The expression for $p$ shows that $p(x, y) \geqq 0$ in $D_1$, $D_2$ and $D_3$, but $p$ can be written

$$p(x, y) = (x^2 - y^2)^2(x^2 + y^2 - 1) + (x^2 - 1)(y^2 - 1)$$

showing that $p(x, y) \geqq 0$ in $D_4$ and $D_5$.

Assume that $q \in C^{(2)}$, i.e. that $q$ may be written

$$q(x, y) = q_1(x, y) + xq_2(x, y) + yq_3(x, y) + xyq_4(x, y),$$

where $q_i \in \Sigma^{(2)}$, hence

$$p(x, y) = p_1(x, y) + (x + 2)p_2(x, y) + (y + 2)p_3(x, y) + (x + 2)(y + 2)p_4(x, y),$$

where $p_i(x, y) = q_i(x + 2, y + 2) \in \Sigma^{(2)}$. Robinson's polynomial $p$ has zeros at $\{1, 0, -1\}^2 \setminus \{(0, 0)\}$ and so have $p_i$, $i = 1, \ldots, 4$. However, $p_1$ is of degree of most 6, and $p_2$, $p_3$ and $p_4$ are of degree at most 4. Furthermore, each $p_i$ is of the form $p_i = \sum p_{ij}^2$ so each $p_{ij}$ is of degree at most 3 and has zeros at $\{1, 0, 1\}^2 \setminus \{(0, 0)\}$. By Lemma 3.10 we see that $p_{ij}(0, 0) = 0$ for all $i, j$, hence $p(0, 0) = 0$, which is a contradiction since $p(0, 0) = 1$. □

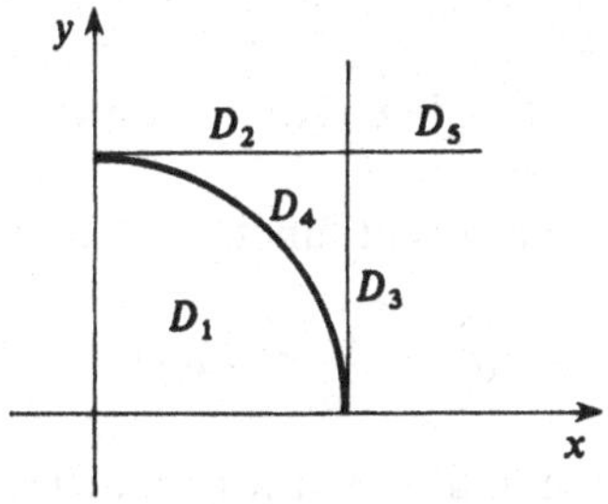


Figure 3

It is worth noticing that even if we replace the cone $\mathscr{P}(\mathbb{N}_0^k)$ by the smaller cone $\mathscr{H}(\mathbb{N}_0^k)$ of moment functions we have no generalization of Theorem 2.5.

**3.12. Theorem.** *For $k \geqq 2$ there exists a function $\varphi\colon \mathbb{N}_0^k \to \mathbb{R}$ such that $E_a\varphi \in \mathscr{H}(\mathbb{N}_0^k)$ for all $a \in \mathbb{N}_0^k$, but $\varphi$ is not a Stieltjes moment function.*

Proof. Let $F = [0, \infty[^k$ and let $A_+^{(k)}(F)$ be the set of polynomials $p \in A^{(k)}$ which are nonnegative on $F$. Let $D^{(k)}$ be the convex cone in $A^{(k)}$ generated by the polynomials $x^a p(x)$ where $a \in \mathbb{N}_0^k$, $p \in A_+^{(k)}$. We have

$$D^{(k)} = \sum_{a \in \{0,1\}^k} x^a A_+^{(k)}$$

and $C^{(k)} \subseteq D^{(k)} \subseteq A_+^{(k)}(F)$. Using the same idea of proof as in Lemma 3.9 it follows that $D^{(k)}$ is closed in $A^{(k)}$. Once we have constructed a polynomial $p_0 \in A_+^{(k)}(F) \setminus D^{(k)}$, it can be separated from $D^{(k)}$ by a (continuous) linear functional $L\colon A^{(k)} \to \mathbb{R}$. There is consequently a function $\varphi\colon \mathbb{N}_0^k \to \mathbb{R}$ such that

$$\langle \varphi, p \rangle \geqq 0 \quad \text{for all} \quad p \in D^{(k)} \qquad \text{and} \qquad \langle \varphi, p_0 \rangle < 0,$$

in particular

$$\langle E_a\varphi, p \rangle \geqq 0 \qquad \text{for all} \quad p \in A_+^{(k)},$$

showing by Theorem 1.10 that $E_a\varphi \in \mathscr{H}(\mathbb{N}_0^k)$ for all $a \in \mathbb{N}_0^k$. However, $\langle \varphi, p_0 \rangle < 0$ is inconsistent with $\varphi$ being a Stieltjes moment function. The existence of $p_0 \in A_+^{(k)}(F) \setminus D^{(k)}$ is established in the following lemma for $k = 2$. □

**3.13. Lemma.** *The polynomial*

$$p_0(x, y) = xy(x + y - 1) + 1$$

*belongs to $A_+^{(2)}(F) \setminus D^{(2)}$ where $F = [0, \infty[^2$.*

Proof. It is easy to see that $p_0(x, y) \geqq 0$ for $x, y \geqq 0$, cf. Lemma 3.1. If we assume that $p_0 \in D^{(2)}$ we have

$$p_0(x, y) = s_0(x, y) + xs_1(x, y) + ys_2(x, y) + xys_3(x, y),$$

where $s_i \in A_+^{(2)}$. Each $s_i$ must be of even degree $\leqq 2$ and $s_3$ must be a nonnegative constant. A nonnegative polynomial of degree 2 in any number of variables is a sum of squares of polynomials as is known from the theory of quadratic forms. It follows that $s_i \in \Sigma^{(2)}$, hence $s_i(x^2, y^2) \in \Sigma^{(2)}$ for $i = 0, 1, 2, 3$, but then $p_0(x^2, y^2) \in \Sigma^{(2)}$, in contradiction with Lemma 3.1. □

**3.14. Exercise.** Show that a moment function

$$\varphi(n) = \int_{\mathbb{R}^k} x^n \, d\mu(x)$$

is strictly positive definite if and only if $\mu$ is not supported by an algebraic variety, i.e. a set of the form $p^{-1}(\{0\})$ for $p \in A^{(k)} \setminus \{0\}$. Show that the set of strictly positive definite functions on $\mathbb{N}_0^k$ is dense in $\mathscr{P}(\mathbb{N}_0^k)$, and conclude that there are strictly positive definite functions which are not moment functions, when $k \geqq 2$.

**3.15. Exercise.** Let $\varphi: \mathbb{N}_0^k \to \mathbb{R}$ be strictly positive definite. Show that

$$\langle p, q \rangle_\varphi := \langle \varphi, p\bar{q} \rangle$$

is a scalar product on the space $A_{\mathbb{C}}^{(k)}$ of polynomials with complex coefficients. Let $H$ be the Hilbert space completion of the pre-Hilbert space $(A_{\mathbb{C}}^{(k)}, \langle \cdot, \cdot \rangle_\varphi)$ and let $T_i: A_{\mathbb{C}}^{(k)} \to A_{\mathbb{C}}^{(k)} \subseteq H$ be the densely defined operator $T_i p(x) = x_i p(x)$, $p \in A_{\mathbb{C}}^{(k)}$, $i = 1, \ldots, k$. Show that each $T_i$ is a symmetric operator and that $T_i T_j = T_j T_i$. However, if $\varphi \notin \mathscr{H}(\mathbb{N}_0^k)$, then it is impossible to find self-adjoint extensions $\tilde{T}_i$ of $T_i$ such that $\tilde{T}_1, \ldots, \tilde{T}_n$ commute in the sense of spectral theory. *Hint*: *If $P$ is the spectral measure on* $\mathbb{R}^k$ *of a commuting family* $\tilde{T}_1, \ldots, \tilde{T}_k$ of self-adjoint extensions, then $\mu = \langle P1, 1 \rangle$ satisfies

$$\varphi(n) = \int x^n \, d\mu(x),$$

cf. Fuglede (1983).

**3.16. Exercise** (Haviland 1935, 1936). Let $\varphi: \mathbb{N}_0^k \to \mathbb{R}$ and let $L: A^{(k)} \to \mathbb{R}$ be the linear functional determined by $L(x^n) = \varphi(n), n \in \mathbb{N}_0^k$. Show that $\varphi$ is the moment function for some $\mu \in M_+(\mathbb{R}^k)$ supported by a closed subset $F \subseteq \mathbb{R}^k$ if and only if $L(p) \geqq 0$ for all $p \in A^{(k)}$ which are nonnegative on $F$. *Hint*: Use Exercise 2.2.12.

**3.17. Exercise.** Let $\varphi: \mathbb{N}_0^k \to \mathbb{R}$ be given. Show that $\varphi \in \mathscr{P}(\mathbb{N}_0^k)$ and

$$(I - (E_1^2 + \cdots + E_k^2))\varphi \in \mathscr{P}(\mathbb{N}_0^k)$$

if and only if $\varphi$ is a moment function for a measure $\mu \in M_+(\mathbb{R}^k)$ supported by the unit ball $\{x \in \mathbb{R}^k \mid \|x\| \leqq 1\}$ where $E_j = E_{e_j}$ and $e_1 = (1, 0, \ldots, 0), \ldots, e_k = (0, \ldots, 0, 1)$. *Hint*: Let $H$ be the RKHS associated with $\varphi$ and $H_0$ the dense subspace spanned by the functions $\varphi_n$, $n \in \mathbb{N}_0^k$, defined by $\varphi_n(m) = \varphi(n + m)$. Show that each $E_j$ is a bounded self-adjoint operator on $H$ and that $E_1, \ldots, E_k$ commute. Show that $\mu = \langle P1, 1 \rangle$ where $P$ is the spectral measure on $\mathbb{R}^k$ of the commuting family $E_1, \ldots, E_k$.

**3.18. Exercise.** Let $S = (\mathbb{N}_0^{(\infty)}, +)$ be the countable direct sum of $\mathbb{N}_0$ consisting of all sequences $n = (n_1, n_2, \ldots)$ of nonnegative integers which become eventually zero. Show that $S^*$ is isomorphic and homeomorphic with $(\mathbb{R}^{\mathbb{N}}, \cdot)$. Show that, in the terminology of §1, $\Sigma$ is closed in $V(\mathbb{R})$ in its finest locally convex topology and a proper subset of $V(\mathbb{R})_+$. Conclude that $\mathscr{P}(S) \neq \mathscr{H}(S)$.

Show that $(\mathbb{N}, \cdot)$ is isomorphic with $(\mathbb{N}_0^{(\infty)}, +)$ and conclude that $\mathscr{P}(\mathbb{N}, \cdot) \neq \mathscr{H}(\mathbb{N}, \cdot)$.

# §4. The Two-Sided Moment Problem

This problem consists in characterizing the *two-sided moment sequences*, i.e. the functions $\varphi: \mathbb{Z} \to \mathbb{R}$ of the form

$$\varphi(n) = \int_{\mathbb{R}\setminus\{0\}} x^n \, d\mu(x) \qquad \text{for} \quad n \in \mathbb{Z}, \tag{1}$$

where $\mu \in M_+(\mathbb{R}\setminus\{0\})$ is such that $x^n$ is $\mu$-integrable for all $n \in \mathbb{Z}$. The problem is called the *strong Hamburger moment problem* in Jones *et al.* (1984). We shall solve it by applying the preceding theory to the semigroup $(\mathbb{Z}, +)$ with the identical involution.

For $x \in \mathbb{R}\setminus\{0\}$ the function $\rho_x: \mathbb{Z} \to \mathbb{R}$ given by $\rho_x(n) = x^n$, $n \in \mathbb{Z}$, is an element of $\mathbb{Z}^*$, and it is easy to see that the mapping $x \mapsto \rho_x$ is a topological semigroup isomorphism of $(\mathbb{R}\setminus\{0\}, \cdot)$ onto $\mathbb{Z}^*$. Under this isomorphism $\hat{\mathbb{Z}}$ corresponds to $(\{-1, 1\}, \cdot)$ so

$$\mathscr{P}^b(\mathbb{Z}) = \{n \mapsto a(-1)^n + b \,|\, a, b \geqq 0\}.$$

The semigroup $(\mathbb{Z}, +)$ is generated by $\{-1, 1\}$. The functions $\chi_n$, $n \in \mathbb{Z}$, cf. (3) in §1, are the functions $x \mapsto x^n$ on $\mathbb{R}\setminus\{0\}$, which are linearly independent, so $V(\mathbb{R})$ is the subspace of $C(\mathbb{R}\setminus\{0\})$ spanned by $x^n$, $n \in \mathbb{Z}$. It follows that $V(\mathbb{R})$ is the set of functions on $\mathbb{R}\setminus\{0\}$ of the form $p(x)/x^n$, where $n \geqq 0$ and $p$ is a polynomial with real coefficients. A representation $f(x) = p(x)/x^n$ of $f \in V(\mathbb{R})$ is of course not unique, and we may always assume that $n$ is even. If $f \in V(\mathbb{R})_+$ we may write $f(x) = p(x)/x^{2n}$, where $p(x) \geqq 0$ for $x \in \mathbb{R}\setminus\{0\}$, hence for all $x \in \mathbb{R}$, and by Lemma 2.1 we may write $p = p_1^2 + p_2^2$ so that

$$f(x) = \left(\frac{p_1(x)}{x^n}\right)^2 + \left(\frac{p_2(x)}{x^n}\right)^2.$$

With the terminology of (6) in §1 we then have $\Sigma = V(\mathbb{R})_+$, hence $\mathscr{P}(\mathbb{Z}) = \mathscr{H}(\mathbb{Z})$, and we have proved the following result:

**4.1. Theorem.** *A function $\varphi: \mathbb{Z} \to \mathbb{R}$ is a two-sided moment function if and only if $\varphi \in \mathscr{P}(\mathbb{Z})$. The set $E_+(\mathbb{R}\setminus\{0\}, \varphi)$ of representing measures is a nonempty compact convex set in $M_+(\mathbb{R}\setminus\{0\})$ with the vague topology.*

**4.2. Remark.** A function $\varphi \in \mathbb{R}^{\mathbb{Z}}$ belongs to $\mathscr{P}(\mathbb{Z})$ if and only if for every $n \geqq 0$ the $(2n + 1) \times (2n + 1)$ matrix

$$(\varphi(j + k)), \qquad j, k \in \{-n, -n + 1, \ldots, 0, 1, \ldots, n\}$$

is positive definite, and $\varphi$ is strictly positive definite if and only if these matrices are strictly positive definite. Introducing the determinants

$$H_k^{(n)} = \begin{vmatrix} \varphi(n) & \varphi(n+1) & \cdots & \varphi(n+k) \\ \varphi(n+1) & \varphi(n+2) & \cdots & \varphi(n+k+1) \\ \vdots & \vdots & & \vdots \\ \varphi(n+k) & \varphi(n+k+1) & \cdots & \varphi(n+2k) \end{vmatrix}$$

for $n \in \mathbb{Z}$, $k = 0, 1, \ldots$ it follows by an easy modification of Theorem 3.1.16 that $\varphi$ is strictly positive definite if and only if

$$H_{2n}^{(-2n)} > 0, \qquad H_{2n+1}^{(-2n)} > 0 \qquad \text{for} \quad n = 0, 1, \ldots.$$

Let $\varphi \in \mathscr{P}(\mathbb{Z})$ and $\mu \in E_+(\mathbb{R}\setminus\{0\}, \varphi)$. For $\{c_{-n}, c_{-n+1}, \ldots, c_0, c_1, \ldots, c_n\} \subseteq \mathbb{R}$ we define

$$p(x) = \sum_{k=-n}^{n} c_k x^k$$

and have

$$\sum_{j,k=-n}^{n} c_j c_k \varphi(j+k) = \int p(x)^2 \, d\mu(x),$$

and it follows that $\varphi$ is strictly positive definite precisely when $\operatorname{supp}(\mu)$ is an infinite set.

We shall now discuss the possibility of extending a positive definite sequence $s: \mathbb{N}_0 \to \mathbb{R}$ to a positive definite function $\varphi: \mathbb{Z} \to \mathbb{R}$.

**4.3. Theorem.** *Let $s: \mathbb{N}_0 \to \mathbb{R}$ be a positive definite sequence and let $E_+(\mathbb{R}, s)$ be the set of representing measures. Then $s$ can be extended to a positive definite function $\varphi: \mathbb{Z} \to \mathbb{R}$ if and only if there exists a measure $\mu \in E_+(\mathbb{R}, s)$ satisfying*

$$\mu(\{0\}) = 0 \qquad \textit{and} \qquad \int_{\mathbb{R}\setminus\{0\}} \frac{1}{|x|^n} \, d\mu(x) < \infty \qquad \textit{for} \quad n \geqq 1. \tag{2}$$

*If such a measure $\mu$ exists then*

$$\varphi(n) = \int_{\mathbb{R}\setminus\{0\}} x^n \, d\mu(x), \qquad n \in \mathbb{Z} \tag{3}$$

*is a positive definite extension of $s$.*

Proof. If $\varphi$ is a positive definite extension of $s$ there is a measure $\mu \in E_+(\mathbb{R}\setminus\{0\}, \varphi)$ such that

$$\varphi(n) = \int_{\mathbb{R}\setminus\{0\}} x^n \, d\mu(x) \qquad \text{for} \quad n \in \mathbb{Z}.$$

Since $\mu(\mathbb{R}\setminus\{0\}) = \varphi(0) < \infty$ we can extend $\mu$ to a Radon measure $\bar{\mu}$ on $\mathbb{R}$ by setting

$$\bar{\mu}(B) = \mu(B\setminus\{0\}) \qquad \text{for} \quad B \in \mathscr{B}(\mathbb{R}),$$

and it is clear that $\bar{\mu} \in E_+(\mathbb{R}, s)$ and that $\bar{\mu}$ satisfies (2). On the other hand if there exists a measure $\mu \in E_+(\mathbb{R}, s)$ satisfying (2), then $\varphi$ defined by (3) is a positive definite extension of $s$. □

If $s$ is determinate we at once get the following:

**4.4. Corollary.** *Let $s\colon \mathbb{N}_0 \to \mathbb{R}$ be a determinate positive definite sequence with representing measure $\mu$. Then $s$ can be extended to a positive definite function $\varphi\colon \mathbb{Z} \to \mathbb{R}$ if and only if $\mu$ satisfies (2). If this is the case then the extension is uniquely determined, it is given by (3) and is determinate.*

Using well-known results from the theory of the one-dimensional moment problem in the indeterminate case we can prove:

**4.5. Theorem.** *Let $s\colon \mathbb{N}_0 \to \mathbb{R}$ be an indeterminate positive definite sequence. The set $F$ of positive definite functions $\varphi\colon \mathbb{Z} \to \mathbb{R}$ extending $s$ is a closed convex set in $\mathscr{P}(\mathbb{Z})$, and there is a continuous injection $t \mapsto \varphi_t$ of $\mathbb{R}$ into $F$ such that $\varphi_t$ is determinate for each $t \in \mathbb{R}$.*

Proof. Without loss of generality we may assume that $s(0) = 1$. It is clear that $F$ is a closed convex set. To see the last assertion we remark that any representing measure $\mu \in E_+(\mathbb{R}, s)$ for which $0 \notin \operatorname{supp}(\mu)$, satisfies (2) and gives an extension of $s$ by the formula (3). That there are such measures in $E_+(\mathbb{R}, s)$ follows from the Nevanlinna parametrization of $E_+(\mathbb{R}, s)$, cf. Akhiezer (1965), according to which the Nevanlinna extremal measures $\sigma_t \in E_+(\mathbb{R}, s)$ for $t \in \mathbb{R} \cup \{\infty\}$ satisfy

$$\int \frac{d\sigma_t(x)}{x - z} = -\frac{A(z)t - C(z)}{B(z)t - D(z)} \qquad \text{for} \quad z \in \mathbb{C}\setminus\mathbb{R}, \tag{4}$$

where $A, B, C, D$ are certain entire functions and $\sigma_t$ is a discrete measure with mass at the zeros of the entire function $z \mapsto B(z)t - D(z)$. (For $t = \infty$ this function shall be interpreted as $B(z)$.) It is known that $A(0) = D(0) = 0$ and $-B(0) = C(0) = 1$. It follows that $\sigma_t$ has mass at 0 if and only if $t = 0$, hence that $0 \notin \operatorname{supp}(\sigma_t)$ for $t \in (\mathbb{R} \cup \{\infty\})\setminus\{0\}$. If we let $z$ tend to zero through imaginary values we get from (4) that

$$\int \frac{d\sigma_t(x)}{x} = -\frac{1}{t}, \qquad t \in (\mathbb{R} \cup \{\infty\})\setminus\{0\},$$

and it follows that

$$\varphi_t(n) := \int_{\mathbb{R}\setminus\{0\}} x^n \, d\sigma_{1/t}(x), \qquad n \in \mathbb{Z},$$

belongs to $F$ for each $t \in \mathbb{R}$. Since $\varphi_t(-1) = -t$ the mapping $t \mapsto \varphi_t$ is one-to-one. The continuity of the mapping $t \mapsto \varphi_t(n)$ is clear for $n \geqq 0$, simply because the function is constant $s(n)$, and follows for $n \leqq -1$ by Proposition 2.4.4. In fact, for fixed $t_0 \in \mathbb{R}$ there exists $\varepsilon > 0$ such that $\operatorname{supp}(\sigma_{1/t}) \subseteq \mathbb{R}\setminus]-\varepsilon, \varepsilon[$ for $t \in ]t_0 - \varepsilon, t_0 + \varepsilon[$, and then

$$x^n \in C^0(\mathbb{R}\setminus]-\varepsilon, \varepsilon[)$$

when $n \leqq -1$.

We shall finally prove that $\varphi_t \in \mathscr{P}(\mathbb{Z})$ is determinate. We first establish that $(1/x^2)\sigma_{1/t}$ generates a determinate Hamburger moment sequence, and for this it suffices to see that

$$\inf\left\{\int p(x)^2 \frac{1}{x^2}\, d\sigma_{1/t}(x) \,|\, p \text{ polynomial}, p(0) = 1\right\} = 0, \tag{5}$$

cf. Akhiezer (1965, pp. 49 and 60). Now, $1/x \in \mathscr{L}^2(\sigma_{1/t})$ and the set of polynomials is dense in $\mathscr{L}^2(\sigma_{1/t})$ because $\sigma_{1/t}$ is Nevanlinna extremal, so for $\varepsilon > 0$ there is a polynomial $q$ such that

$$\varepsilon > \int \left(\frac{1}{x} - q(x)\right)^2 d\sigma_{1/t}(x) = \int (1 - xq(x))^2 \frac{1}{x^2}\, d\sigma_{1/t}(x),$$

and (5) follows because $p(x) = 1 - xq(x)$ is a polynomial verifying $p(0) = 1$.

To see that $\varphi_t$ is determinate we consider $\mu \in E_+(\mathbb{R}\setminus\{0\}, \varphi_t)$ and will prove that $\mu = \sigma_{1/t}$. By assumption

$$\int_{\mathbb{R}\setminus\{0\}} x^n\, d\mu(x) = \int_{\mathbb{R}\setminus\{0\}} x^n\, d\sigma_{1/t}(x) \qquad \text{for} \quad n \in \mathbb{Z},$$

in particular

$$\int_{\mathbb{R}\setminus\{0\}} x^n \frac{1}{x^2}\, d\mu(x) = \int_{\mathbb{R}\setminus\{0\}} x^n \frac{1}{x^2}\, d\sigma_{1/t}(x) \qquad \text{for} \quad n \geqq 0,$$

and we conclude that $(1/x^2)\mu = (1/x^2)\sigma_{1/t}$, whence $\mu = \sigma_{1/t}$. □

**4.6. Example.** We will give an example of an indeterminate two-sided moment sequence.

The log-normal distribution has the following density function on $]0, \infty[$

$$d(x) = \frac{1}{\sqrt{2\pi}} x^{-1} e^{-(1/2)(\log x)^2},$$

which generates $\varphi \in \mathscr{P}(\mathbb{Z})$ given as

$$\varphi(k) = \int_0^\infty \frac{1}{\sqrt{2\pi}} x^{k-1} e^{-(1/2)(\log x)^2}\, dx, \qquad k \in \mathbb{Z}.$$

Substituting $t = \log x$ yields

$$\varphi(k) = \frac{1}{\sqrt{2\pi}} \int_{-\infty}^{\infty} e^{-(1/2)t^2 + kt}\, dt = e^{(1/2)k^2} \frac{1}{\sqrt{2\pi}} \int_{-\infty}^{\infty} e^{-(1/2)(t-k)^2}\, dt = e^{(1/2)k^2}.$$

(This should be compared with the fact that $\psi(k) = -k^2$ is negative definite on $(\mathbb{Z}, +)$ in the semigroup sense.) That $\varphi$ is indeterminate follows from the fact that all the density functions

$$d_a(x) = d(x)(1 + a \sin(2\pi \log x)), \qquad x \in\, ]0, \infty[,$$

where $a \in [-1, 1]$, generate the same moment sequence $\varphi$ because for $k \in \mathbb{Z}$

$$\int_0^\infty x^k \sin(2\pi \log x)\, d(x)\, dx = \frac{1}{\sqrt{2\pi}} \int_{-\infty}^\infty e^{-(1/2)t^2 + kt} \sin(2\pi t)\, dt$$

$$= e^{(1/2)k^2} \frac{1}{\sqrt{2\pi}} \int_{-\infty}^\infty e^{-(1/2)y^2} \sin(2\pi y)\, dy = 0.$$

**4.7. Exercise.** A function $\varphi \in \mathbb{R}^{\mathbb{Z}}$ is called a *two-sided Stieltjes moment sequence* if there is a measure $\mu \in M_+(]0, \infty[)$ such that

$$\varphi(k) = \int_0^\infty x^k\, d\mu(x) \qquad \text{for} \quad k \in \mathbb{Z}.$$

Show that the following conditions are equivalent for $\varphi \in \mathbb{R}^{\mathbb{Z}}$:

(i) $\varphi$ is a two-sided Stieltjes moment sequence.
(ii) $f: \mathbb{Z} \to \mathbb{R}$ defined by

$$f(k) = \begin{cases} \varphi\left(\dfrac{k}{2}\right), & k \text{ even,} \\ 0, & k \text{ odd,} \end{cases}$$

belongs to $\mathscr{P}(\mathbb{Z})$.
(iii) $\varphi, E_1\varphi \in \mathscr{P}(\mathbb{Z})$.

**4.8. Exercise.** Let $S = (\mathbb{Z}^k, +)$. Show that $S^*$ is isomorphic and homeomorphic with $((\mathbb{R}\setminus\{0\})^k, \cdot)$ under the mapping $x \mapsto \rho_x$ where $\rho_x(n) = x^n = x_1^{n_1} \cdots x_k^{n_k}$ for $x = (x_1, \ldots, x_k) \in (\mathbb{R}\setminus\{0\})^k$ and $n = (n_1, \ldots, n_k) \in \mathbb{Z}^k$.

Show that, in the terminology of §1, $\Sigma$ is closed in $V(\mathbb{R})$ and a proper subset of $V(\mathbb{R})_+$ when $k \geqq 2$. Conclude that there exist functions $\varphi \in \mathscr{P}(\mathbb{Z}^k)\setminus\mathscr{H}(\mathbb{Z}^k)$ when $k \geqq 2$.

**4.9. Exercise.** Show that a function $\psi: \mathbb{Z} \to \mathbb{R}$ is negative definite in the semigroup sense on $(\mathbb{Z}, +)$ if and only if it has a representation of the form

$$\psi(n) = a + bn - cn^2 + \int_{\mathbb{R}\setminus\{0,1\}} (1 - x^n - n(1 - x))\, d\mu(x), \qquad n \in \mathbb{Z},$$

where $a, b \in \mathbb{R}$, $c \geqq 0$ and $\mu \in M_+(\mathbb{R}\setminus\{0, 1\})$ satisfies

$$\int_{0<|1-x|<1/2} (1 - x)^2\, d\mu(x) < \infty, \qquad \int_{0<|x|<1/2} \frac{1}{|x|^n}\, d\mu(x) < \infty,$$

$$\int_{|x| \geqq 2} |x|^n\, d\mu(x) < \infty$$

for all $n \geqq 0$.

**4.10. Exercise.** Let $S = (\mathbb{Z}^{(\infty)}, +)$ be the countable direct sum of $\mathbb{Z}$. Show that $\mathscr{P}(S) \neq \mathscr{H}(S)$. Show that $S$ is isomorphic with $(\mathbb{Q}_+\setminus\{0\}, \cdot)$. *Hint*: See Exercise 3.18.

## §5. Perfect Semigroups

**5.1. Definition.** An abelian semigroup $(S, +, *)$ with involution is called *perfect* if every $\varphi \in \mathscr{P}(S)$ is a determinate moment function.

Every semigroup with involution for which $\mathscr{P}(S) = \mathscr{P}^b(S)$ is perfect, cf. 1.5, so, in particular, finite semigroups, idempotent semigroups and abelian groups with the involution $x^* = -x$ are perfect semigroups. Abelian groups with the identical involution need not be perfect as the example of $(\mathbb{Z}, +)$ shows, cf. 4.6. Although $\mathscr{P}(\mathbb{N}_0) = \mathscr{H}(\mathbb{N}_0)$ the semigroup $(\mathbb{N}_0, +)$ is not perfect since there exist indeterminate moment sequences. By the results of §§3 and 4 $(\mathbb{N}_0^k, +)$ and $(\mathbb{Z}^k, +)$ are not perfect. Further examples of perfect semigroups are given below.

Let $S$ be a perfect semigroup. The set of signed measures of the form $\mu_1 - \mu_2 + i(\mu_3 - \mu_4)$ with $\mu_j \in E_+(S^*)$ will be denoted $E(S^*)$. Extending the definition of convolution of measures in $E_+(S^*)$ to $E(S^*)$ by bilinearity, $E(S^*)$ becomes an algebra. The generalized Laplace transformation defined in 4.2.10 can be extended from $M^c(S^*)$ to $E(S^*)$ in the following way:

For $\mu \in E(S^*)$ we denote by $\hat{\mu}$ the function

$$\hat{\mu}(s) = \int_{S^*} \rho(s)\, d\mu(\rho), \qquad s \in S,$$

and the properties (i), (ii) and (iii) of 4.2.10 remain valid:

**5.2. Proposition.** *Let $S$ be a perfect semigroup. The generalized Laplace transformation $\mu \mapsto \hat{\mu}$ is an injective algebra homomorphism of $E(S^*)$ into $\mathbb{C}^S$.*

Proof. We restrict attention to the injectivity, the rest of the proposition being obvious. Suppose that

$$\hat{\mu}_1(s) - \hat{\mu}_2(s) + i(\hat{\mu}_3(s) - \hat{\mu}_4(s)) = 0 \qquad \text{for all} \quad s \in S,$$

where $\mu_j \in E_+(S^*)$ for $j = 1, \ldots, 4$. Since $\hat{\mu}_j(s^*) = \overline{\hat{\mu}_j(s)}$, $j = 1, \ldots, 4$ we also have that

$$\hat{\mu}_1(s) - \hat{\mu}_2(s) - i(\hat{\mu}_3(s) - \hat{\mu}_4(s)) = 0 \qquad \text{for all} \quad s \in S,$$

hence

$$\hat{\mu}_1(s) = \hat{\mu}_2(s) \qquad \text{and} \qquad \hat{\mu}_3(s) = \hat{\mu}_4(s) \qquad \text{for} \quad s \in S.$$

Since $S$ is perfect we conclude that $\mu_1 = \mu_2$ because they represent the same moment function. Similarly $\mu_3 = \mu_4$. □

**5.3. Remark.** Let $(\varphi_\alpha)_{\alpha\in A}$ be a net from $\mathscr{P}(S)$ converging pointwise to a function $\varphi\in\mathscr{P}(S)$ on a perfect semigroup, and let $(\mu_\alpha)_{\alpha\in A}$ and $\mu$ be the uniquely determined measures in $E_+(S^*)$ such that $\hat{\mu}_\alpha = \varphi_\alpha$, $\hat{\mu} = \varphi$. Then

$$\lim_\alpha \int_{S^*} f\,d\mu_\alpha = \int_{S^*} f\,d\mu$$

for all $f\in V(\mathbb{C})$, the subspace of $\mathbb{C}^{S^*}$ spanned by the functions $\chi_s$, $s\in S$. It is natural to believe that $\lim_\alpha \mu_\alpha = \mu$ weakly, but we have not been able to settle this question. However, if $S$ in addition is finitely generated, or if more generally $S^*$ is locally compact and $V(\mathbb{R})$ is an adapted space, then $\lim_\alpha \mu_\alpha = \mu$ vaguely, and hence weakly, by 2.4.2. In fact, a modification of the proof in Proposition 1.7 will show that any vague accumulation point $\sigma$ of the net $(\mu_\alpha)_{\alpha\in A}$ satisfies $\hat{\sigma} = \varphi$, and since $\varphi$ has a unique representing measure by definition, every vague accumulation point of $(\mu_\alpha)_{\alpha\in A}$ is equal to $\mu$, hence $\lim_\alpha \mu_\alpha = \mu$ vaguely. This remark should be compared with Exercise 1.12.

As an application of Theorem 2.1.10 on Radon bimeasures we can prove

**5.4. Theorem.** *The product $S\times T$ of two perfect semigroups $S$ and $T$ is perfect.*

Proof. For $(\rho,\zeta)\in S^*\times T^*$ the mapping $(s,t)\mapsto \rho(s)\zeta(t)$ is a semicharacter on $S\times T$, and every semicharacter $\xi$ on $S\times T$ is of this form with $\rho(s) = \xi(s,0)$ and $\zeta(t) = \xi(0,t)$. This makes it possible to identify $(S\times T)^*$ with $S^*\times T^*$, cf. 4.2.13.

Let $\varphi\in\mathscr{P}(S\times T)$. For each $t\in T$ we have

$$\varphi(\cdot, t^* + t)\in\mathscr{P}(S), \tag{1}$$

$$\varphi(\cdot, t^* + t) + \varphi(\cdot, 0) - \varphi(\cdot, t^*) - \varphi(\cdot, t)\in\mathscr{P}(S), \tag{2}$$

$$\varphi(\cdot, t^* + t) + \varphi(\cdot, 0) - i\varphi(\cdot, t^*) + i\varphi(\cdot, t)\in\mathscr{P}(S). \tag{3}$$

Here (1) follows simply because $(s_1^* + s_2, t^* + t) = (s_1, t)^* + (s_2, t)$. To see (2) and (3) let $\{s_1,\dots,s_n\}\subseteq S$ and $\{c_1,\dots,c_n\}\subseteq\mathbb{C}$ be given. By expressing the defining property of $\varphi\in\mathscr{P}(S\times T)$ for the sets

$$\{(s_1,t),\dots,(s_n,t),(s_1,0),\dots,(s_n,0)\}\subseteq S\times T$$

and $\{c_1,\dots,c_n,-c_1,\dots,-c_n\}\subseteq\mathbb{C}$ (resp. $\{c_1,\dots,c_n,ic_1,\dots,ic_n\}\subseteq\mathbb{C}$) we find

$$\sum_{j,k=1}^n c_j\bar{c}_k(\varphi(s_j^* + s_k, t^* + t) + \varphi(s_j^* + s_k, 0) - \varphi(s_j^* + s_k, t^*) - \varphi(s_j^* + s_k, t)) \geq 0,$$

(resp.

$$\sum_{j,k=1}^n c_j\bar{c}_k(\varphi(s_j^* + s_k, t^* + t) + \varphi(s_j^* + s_k, 0) - i\varphi(s_j^* + s_k, t^*) + i\varphi(s_j^* + s_k, t)) \geq 0),$$

which shows (2) and (3). For $t \in T$ we denote by $\sigma_t$, $\tau_t$ and $\nu_t$ the uniquely determined representing measures in $E_+(S^*)$ for the functions in (1), (2) and (3), and we define

$$\mu_t := \tfrac{1}{2}((\sigma_t + \sigma_0 - \tau_t) + i(\sigma_t + \sigma_0 - \nu_t)).$$

Then $\mu_t$ is the unique signed measure in $E(S^*)$ such that

$$\varphi(s, t) = \int_{S^*} \rho(s)\, d\mu_t(\rho) \qquad \text{for} \quad (s, t) \in S \times T.$$

The mapping $t \mapsto \mu_t$ of $T$ into $E(S^*)$ is positive definite, i.e. for $\{t_1, \ldots, t_n\} \subseteq T$ and $\{c_1, \ldots, c_n\} \subseteq \mathbb{C}$, we have

$$\sum_{j,k=1}^{n} c_j \bar{c}_k \mu_{t_j^* + t_k} \in E_+(S^*).$$

To see this we consider the function $F: S \to \mathbb{C}$ defined by

$$F(s) = \int_{S^*} \rho(s) d\left( \sum_{j,k=1}^{n} c_j \bar{c}_k \mu_{t_j^* + t_k} \right) = \sum_{j,k=1}^{n} c_j \bar{c}_k \varphi(s, t_j^* + t_k)$$

and claim that $F \in \mathscr{P}(S)$. In fact, if $\{s_1, \ldots, s_m\} \subseteq S$ and $\{d_1, \ldots, d_m\} \subseteq \mathbb{C}$ we find

$$\sum_{p,q=1}^{m} d_p \bar{d}_q F(s_p^* + s_q) = \sum_{p,q=1}^{m} \sum_{j,k=1}^{n} d_p c_j \overline{d_q c_k} \varphi((s_p, t_j)^* + (s_q, t_k)) \geqq 0.$$

Since $S$ is perfect there is a measure $\mu \in E_+(S^*)$ such that

$$F(s) = \int_{S^*} \rho(s)\, d\mu(\rho) \qquad \text{for} \quad s \in S$$

and by Proposition 5.2 it follows that

$$\sum_{j,k=1}^{n} c_j \bar{c}_k \mu_{t_j^* + t_k} = \mu \in E_+(S^*).$$

In particular, for any set $A \in \mathscr{B}(S^*)$ the function $t \mapsto \mu_t(A)$ belongs to $\mathscr{P}(T)$ and is therefore of the form

$$\mu_t(A) = \int_{T^*} \zeta(t)\, d\tau_A(\zeta) \qquad \text{for} \quad t \in T$$

for a uniquely determined measure $\tau_A \in E_+(T^*)$. The mapping $A \mapsto \tau_A$ of $\mathscr{B}(S^*)$ into $E_+(T^*)$ is a "Radon vector measure", i.e.

(i) $\tau_\varnothing = 0$;
(ii) $\tau_{\bigcup_1^\infty A_n} = \sum_{n=1}^\infty \tau_{A_n}$ when $(A_n)$ is a sequence of disjoint sets in $\mathscr{B}(S^*)$;
(iii) $\tau_A = \sup\{\tau_K \mid K \in \mathscr{K}(S^*), K \subseteq A\}$ for $A \in \mathscr{B}(S^*)$.

Here (i) is clear. Concerning (ii) we have for $t \in T$:

$$\int_{T^*} \zeta(t)\, d\tau_{\bigcup_1^\infty A_n}(\zeta) = \mu_t\left(\bigcup_1^\infty A_n\right) = \sum_{n=1}^\infty \mu_t(A_n) = \sum_{n=1}^\infty \int_{T^*} \zeta(t)\, d\tau_{A_n}(\zeta),$$

in particular for $t = 0$

$$\sum_{n=1}^{\infty} \tau_{A_n}(T^*) = \mu_0\left(\bigcup_1^{\infty} A_n\right) < \infty \tag{4}$$

and for $t$ replaced by $t^* + t$

$$\sum_{n=1}^{\infty} \int_{T^*} |\zeta(t)|^2 \, d\tau_{A_n}(\zeta) = \mu_{t^*+t}\left(\bigcup_1^{\infty} A_n\right) < \infty. \tag{5}$$

Equation (4) shows that $\sum_{n=1}^{\infty} \tau_{A_n}$ is a Radon measure, cf. Exercise 2.1.28, and since $|\zeta(t)| \leqq 1 + |\zeta(t)|^2$ for $\zeta \in T^*$, (4) and (5) imply that $\sum_{n=1}^{\infty} \tau_{A_n} \in E_+(T^*)$ and (ii) follows by Proposition 5.2.

From (i) and (ii) we see that $\tau$ is increasing: If $A_1, A_2 \in \mathscr{B}(S^*)$, $A_1 \subseteq A_2$ then $\tau_{A_1} \leqq \tau_{A_2}$. For each $A \in \mathscr{B}(S^*)$ the net $\{\tau_K \mid K \in \mathscr{K}(S^*), K \subseteq A\}$ is increasing if the index set is ordered by inclusion, so by Exercise 2.1.29

$$\tilde{\tau}_A := \sup\{\tau_K \mid K \in \mathscr{K}(S^*), K \subseteq A\}$$

is a Radon measure on $T^*$ and $\tilde{\tau}_A \leqq \tau_A$ so in particular $\tilde{\tau}_A \in E_+(T^*)$. For $t \in T$ we find

$$\int_{T^*} \zeta(t) \, d\tilde{\tau}_A(\zeta) = \lim_K \int_{T^*} \zeta(t) \, d\tau_K(\zeta) = \lim_K \mu_t(K) = \mu_t(A)$$

$$= \int_{T^*} \zeta(t) \, d\tau_A(\zeta),$$

where the limit is along the index set $\{K \in \mathscr{K}(S^*), K \subseteq A\}$, showing by 5.2 that $\tau_A = \tilde{\tau}_A$ for every $A \in \mathscr{B}(S^*)$.

The function $\Phi\colon \mathscr{B}(S^*) \times \mathscr{B}(T^*) \to [0, \infty[$ defined by

$$\Phi(A, B) = \tau_A(B) \qquad \text{for} \quad A \in \mathscr{B}(S^*), \quad B \in \mathscr{B}(T^*)$$

is a Radon bimeasure by (i), (ii) and (iii), so Theorem 2.1.10 implies the existence of a Radon measure $\kappa$ on $S^* \times T^*$ such that

$$\Phi(A, B) = \int_{S^* \times T^*} 1_A \otimes 1_B \, d\kappa = \int_{T^*} 1_B \, d\tau_A \qquad \text{for} \quad A \in \mathscr{B}(S^*), \quad B \in \mathscr{B}(T^*),$$

where for functions $f\colon S^* \to \mathbb{C}$, $g\colon T^* \to \mathbb{C}$ the tensor product $f \otimes g$ denotes the function on $S^* \times T^*$ defined by $f \otimes g(\rho, \zeta) = f(\rho)g(\zeta)$ for $(\rho, \zeta) \in S^* \times T^*$.

By standard arguments from integration theory we then get

$$\int_{S^* \times T^*} 1_A \otimes h \, d\kappa = \int_{T^*} h \, d\tau_A$$

for $A \in \mathscr{B}(S^*)$ and any $\tau_A$-integrable function $h\colon T^* \to \mathbb{C}$, in particular

$$\mu_t(A) = \int_{T^*} \zeta(t) \, d\tau_A(\zeta) = \int_{S^* \times T^*} 1_A(\rho)\zeta(t) \, d\kappa(\rho, \zeta) \qquad \text{for} \quad t \in T.$$

Replacing $t$ by $t^* + t$ and using similar standard arguments we find

$$\int_{S^*} g\, d\mu_{t^*+t} = \int_{S^* \times T^*} g(\rho)|\zeta(t)|^2\, d\kappa(\rho, \zeta)$$

for $t \in T$ and any Borel measurable function $g: S^* \to [0, \infty]$, in particular

$$\int_{S^* \times T^*} |\rho(s)|^2 |\zeta(t)|^2\, d\kappa(\rho, \zeta) = \int_{S^*} |\rho(s)|^2\, d\mu_{t^*+t}(\rho) < \infty,$$

showing that $\kappa \in E_+(S^* \times T^*)$. Finally we get

$$\int_{S^* \times T^*} \rho(s)\zeta(t)\, d\kappa(\rho, \zeta) = \int_{S^*} \rho(s)\, d\mu_t(\rho) = \varphi(s, t) \qquad \text{for} \quad (s, t) \in S \times T$$

showing that $\varphi$ is a moment function with representing measure $\kappa$.

We still have to show the uniqueness of the representing measure. Suppose $\kappa_1, \kappa_2 \in E_+(S^* \times T^*)$ satisfy

$$\varphi(s, t) = \int \chi_{(s,t)}\, d\kappa_1 = \int \chi_{(s,t)}\, d\kappa_2 \qquad \text{for} \quad (s, t) \in S \times T,$$

where $\chi_{(s,t)}: S^* \times T^* \to \mathbb{C}$ is given by

$$\chi_{(s,t)}(\rho, \zeta) = \rho(s)\zeta(t) \qquad \text{for} \quad \rho \in S^*, \quad \zeta \in T^*.$$

Denoting $\pi_1$ and $\pi_2$ the projections of $S^* \times T^*$ onto $S^*$ and $T^*$, the image measures $(\chi_{(0,t)}\kappa_1)^{\pi_1}$ and $(\chi_{(0,t)}\kappa_2)^{\pi_1}$ belong to $E(S^*)$ and have the same generalized Laplace transform. (The notion of image measures is extended in an obvious way from positive finite Radon measures to finite signed Radon measures.) They are therefore equal and we find

$$\int 1_A \otimes \chi_t\, d\kappa_1 = \int 1_A \otimes \chi_t\, d\kappa_2 \qquad \text{for} \quad t \in T \quad \text{and} \quad A \in \mathscr{B}(S^*),$$

showing that the image measures $(1_{A \times T^*}\kappa_1)^{\pi_2}$ and $(1_{A \times T^*}\kappa_2)^{\pi_2}$ belong to $E_+(T^*)$ and represent the same function in $\mathscr{P}(T)$. This yields $\kappa_1(A \times B) = \kappa_2(A \times B)$ for all $A \in \mathscr{B}(S^*)$, $B \in \mathscr{B}(T^*)$ and hence $\kappa_1 = \kappa_2$. □

Another way to preserve perfectness is given by the following result.

**5.5. Theorem.** *Let $S$ and $T$ be $*$-semigroups and let $h: S \to T$ be a surjective $*$-homomorphism. Then if $S$ is perfect, so is $T$.*

Proof. The dual homomorphism $h^*: T^* \to S^*$, defined by $h^*(\tau) := \tau \circ h$ is injective and its image $F = h^*(T^*)$, consisting of all $\rho \in S^*$ which can be factorized over $h$, is a closed subset of $S^*$. Furthermore $h^*$ is a homeomorphism from $T^*$ to $F$.

Let $\varphi \in \mathscr{P}(T)$ be given, then $\varphi \circ h \in \mathscr{P}(S)$ has the unique integral representation

$$\varphi(h(s)) = \int_{S^*} \rho(s)\, d\mu(\rho), \qquad s \in S.$$

We shall prove that $\mu$ is concentrated on $F$. *For* $a, b \in S$ we put

$$G_{a,b} := \{\rho \in S^* \mid \rho(a) \neq \rho(b)\};$$

these sets $G_{a,b}$ are open and

$$F^c = \bigcup_{\substack{a,b \in S \\ h(a) = h(b)}} G_{a,b},$$

hence it suffices to show that $\mu(G_{a,b}) = 0$ if $h(a) = h(b)$. But under this assumption we get for $s \in S$

$$\begin{aligned} \int \rho(s)\rho(a)\, d\mu(\rho) &= \int \rho(s+a)\, d\mu(\rho) = \varphi(h(s+a)) \\ &= \varphi(h(s) + h(a)) = \varphi(h(s) + h(b)) = \varphi(h(s+b)) \\ &= \int \rho(s)\rho(b)\, d\mu(\rho), \end{aligned}$$

and therefore

$$\rho(a)\mu = \rho(b)\mu$$

by Proposition 5.2, i.e. $\operatorname{Re} \rho(a)\mu = \operatorname{Re} \rho(b)\mu$ and $\operatorname{Im} \rho(a)\mu = \operatorname{Im} \rho(b)\mu$. This implies

$$\int_{\{\operatorname{Re} \rho(b) > \operatorname{Re} \rho(a)\}} [\operatorname{Re} \rho(b) - \operatorname{Re} \rho(a)]\, d\mu = 0$$

so that $\mu(\{\rho \mid \operatorname{Re} \rho(b) > \operatorname{Re} \rho(a)\}) = 0$ and similarly $\mu(\{\rho \mid \operatorname{Re} \rho(b) < \operatorname{Re} \rho(a)\}) = 0$ as well as the corresponding result for the imaginary part, i.e. we have $\mu(G_{a,b}) = 0$.

Let $\nu \in M_+^b(T^*)$ be the image of $\mu$ under $(h^*)^{-1}$, then $\mu = \nu^{h^*}$ and for $t = h(s) \in T$ we get

$$\begin{aligned} \varphi(t) = \varphi(h(s)) &= \int \rho(s)\, d\mu(\rho) = \int_{h^*(T^*)} \rho(s)\, d\nu^{h^*}(\rho) \\ &= \int_{T^*} (h^*(\tau))(s)\, d\nu(\tau) = \int \tau(h(s))\, d\nu(\tau) \\ &= \int \tau(t)\, d\nu(\tau), \end{aligned}$$

which shows existence of a representing measure for $\varphi$. If $\nu'$ is another measure with this property then by assumption $\nu^{h^*} = (\nu')^{h^*}$ and therefore $\nu = \nu'$, $h^*$ being a homeomorphism. □

We will now give an important example of a perfect semigroup.

**5.6. Proposition.** *The semigroup $S = (\mathbb{Q}_+, +)$ of nonnegative rational numbers is perfect. Every $\varphi \in \mathscr{P}(S)$ has a unique representation*

$$\varphi(r) = a1_{\{0\}}(r) + \int_{-\infty}^{\infty} e^{rx}\, d\mu(x) \qquad \text{for} \quad r \in \mathbb{Q}_+,$$

*where $a \geqq 0$, and $\mu \in M_+(\mathbb{R})$ is such that $e^{rx}$ is $\mu$-integrable for each $r \in \mathbb{Q}_+$.*

Proof. For $x \in \mathbb{R}$ the function $\rho_x\colon \mathbb{Q}_+ \to \mathbb{R}$ defined by $\rho_x(r) = e^{rx}$ is a semicharacter and so is $\rho_{-\infty} := 1_{\{0\}}$. Conversely, if $\rho \in S^*$ then $\rho(1) \geqq 0$ and defining $x = \log \rho(1) \in \mathbb{R} \cup \{-\infty\}$ we have $\rho = \rho_x$. It is easy to see that $x \mapsto \rho_x$ is a topological semigroup isomorphism of $([-\infty, \infty[, +)$ onto $S^*$, the topology on $[-\infty, \infty[$ being the obvious with $\{[-\infty, -n[ \,|\, n \in \mathbb{N}\}$ as a base for the filter of neighbourhoods of $-\infty$.

Let $\varphi \in \mathscr{P}(\mathbb{Q}_+)$ and assume $\varphi(0) = 1$. For each $n \in \mathbb{N}$ we have that $(\varphi(k/n))_{k \geqq 0}$ is a Stieltjes moment sequence because

$$k \mapsto \varphi\left(\frac{k}{n}\right) \qquad \text{and} \qquad k \mapsto \varphi\left(\frac{k+1}{n}\right) = \varphi\left(\frac{k}{n} + 2\frac{1}{2n}\right)$$

are positive definite on $(\mathbb{N}_0, +)$, cf. Theorem 2.5. Therefore there exists $\mu_n \in M^1_+([0, \infty[)$ such that

$$\varphi\left(\frac{k}{n}\right) = \int_0^{\infty} x^k\, d\mu_n(x) \qquad \text{for} \quad k \geqq 0, \quad n \in \mathbb{N}.$$

The mapping $f_n\colon [-\infty, \infty[ \to [0, \infty[$ defined by

$$f_n(x) = e^{x/n} \qquad \text{for} \quad x \in [-\infty, \infty[$$

is a homeomorphism, so there exists $\tau_n \in M^1_+([-\infty, \infty[)$ such that $\tau_n^{f_n} = \mu_n$ (cf. 2.1.14), hence

$$\varphi\left(\frac{k}{n}\right) = \int_{[-\infty, \infty[} e^{(k/n)x}\, d\tau_n(x).$$

By 2.4.6 and 2.4.10 there is an increasing sequence $n_1 < n_2 < \cdots$ in $\mathbb{N}$ such that $\tau_{(n_k)!}$ converges vaguely to a measure $\tau \in M_+([-\infty, \infty[)$ of total mass $\leqq 1$. Let $r = p/q \in \mathbb{Q}_+$ be fixed and assume that $p, q \in \mathbb{N}_0$ have no common prime factors, $q \neq 0$. For $k \in \mathbb{N}$ such that $n_k \geqq q$ we find

$$\varphi(r) = \varphi\left(\frac{r(n_k)!}{(n_k)!}\right) = \int_{[-\infty, \infty[} e^{rx}\, d\tau_{(n_k)!}(x),$$

and using that for each nonnegative continuous function $f$ on a locally compact space the integral $\int f\, d\mu$ is lower semicontinuous in $\mu$ with respect to the vague topology (cf. 2.4.1), we find

$$\int_{[-\infty, \infty[} e^{rx}\, d\tau(x) \leqq \varphi(r),$$

which shows that $e^{rx}$ is $\tau$-integrable for all $r \in \mathbb{Q}_+$. Let $h \in C^c_+([-\infty, \infty[)$ satisfy $0 \leqq h \leqq 1$, $h(x) = 1$ for $x \in [-\infty, 0]$, $h(x) = 0$ for $x \geqq 1$. Then

$$\varphi(r) = \int_{[-\infty, \infty[} e^{rx}h(x)\, d\tau_{(n_k)!}(x) + \int_{[-\infty, \infty[} e^{(r+1)x}(1 - h(x))e^{-x}\, d\tau_{(n_k)!}(x), \tag{6}$$

and since $e^{rx}h(x)$ has compact support in $[-\infty, \infty[$ the first term converges to

$$\int_{[-\infty, \infty[} e^{rx}h(x)\, d\tau(x).$$

Furthermore, $(1 - h(x))e^{-x} \in C^0([-\infty, \infty[)$ and the sequence

$$(e^{(r+1)x}\tau_{(n_k)!})_{k \geqq 1}$$

converges vaguely to $e^{(r+1)x}\tau$ with the total masses bounded by $\varphi(r + 1)$. By Proposition 2.4.4 it follows that the second term of (6) converges to

$$\int_{[-\infty, \infty[} e^{(r+1)x}(1 - h(x))e^{-x}\, d\tau(x) = \int_{[-\infty, \infty[} e^{rx}(1 - h(x))\, d\tau(x),$$

hence

$$\varphi(r) = \int_{[-\infty, \infty[} e^{rx}\, d\tau(x) \qquad \text{for} \quad r \in \mathbb{Q}_+,$$

showing that $\varphi \in \mathscr{H}(\mathbb{Q}_+)$. Putting $a = \tau(\{-\infty\})$ and $\mu = \tau|\mathbb{R}$ we have

$$\varphi(r) = a1_{\{0\}}(r) + \int_{-\infty}^{\infty} e^{rx}\, d\mu(x),$$

hence

$$\lim_{r \to 0} \varphi(r) = \mu(\mathbb{R}) = \varphi(0) - a.$$

The function

$$\Phi(z) = \int_{-\infty}^{\infty} e^{zx}\, d\mu(x)$$

is well defined and continuous on the closed half-plane $\{z \in \mathbb{C} | \operatorname{Re} z \geqq 0\}$ and holomorphic in its interior, hence uniquely determined by $\varphi$. On the imaginary axis $\Phi$ is the Fourier transform of $\mu$ which shows that $\mu$ is uniquely determined by $\varphi$. □

By Proposition 5.6 and Theorem 5.4 we get that $(\mathbb{Q}^k_+, +)$ is a perfect semigroup for $k \geqq 1$. For $x = (x_1, \ldots, x_k) \in [-\infty, \infty[^k$ and $r = (r_1, \ldots, r_k) \in \mathbb{Q}^k_+$ we define $\langle x, r\rangle = \sum_{j=1}^k x_j r_j$ and $\rho_x(r) = e^{\langle x, r\rangle}$ and so we have:

**5.7. Proposition.** *For every $\varphi \in \mathscr{P}(\mathbb{Q}_+^k)$ there is a unique measure $\mu \in M_+([-\infty, \infty[^k)$ such that*

$$\varphi(r) = \int_{[-\infty, \infty[^k} e^{\langle x, r\rangle}\, d\mu(x) \qquad \textit{for} \quad r \in \mathbb{Q}_+^k.$$

As an application of this result we prove:

**5.8. Theorem.** *A necessary and sufficient condition that a function $\varphi\colon \mathbb{R}_+^k \to \mathbb{R}$ can be represented as*

$$\varphi(y) = \int_{\mathbb{R}^k} e^{\langle x, y\rangle}\, d\mu(x) \qquad \textit{for} \quad y \in \mathbb{R}_+^k \tag{7}$$

*with $\mu \in M_+(\mathbb{R}^k)$, is that $\varphi$ is continuous and positive definite on $(\mathbb{R}_+^k, +)$. The measure $\mu$ is uniquely determined by $\varphi$.*

Proof. It is clear that $\varphi$ given by (7) is continuous on $\mathbb{R}_+^k$ and positive definite on $(\mathbb{R}_+^k, +)$. Conversely, if $\varphi$ has these properties the restriction of $\varphi$ to $\mathbb{Q}_+^k$ belongs to $\mathscr{P}(\mathbb{Q}_+^k)$ and is hence of the form

$$\varphi(r) = \int_{[-\infty, \infty[^k} e^{\langle x, r\rangle}\, d\mu(x) \qquad \text{for} \quad r \in \mathbb{Q}_+^k$$

with $\mu \in M_+([-\infty, \infty[^k)$. As in the proof of Proposition 4.4.7 the continuity of $\varphi$ at $(0, \ldots, 0)$ implies that $\mu$ is concentrated on $\mathbb{R}^k$. The two continuous functions $\varphi$ and

$$y \mapsto \int_{\mathbb{R}^k} e^{\langle x, y\rangle}\, d\mu(x)$$

agree on $\mathbb{Q}_+^k$ and (7) follows. The uniqueness follows from Proposition 5.7. □

**5.9. Remark.** The function $\varphi$ in Theorem 5.8 has a unique continuous extension $\tilde{\varphi}$ to $\{z \in \mathbb{C}^k \mid \operatorname{Re} z_j \geqq 0, j = 1, \ldots, k\}$ which is holomorphic in the interior. It is given as

$$\tilde{\varphi}(z) = \int_{\mathbb{R}^k} e^{\langle x, z\rangle}\, d\mu(x)$$

and $\tilde{\varphi}(iy)$, $y \in \mathbb{R}^k$, is the Fourier transform of $\mu$.

The results about $\mathbb{Q}_+$ can be extended to $(\mathbb{Q}, +)$ with the identical involution.

**5.10. Proposition.** *The semigroup $(\mathbb{Q}, +)$ with the identical involution is perfect. Every $\varphi \in \mathscr{P}(\mathbb{Q})$ has a unique representation as*

$$\varphi(r) = \int_{-\infty}^{\infty} e^{rx}\, d\mu(x) \qquad \textit{for} \quad r \in \mathbb{Q},$$

*where $\mu \in M_+(\mathbb{R})$.*

PROOF. Since $h: \mathbb{Q}_+ \times \mathbb{Q}_+ \to \mathbb{Q}$ defined by $h(s, t) = s - t$ is a homomorphism and onto, it follows from 5.4, 5.5 and 5.6 that $(\mathbb{Q}, +)$ is perfect, too. Furthermore, the proof of Theorem 5.5 shows that $\mathbb{Q}^*$ is isomorphic (and homeomorphic) to $h^*(\mathbb{Q}^*)$. Noting that $(\mathbb{Q}_+^2)^* \simeq ([-\infty, \infty[^2, +)$, cf. Proposition 5.7, we get for $\tau \in \mathbb{Q}^*$, $h^*(\tau) = (x, y) \in [-\infty, \infty[^2$, and $s, t \in \mathbb{Q}_+$

$$e^{xs+yt} = (h^*(\tau))(s, t) = \tau(h(s, t)) = \tau(s - t),$$

implying $x > -\infty$, $y > -\infty$ and $y = -x$, i.e. $\tau(s) = e^{xs}$. Hence $\mathbb{Q}^*$ is isomorphic with $(\mathbb{R}, +)$ and a given $\varphi \in \mathscr{P}(\mathbb{Q})$ has the unique integral representation

$$\varphi(r) = \int_{-\infty}^{\infty} e^{rx}\, d\mu(x) \qquad \text{for} \quad r \in \mathbb{Q},$$

where $\mu \in M_+^b(\mathbb{R})$. □

Also $(\mathbb{Q}^k, +)$ is perfect and in analogy with 5.8 we find:

**5.11. Theorem.** *A necessary and sufficient condition that a function $\varphi: \mathbb{R}^k \to \mathbb{R}$ can be represented as*

$$\varphi(y) = \int_{\mathbb{R}^k} e^{\langle x, y\rangle}\, d\mu(x) \qquad \textit{for} \quad y \in \mathbb{R}^k$$

*with $\mu \in M_+(\mathbb{R}^k)$ is that $\varphi$ is continuous and positive definite in the semigroup sense on $(\mathbb{R}^k, +)$. The measure $\mu$ is uniquely determined and $\varphi$ can be extended to an entire holomorphic function on $\mathbb{C}^k$.*

Theorems 5.8 and 5.11 can be generalized to a non-semigroup setting, cf., Widder (1941, p. 273) and Akhiezer (1965, pp. 211 and 229).

**5.12. Theorem.** *Let $H \subseteq \mathbb{R}^k$ be an interval. A necessary and sufficient condition that a function $\varphi: H \to \mathbb{R}$ can be represented as*

$$\varphi(y) = \int_{\mathbb{R}^k} e^{\langle x, y\rangle}\, d\mu(x) \qquad \textit{for} \quad y \in H$$

*with $\mu \in M_+(\mathbb{R}^k)$ is that $\varphi$ is continuous and the kernel $(x, y) \mapsto \varphi(x + y)$ is positive definite on $(\frac{1}{2}H) \times (\frac{1}{2}H)$.*

The proof will not be given here, but let us indicate how a proof may run. The result is first established for $k = 1$ using the truncated moment problem, cf. Akhiezer (1965), and then extended to $k$ dimensions by the technique of Theorem 5.4.

We will now give an integral representation of the cone $\mathscr{N}(\mathbb{Q}_+)$ of negative definite functions on the semigroup $\mathbb{Q}_+$.

**5.13. Proposition.** *A function $\psi: \mathbb{Q}_+ \to \mathbb{R}$ is negative definite if and only if it has a representation of the form*

$$\psi(r) = a + br - cr^2 + d1_{\mathbb{Q}_+\setminus\{0\}}(r) + \int_{\mathbb{R}\setminus\{0\}} (1 - e^{rx} - r(1 - e^x))\, d\mu(x)$$

*for $r \in \mathbb{Q}_+$, where $a, b \in \mathbb{R}$, $c, d \geqq 0$ and $\mu \in M_+(\mathbb{R}\setminus\{0\})$ satisfies*

$$\int_{0<|x|\leqq 1} x^2\, d\mu(x) < \infty, \qquad \int_{1<|x|} e^{rx}\, d\mu(x) < \infty \qquad \textit{for all} \quad r \in \mathbb{Q}_+.$$

*The quintuple $(a, b, c, d, \mu)$ is uniquely determined by $\psi$.*

Proof. We note that

$$\lim_{x\to 0} \frac{1 - e^{rx} - r(1 - e^x)}{(1 - e^x)^2} = \tfrac{1}{2}(r - r^2),$$

and it follows easily that any function of the above form is negative definite.

Conversely, it suffices to prove the above integral representation with $a = 0$ for $\psi \in \mathcal{N}(\mathbb{Q}_+)$ such that $\psi(0) = 0$. By Theorem 3.2.2 and Proposition 5.6 there exists a unique measure $\mu_t \in E_+(\underline{\mathbb{R}})$ on the locally compact space $\underline{\mathbb{R}} := [-\infty, \infty[$ such that

$$e^{-t\psi(r)} = \int_{\underline{\mathbb{R}}} e^{rx}\, d\mu_t(x) \qquad \text{for} \quad r \in \mathbb{Q}_+, \quad t > 0,$$

and $\mu_t$ is a probability measure. We have

$$\frac{1}{t}\int_{\underline{\mathbb{R}}} (1 - e^{rx} - r(1 - e^x))\, d\mu_t(x) = \frac{1}{t}(1 - e^{-t\psi(r)}) - \frac{r}{t}(1 - e^{-t\psi(1)})$$

and

$$0 \leqq \frac{1}{t}\int_{\underline{\mathbb{R}}} (1 - e^{rx})^2\, d\mu_t(x) = \frac{1}{t}(1 - 2e^{-t\psi(r)} + e^{-t\psi(2r)}),$$

which show that

$$\lim_{t\to 0} \frac{1}{t}\int_{\underline{\mathbb{R}}} (1 - e^{rx} - r(1 - e^x))\, d\mu_t(x) = \psi(r) - r\psi(1)$$

and

$$0 \leqq \lim_{t\to 0} \frac{1}{t}\int_{\underline{\mathbb{R}}} (1 - e^{rx})^2\, d\mu_t(x) = 2\psi(r) - \psi(2r).$$

The last equation implies that there is a constant $A_r$ depending on $r \in \mathbb{Q}_+$ such that

$$\frac{1}{t}\int_{\underline{\mathbb{R}}} (1 - e^{rx})^2\, d\mu_t(x) \leqq A_r \qquad \text{for} \quad 0 < t \leqq 1,$$

so by Proposition 2.4.6 there is a measure $\sigma \in M_+(\overline{\mathbb{R}})$ and a sequence $(t_j)$ tending to zero such that

$$\lim_{j\to\infty} \frac{1}{t_j}(1-e^x)^2\mu_{t_j} = \sigma$$

vaguely. Since $(1-e^{rx})^2(1-e^x)^{-2}$ is continuous on $\overline{\mathbb{R}}$ we get

$$\lim_{j\to\infty} \frac{1}{t_j}\int_{\overline{\mathbb{R}}}(1-e^{rx})^2 f(x)\,d\mu_{t_j} = \int_{\overline{\mathbb{R}}}\left(\frac{1-e^{rx}}{1-e^x}\right)^2 f(x)\,d\sigma(x)$$

for all $f \in C^c(\overline{\mathbb{R}})$, and by Proposition 2.4.4 this holds even for $f \in C^0(\overline{\mathbb{R}})$.

Choosing a function $\varphi \in C^c_+(\overline{\mathbb{R}})$ with $\varphi(x) = 1$ for $x \in [-1, 1]$ we find for $r \in \mathbb{Q}_+\setminus\{0\}$

$$\begin{aligned}
&\frac{1}{t_j}\int_{\overline{\mathbb{R}}}(1-e^{rx}-r(1-e^x))\,d\mu_{t_j}(x)\\
&\quad= \frac{1}{t_j}\int_{\overline{\mathbb{R}}}\frac{1-e^{rx}-r(1-e^x)}{(1-e^x)^2}\varphi(x)(1-e^x)^2\,d\mu_{t_j}(x)\\
&\qquad+\frac{1}{t_j}\int_{\overline{\mathbb{R}}} f(x)(1-e^{(r+1)x})^2\,d\mu_{t_j}(x),
\end{aligned}$$

where

$$f(x) := \frac{1-e^{rx}-r(1-e^x)}{(1-e^{(r+1)x})^2}(1-\varphi(x)) \in C^0(\overline{\mathbb{R}}).$$

Letting $j \to \infty$ we therefore get

$$\begin{aligned}
\psi(r) - r\psi(1) &= \int_{\overline{\mathbb{R}}}\frac{1-e^{rx}-r(1-e^x)}{(1-e^x)^2}\varphi(x)\,d\sigma(x)\\
&\quad+\int_{\overline{\mathbb{R}}}\left(\frac{1-e^{(r+1)x}}{1-e^x}\right)^2 f(x)\,d\sigma(x)\\
&= \int_{\overline{\mathbb{R}}}\frac{1-e^{rx}-r(1-e^x)}{(1-e^x)^2}\,d\sigma(x)\\
&= \tfrac{1}{2}(r-r^2)\sigma(\{0\}) + (1-r)\sigma(\{-\infty\})\\
&\quad+\int_{\mathbb{R}\setminus\{0\}}(1-e^{rx}-r(1-e^x))\,d\mu(x),
\end{aligned}$$

where $\mu = (1-e^x)^{-2}(\sigma|\mathbb{R}\setminus\{0\})$, hence

$$\psi(r) = br - cr^2 + d1_{\mathbb{Q}_+\setminus\{0\}}(r) + \int_{\mathbb{R}\setminus\{0\}}(1-e^{rx}-r(1-e^x))\,d\mu(x),$$

with $b = \psi(1) + \frac{1}{2}\sigma(\{0\}) - \sigma(\{-\infty\})$, $c = \frac{1}{2}\sigma(\{0\})$, $d = \sigma(\{-\infty\})$.

By Proposition 2.4.6

$$\int_{\mathbb{R}} \left(\frac{1 - e^{rx}}{1 - e^x}\right)^2 d\sigma(x) \leqq A_r \qquad \text{for} \quad r \in \mathbb{Q}_+,$$

and it follows that $\mu$ has the asserted properties.

To see unicity of the representation we first remark that $a = \psi(0)$ and $a + d = \psi(0+)$, which show that $a$, $d$ are uniquely determined. For $\psi \in \mathcal{N}(\mathbb{Q}_+)$ of the form

$$\psi(r) = br - cr^2 + \int_{\mathbb{R}\setminus\{0\}} (1 - e^{rx} - r(1 - e^x))\, d\mu(x)$$

we find

$$2\psi(r + 1) - \psi(r) - \psi(r + 2) = 2c + \int_{\mathbb{R}\setminus\{0\}} e^{rx}(1 - e^x)^2\, d\mu(x)$$

$$= \int_{\mathbb{R}} e^{rx}\, d\tau(x),$$

where $\tau = 2c\varepsilon_0 + (1 - e^x)^2\mu$. By the uniqueness assertion in Proposition 5.6 it follows that $\tau$ and hence $c$ and $\mu$ are uniquely determined. Finally, $b$ is also uniquely determined. □

In the above result $\psi$ is continuous if and only if $d = 0$, and if $d = 0$ the representation shows that $\psi$ can be extended to a continuous negative definite function on $\mathbb{R}_+$. The term $-r(1 - e^x)$ is a correction term analogous to a Lévy function, cf. 4.3.17, and it can be replaced by any term of the form $r\varphi(x)$, where $\varphi$ is a locally bounded Borel function such that $\varphi(x) \sim x$ for $x \to 0$ and which is integrable at $\pm\infty$ with respect to all the measures $\mu$ appearing in the representation. As examples we have $\varphi(x) = xe^x$ and $\varphi(x) = x(1 + x^2)^{-1}$.

This leads to the following result only formulated for continuous negative definite functions:

**5.14. Theorem.** *A function $\psi\colon \mathbb{R}_+ \to \mathbb{R}$ is continuous and negative definite if and only if it has a representation of the form*

$$\psi(y) = a + by - cy^2 + \int_{\mathbb{R}\setminus\{0\}} \left(1 - e^{xy} + \frac{xy}{1 + x^2}\right) d\mu(x),$$

*where $a, b \in \mathbb{R}$, $c \geqq 0$ and $\mu \in M_+(\mathbb{R}\setminus\{0\})$ satisfies*

$$\int_{0<|x|\leqq 1} x^2\, d\mu(x) < \infty, \qquad \int_{1<|x|} e^{xy}\, d\mu(x) < \infty \qquad \text{for all} \quad y \in \mathbb{R}_+$$

*The representation is unique.*

Any function $\psi$ as above has an extension to the closed half-plane $\mathbb{C}_+$ given by

$$\psi(z) = a + bz - cz^2 + \int_{\mathbb{R}\setminus\{0\}} \left(1 - e^{xz} + \frac{xz}{1 + x^2}\right) d\mu(x). \tag{8}$$

This extension is continuous in $\mathbb{C}_+$ and holomorphic in the open half-plane and uniquely determined by these two properties. In particular, we find

$$\psi(-iy) = a - iby + cy^2 + \int_{\mathbb{R}\setminus\{0\}} \left(1 - e^{-ixy} - \frac{ixy}{1 + x^2}\right) d\mu(x) \quad \text{for } y \in \mathbb{R}, \tag{9}$$

which is easily seen to be a continuous negative definite function on $\mathbb{R}$ in the group sense. In fact, formula (9) is the classical Lévy–Khinchin representation, cf. Loève (1963), with the exception that it is normally assumed that $a = 0$. Note also that (9) does not give all continuous negative definite functions because of the integrability conditions on $\mu$ given in Theorem 5.14. Formula (9) gives all continuous negative definite functions if the only restriction on $\mu \in M_+(\mathbb{R}\setminus\{0\})$ is

$$\int_{\mathbb{R}\setminus\{0\}} \frac{x^2}{1 + x^2}\, d\mu(x) < \infty.$$

With each continuous negative definite function $\psi: \mathbb{R}_+ \to \mathbb{R}$ can be associated a convolution semigroup $(\mu_t)_{t \geqq 0}$ on $\mathbb{R}$ such that

$$e^{-t\psi(y)} = \int e^{xy}\, d\mu_t(x) \qquad \text{for} \quad y \geqq 0, \quad t \geqq 0. \tag{10}$$

In fact, by Theorem 3.2.2 $\exp(-t\psi)$ is continuous and positive definite, so the results in 5.6 and 5.8 ensure the existence of a unique measure $\mu_t \in M^b_+(\underline{\mathbb{R}})$ such that (10) holds, and $\mu_t$ is concentrated on $\mathbb{R}$. By the uniqueness assertion in 5.8 it also follows that $\mu_0 = \varepsilon_0$ and $\mu_t * \mu_s = \mu_{t+s}$. The function $x \mapsto \exp(xy)$ is continuous on $\underline{\mathbb{R}}$ for $y \geqq 0$, and we claim that the subspace of $C(\underline{\mathbb{R}})$ spanned by these functions is adapted. This follows by the fact that

$$x \mapsto e^{xy}(1 + e^{x(y+1)})^{-1}$$

belongs to $C^0(\underline{\mathbb{R}})$ for $y \geqq 0$. Using the adaptedness we see by Remark 5.3 that $t \mapsto \mu_t$ is weakly continuous. The measure $\mu_t$ decreases exponentially on $\mathbb{R}_+$ in the sense that

$$\int_{\mathbb{R}} e^{xy}\, d\mu_t(x) < \infty \qquad \text{for all} \quad y \geqq 0,$$

and it follows that

$$z \mapsto \int e^{xz}\, d\mu_t(x)$$

is continuous on the closed half-plane $\{z \in \mathbb{C} \mid \operatorname{Re} z \geqq 0\}$ and holomorphic in its interior. By the uniqueness of holomorphic extensions we have by (8) that

$$e^{-t\psi(z)} = \int e^{xz}\, d\mu_t(x) \qquad \text{for} \quad \operatorname{Re} z \geqq 0, \quad t \geqq 0,$$

so $y \mapsto \psi(-iy)$ is the continuous negative definite function in the group sense associated with $(\mu_t)_{t \geqq 0}$, cf. Berg and Forst (1975).

**5.15. Example.** For $0 < \alpha \leqq 2$ we define $\psi_\alpha: \mathbb{R}_+ \to \mathbb{R}$ by

$$\psi_\alpha(y) = \begin{cases} y^\alpha, & 0 < \alpha \leqq 1, \\ -y^\alpha, & 1 < \alpha \leqq 2. \end{cases}$$

Then $\psi_\alpha$ is continuous and negative definite on $\mathbb{R}_+$. We first remark that $\psi_1$, $-\psi_1$ and $\psi_2$ are all negative definite because if $y_1, \ldots, y_n \in \mathbb{R}_+$ and $c_1, \ldots, c_n \in \mathbb{R}$ with $\sum_{j=1}^n c_j = 0$ we have

$$\sum_{j,k=1}^n c_j c_k (y_j + y_k) = 0,$$

$$\sum_{j,k=1}^n c_j c_k (y_j + y_k)^2 = 2\left(\sum_{j=1}^n c_j y_j\right)^2 \geqq 0.$$

That $\psi_\alpha$ is negative definite follows now by Corollary 3.2.10 for $0 < \alpha < 1$ and by Corollary 3.2.11 for $1 < \alpha < 2$.

Let $(\mu_t^{(\alpha)})_{t \geqq 0}$ be the corresponding convolution semigroup on $\mathbb{R}$, determined by the formula

$$e^{-t\psi_\alpha(y)} = \int_{-\infty}^{\infty} e^{xy}\, d\mu_t^{(\alpha)}(x), \qquad y \geqq 0, \quad t \geqq 0.$$

Since $\psi_\alpha(0) = 0$ each $\mu_t^{(\alpha)}$ is a probability measure, and $\psi_\alpha$ being nonnegative for $0 < \alpha \leqq 1$ we see by Corollary 4.4.5 that $\mu_t^{(\alpha)}$ is concentrated on $]-\infty, 0]$ for $0 < \alpha \leqq 1$ but not for $1 < \alpha \leqq 2$. For $0 < \alpha < 1$ the measure $\mu_t^{(\alpha)}$ (or its reflection in the origin) is a so-called one-sided stable measure of index $\alpha$, cf. Hall (1981). Clearly $\mu_t^{(1)} = \varepsilon_{-t}$.

For $\alpha \in \,]0, 2] \setminus \{1\}$, we have

$$e^{tz^\alpha \operatorname{sgn}(\alpha - 1)} = \int_{-\infty}^{\infty} e^{xz}\, d\mu_t^{(\alpha)}(x), \qquad \operatorname{Re} z \geqq 0,$$

where $z^\alpha$ is the holomorphic extension of $y^\alpha$ to the right half-plane, i.e.

$$z^\alpha = |z|^\alpha \exp(i\alpha \arg z)$$

with $\arg z \in [-\pi/2, \pi/2]$. In particular

$$(-iy)^\alpha = |y|^\alpha \exp\left(-i\alpha \frac{\pi}{2} \operatorname{sgn}(y)\right)$$

$$= \cos\left(\alpha \frac{\pi}{2}\right) |y|^\alpha \left(1 - i \operatorname{sgn}(y) \tan\left(\alpha \frac{\pi}{2}\right)\right),$$

hence

$$|\exp(t(-iy)^\alpha \operatorname{sgn}(\alpha - 1))| = \exp\left(t \operatorname{sgn}(\alpha - 1)\cos\left(\alpha\frac{\pi}{2}\right)|y|^\alpha\right), \qquad (11)$$

which is Lebesgue integrable since $\operatorname{sgn}(\alpha - 1)\cos(\alpha(\pi/2)) < 0$. By the Fourier inversion theorem $\mu_t^{(\alpha)}$ has a density $g_t^{(\alpha)} \in C^0(\mathbb{R})$ given by

$$g_t^{(\alpha)}(x) = \frac{1}{2\pi}\int_{-\infty}^{\infty} e^{ixy}e^{t(-iy)^\alpha \operatorname{sgn}(\alpha-1)}\,dy.$$

Since the function in (11) remains integrable after multiplication with any power of $y$ we see that $g_t^{(\alpha)} \in C^\infty(\mathbb{R})$. The measure $\mu_t^{(\alpha)}$ is a stable measure of index $\alpha$, cf. Hall (1981). For $\alpha = 2$ we have

$$g_t^{(2)}(x) = \frac{1}{2\pi}\int_{-\infty}^{\infty} e^{ixy}e^{-ty^2}\,dt = \frac{1}{\sqrt{4\pi t}}e^{-x^2/4t},$$

which is the density of a normal distribution and $(\mu_t^{(2)})_{t\geqq 0}$ is the so-called Gaussian semigroup. For an explicit formula for $g_t^{(1/2)}$ see, for example, Berg and Forst (1975, p. 71).

The restriction of $\psi_\alpha$ to $\mathbb{N}_0$ is negative definite on $(\mathbb{N}_0, +)$ so

$$s_\alpha(n) = e^{-\psi_\alpha(n)} = \begin{cases} e^{-n^\alpha}, & 0 < \alpha \leqq 1, \\ e^{n^\alpha}, & 1 < \alpha \leqq 2, \end{cases}$$

is a Hamburger moment sequence. For $\alpha \neq 1$ we have

$$s_\alpha(n) = \int_{-\infty}^{\infty} e^{xn}g_1^{(\alpha)}(x)\,dx = \int_0^{\infty} t^n g_1^{(\alpha)}(\log t)t^{-1}\,dt, \qquad n \geqq 0,$$

showing that $s_\alpha$ is a Stieltjes moment sequence. For $0 < \alpha < 1$ we have that $g_1^{(\alpha)}(\log t) = 0$ for $t \geqq 1$ so that $s_\alpha$ is completely monotone, and $n^\alpha$ is completely alternating. By comparison with Example 4.6 it can be seen that $s_2$ is indeterminate.

**5.16. Example.** The important function from information theory $\psi(y) = -y\log y$ is continuous and negative definite on $\mathbb{R}_+$. In fact, from Example 5.15 we get that $(-y^\alpha + y)(\alpha - 1)^{-1}$ is negative definite for $\alpha \in ]0, 2]\setminus\{1\}$, and letting $\alpha \to 1$ we get $\psi(y)$ in the limit.

Let $(\mu_t)_{t\geqq 0}$ be the corresponding convolution semigroup of probabilities on $\mathbb{R}$ determined by

$$e^{-t\psi(y)} = y^{ty} = \int_{-\infty}^{\infty} e^{xy}\,d\mu_t(x), \qquad y \geqq 0, \quad t \geqq 0.$$

In particular $y^y$ is continuous and positive definite on $\mathbb{R}_+$. The holomorphic extension of $\psi$ to the right half-plane is

$$\psi(z) = -z\log z = -z(\log|z| + i\arg z), \qquad \operatorname{Re} z \geqq 0,$$

where $\arg z \in [-\pi/2, \pi/2]$. In particular

$$\psi(-iy) = iy\left(\log|y| - i\frac{\pi}{2}\operatorname{sgn}(y)\right) = \frac{\pi}{2}|y| + iy\log|y|$$

so that

$$\exp\left\{-t\left(\frac{\pi}{2}|y| + iy\log|y|\right)\right\} = \int_{-\infty}^{\infty} e^{-ixy}\,d\mu_t(x).$$

Since $y^n \exp\{-t(\pi/2)|y|\}$ is Lebesgue integrable for all $n \in \mathbb{N}$, we see that $\mu_t$ has a $C^\infty$-density $g_t$ given by

$$g_t(x) = \frac{1}{2\pi}\int_{-\infty}^{\infty} e^{ixy}\exp\left\{-t\left(\frac{\pi}{2}|y| + iy\log|y|\right)\right\}dy.$$

The measure $\mu_t$ is a stable measure of index 1, cf. Hall (1981).

The sequence $(n^n)_{n \geq 0}$ is positive definite on $\mathbb{N}_0$, and we have

$$n^n = \int_{-\infty}^{\infty} e^{nx} g_1(x)\,dx = \int_0^{\infty} t^n g_1(\log t)t^{-1}\,dt,$$

showing that $n^n$ is a Stieltjes moment sequence. Using Carleman's criterion, cf. Akhiezer (1965), it follows that $n^n$ is determinate.

The result of 5.14 can be extended from $\mathbb{R}_+$ to $\mathbb{R}_+^k$ and reads as follows:

**5.17. Theorem.** *A function $\psi: \mathbb{R}_+^k \to \mathbb{R}$ is continuous and negative definite if and only if it has a representation of the form*

$$\psi(y) = a + \langle b, y\rangle - q(y) + \int_{\mathbb{R}^k\setminus\{0\}}\left(1 - e^{\langle x, y\rangle} + \frac{\langle x, y\rangle}{1 + \|x\|^2}\right)d\mu(x), \qquad y \in \mathbb{R}_+^k, \quad (12)$$

*where $a \in \mathbb{R}$, $b \in \mathbb{R}^k$, $q(y) = \sum_{n,m=1}^k a_{nm} y_n y_m$ is a nonnegative quadratic form on $\mathbb{R}^k$ and $\mu \in M_+(\mathbb{R}^k\setminus\{0\})$ is such that*

$$\int_{0 < \|x\| \leq 1} \|x\|^2\,d\mu(x) < \infty,$$

$$\int_{\|x\| > 1} e^{\langle x, y\rangle}\,d\mu(x) < \infty \qquad \textit{for all} \quad y \in \mathbb{R}_+^k.$$

*The quadruple $(a, b, q, \mu)$ is uniquely determined.*

The proof is a modification of the proof of Theorem 5.18 below and will be left as an exercise for the reader. It should be noted that the continuous negative definite functions on $\mathbb{R}_+^k$ are in one-to-one correspondence with the convolution semigroups $(\mu_t)_{t \geq 0}$ on $\mathbb{R}^k$ satisfying

$$\int_{\mathbb{R}^k} e^{\langle x, y\rangle}\,d\mu_t(x) < \infty \qquad \text{for} \quad t \geq 0, \quad y \in \mathbb{R}_+^k,$$

the correspondence being established by

$$e^{-t\psi(y)} = \int_{\mathbb{R}^k} e^{\langle x, y\rangle}\, d\mu_t(x) \qquad \text{for} \quad t \geqq 0, \quad y \in \mathbb{R}_+^k.$$

Formula (12) shows that $\psi$ has a continuous extension to $\mathbb{C}_+^k$ which is holomorphic in its interior, and $\exp(-t\psi(-iy))$ is the Fourier transform of $\mu_t$.

Just as functions $\varphi \in \mathscr{P}(\mathbb{Q}^k)$ automatically are continuous, so are functions which are negative definite in the semigroup sense on $\mathbb{Q}^k$. The following result can therefore be regarded as an integral representation of the cone $\mathscr{N}(\mathbb{Q}^k)$ or as an integral representation of the continuous functions $\psi: \mathbb{R}^k \to \mathbb{R}$ which are negative definite on $(\mathbb{R}^k, +)$ in the semigroup sense. The result is due to Bickel and van Zwet (1980).

**5.18. Theorem.** *A function $\psi: \mathbb{R}^k \to \mathbb{R}$ is continuous and negative definite in the semigroup sense if and only if it has a representation of the form*

$$\psi(y) = a + \langle b, y\rangle - q(y) + \int_{\mathbb{R}^k\setminus\{0\}} \left(1 - e^{\langle x, y\rangle} + \frac{\langle x, y\rangle}{1 + \|x\|^2}\right) d\mu(x), \qquad y \in \mathbb{R}^k,$$

*where $a \in \mathbb{R}$, $b \in \mathbb{R}^k$, $q(y) = \sum_{n,m=1}^k a_{nm} y_n y_m$ is a nonnegative quadratic form on $\mathbb{R}^k$ and $\mu \in M_+(\mathbb{R}^k\setminus\{0\})$ is such that*

$$\int_{0 < \|x\| \leqq 1} \|x\|^2\, d\mu(x) < \infty,$$

$$\int_{1 < \|x\|} e^{\langle x, y\rangle}\, d\mu(x) < \infty \qquad \textit{for} \quad y \in \mathbb{R}^k.$$

*The quadruple $(a, b, q, \mu)$ is uniquely determined by $\psi$.*

Proof. It is easy to see that any function of the above form is continuous and negative definite. Conversely, it suffices to prove that any continuous negative definite function $\psi: \mathbb{R}^k \to \mathbb{R}$ with $\psi(0) = 0$ has a representation as above with $a = 0$. By Theorem 5.11 there is a unique $\mu_t \in M_+^1(\mathbb{R}^k)$ such that

$$\exp(-t\psi(y)) = \int_{\mathbb{R}^k} e^{\langle x, y\rangle}\, d\mu_t(x) \qquad \text{for} \quad t > 0, y \in \mathbb{R}^k.$$

From this equation together with the unicity of $\mu_t$ we get $\mu_t * \mu_s = \mu_{t+s}$ for $t, s > 0$, and using the fact that the subspace of $C(\mathbb{R}^k)$ spanned by the functions $x \mapsto \exp(\langle x, y\rangle)$, $y \in \mathbb{R}^k$ is adapted, we see that $\lim_{t\to 0} \mu_t = \varepsilon_0$ weakly, cf. Remark 5.3, i.e. $(\mu_t)_{t \geqq 0}$ is a convolution semigroup of probability measures on $\mathbb{R}^k$ where we put $\mu_0 = \varepsilon_0$. Defining

$$F(t, z) = \int_{\mathbb{R}^k} e^{\langle x, z\rangle}\, d\mu_t(x) \qquad \text{for} \quad t \geqq 0, \quad z \in \mathbb{C}^k,$$

then $F: [0, \infty[ \times \mathbb{C}^k \to \mathbb{C}$ is continuous and $F(t, \cdot)$ is holomorphic on $\mathbb{C}^k$ for fixed $t \geqq 0$. Since $F(t + s, z) = F(t, z)F(s, z)$ and $F(0, z) = 1$ we see that $F(t, z) \neq 0$, so there exists a unique holomorphic function $\tilde{\psi}: \mathbb{C}^k \to \mathbb{C}$ such that $F(t, z) = \exp(-t\tilde{\psi}(z))$, and $\tilde{\psi}$ is clearly an extension of $\psi$. In particular

$$\exp(-t\tilde{\psi}(-iy)) = \int_{\mathbb{R}^k} e^{-i\langle x, y\rangle}\, d\mu_t(x),$$

which shows that $y \mapsto \tilde{\psi}(-iy)$ is continuous and negative definite on $\mathbb{R}^k$ in the group sense. By the classical Lévy–Khinchin formula in $\mathbb{R}^k$, cf. Courrège (1964), we have

$$\tilde{\psi}(-iy) = -i\langle b, y\rangle + q(y) + \int_{\mathbb{R}^k\setminus\{0\}} \left(1 - e^{-i\langle x, y\rangle} - \frac{i\langle x, y\rangle}{1 + \|x\|^2}\right) d\mu(x),$$

where $b \in \mathbb{R}^k$, $q(y) = \sum_{n,m=1}^k a_{nm} y_n y_m$ is a nonnegative quadratic form and $\mu \in M_+(\mathbb{R}^k\setminus\{0\})$ satisfies

$$\int_{0 < \|x\| \leqq 1} \|x\|^2\, d\mu(x) < \infty \qquad \text{and} \qquad \int_{1 < \|x\|} d\mu(x) < \infty.$$

The Lévy measure $\mu$ is the vague limit as $t \to 0$ of $(1/t)\mu_t|(\mathbb{R}^k\setminus\{0\})$. For each $y \in \mathbb{R}^k$ the family

$$(e^{t\psi(y)} e^{\langle x, y\rangle}\, d\mu_t(x))_{t \geqq 0}$$

is a convolution semigroup of probabilities on $\mathbb{R}^k$ and its Lévy measure is $\exp\langle x, y\rangle\, d\mu(x)$, and therefore

$$\int_{\|x\| > 1} e^{\langle x, y\rangle}\, d\mu(x) < \infty$$

for each $y \in \mathbb{R}^k$. It follows easily that

$$\Phi(z) = \langle b, z\rangle - q(z) + \int_{\mathbb{R}^k\setminus\{0\}} \left(1 - e^{\langle x, z\rangle} + \frac{\langle x, z\rangle}{1 + \|x\|^2}\right) d\mu(x)$$

is well defined and holomorphic for $z \in \mathbb{C}^k$, and since $\Phi(-iy) = \tilde{\psi}(-iy)$ for all $y \in \mathbb{R}^k$ we get $\Phi(z) = \tilde{\psi}(z)$ for $z \in \mathbb{C}^k$, in particular, $\Phi(y) = \psi(y)$ for $y \in \mathbb{R}^k$, which establishes the desired integral representation.

The uniqueness assertion follows by the uniqueness statement in the classical Lévy–Khinchin formula. □

**5.19. Exercise.** For $\mu \in M_+(\mathbb{R}^k)$ let $A_\mu$ be the set of points $y \in \mathbb{R}^k$ for which $x \mapsto \exp\langle x, y\rangle$ is $\mu$-integrable and define

$$f_\mu(y) = \int e^{\langle x, y\rangle}\, d\mu(x) \qquad \text{for} \quad y \in A_\mu.$$

Show that $A_\mu$ is a convex set and that $f_\mu$ is a lower semicontinuous logarithmically convex function. Show that $(u, v) \mapsto f_\mu(u + v)$ is a positive definite kernel on $(\frac{1}{2}A_\mu) \times (\frac{1}{2}A_\mu)$.

Show finally that $f_\mu$ is $C^\infty$ in the interior of $A_\mu$ and that

$$D^\alpha f_\mu(0) = \int x^\alpha \, d\mu(x) \qquad \text{for} \quad \alpha \in \mathbb{N}_0^k$$

if 0 is an interior point of $A_\mu$.

The function $f_\mu$ is called the *moment generating function* of $\mu$.

**5.20. Exercise.** Let $\psi\colon \mathbb{R}_+ \to \mathbb{R}$ be a continuous negative definite function and $(\mu_t)_{t\geqq 0}$ the convolution semigroup such that (10) holds. Show that $\lim_{t\to 0}(1/t)\mu_t|(\mathbb{R}\setminus\{0\}) = \mu$ vaguely, where $\mu$ is the measure in the representation of $\psi$ in Theorem 5.14.

**5.21. Exercise.** Let $S$ be a perfect semigroup and add to $S$ an absorbing element $\omega$, cf. 4.1.3. Then $S \cup \{\omega\}$ is again perfect.

**5.22. Exercise.** Let $S$ be the subsemigroup of $(\mathbb{R}_+, +)$ generated by $\mathbb{Q}_+$ and an irrational number $\alpha > 0$. Show that $S$ is not perfect.

**5.23. Exercise.** Let $\varphi\colon \mathbb{R} \to \mathbb{R}$ be a continuous function which is positive definite on the multiplicative semigroup $(\mathbb{R}, \cdot)$. Show that there are measures $\mu \in M_+(\mathbb{R}_+)$, $\nu \in M_+(]0, \infty[)$ such that

$$\varphi(s) = \int_{\mathbb{R}_+} |s|^y \, d\mu(y) + \int_{]0,\infty[} |s|^y \cdot \operatorname{sgn}(s) \, d\nu(y), \qquad s \in \mathbb{R}.$$

**5.24. Exercise.** Show that $(\mathbb{Q}_+, \cdot)$ and $(\mathbb{Q}, \cdot)$ are not perfect semigroups. *Hint*: Use that $(\mathbb{Q}_+\setminus\{0\}, \cdot)$ is isomorphic with $\mathbb{Z}^{(\infty)}$.

## Notes and Remarks

In his fundamental paper "Recherches sur les fractions continues", Stieltjes formulated and solved the moment problem which bears his name. Later Hamburger generalized Stieltjes' result to moment sequences of measures on the whole real line. Moments of a measure have been studied, before Stieltjes, by Tchebycheff and others, but concerning the history of the moment problem we refer to Shohat and Tamarkin (1943). Our choice of the name "moment function" on a semigroup $S$ is motivated by this classical theory, which corresponds to the semigroup $(\mathbb{N}_0, +)$. Likewise the symbol $\mathscr{H}(S)$ reflects the name of Hamburger. The results in 1.6–1.11 seem to be new but are, of course, known for some concrete semigroups.

For a detailed study of Hamburger's moment problem, in particular of the set $E_+(\mathbb{R}, s)$, we refer the reader to the classical monographs by Akhiezer (1965) and Shohat and Tamarkin (1943). Results about denseness of the set of polynomials in $\mathscr{L}^p(\mathbb{R}, \mu)$ can be found in Berg and Christensen (1981,

1983a). The $F$-moment problem in the case $F = \{x \in \mathbb{R} \mid p(x) \geqq 0\}$, where $p$ is a fixed polynomial, is studied in Berg and Maserick (1982). It contains a characterization of the polynomials $p$ for which the set of $\{p \geqq 0\}$-moment sequences is equal to $\{s \mid s, p(E)s \in \mathscr{P}(\mathbb{N}_0)\}$. Here

$$(p(E)s)_n = \sum_{k=0}^{N} a_k s_{n+k} \qquad \text{if} \quad p(x) = \sum_{k=0}^{N} a_k x^k.$$

The $F$-moment problem where $F = \mathbb{R} \setminus \bigcup_{i=1}^{n} ]a_i, b_i[$ and $a_1 < b_1 < a_2 < b_2 < \cdots < a_n < b_n$ is solved in Švecov (1939) for $n = 1$ and in Fil'štinskiĭ (1964) for arbitrary $n$.

Horn (1969b) showed that a nonzero positive definite sequence $\varphi$: $\mathbb{N}_0 \to \mathbb{R}$ is infinitely divisible if and only if $\psi := -\log \varphi$ has the representation stated in Theorem 2.6.

Already in 1888 Hilbert remarked that there exist nonnegative homogeneous polynomials in three variables which are not a sum of squares of homogeneous polynomials, cf. Hilbert (1888). His method is, however, rather involved and does not lead to a simple explicit example, cf. Gelfand and Vilenkin (1964, pp. 233–234) where Hilbert's construction is indicated. An example along these lines was found by Robinson (1969, 1973) and by Schmüdgen (1979). The polynomial $x_1^2 x_2^2 (x_1^2 + x_2^2 - x_3^2) + x_3^6$ obtained by making the polynomial of Lemma 3.1 homogeneous is probably as simple as possible since the degree must be at least 6, as shown by Hilbert. The paper by Motzkin (1967) contains a similar explicit example. A nonnegative homogeneous polynomial in two variables is a sum of two squares of homogeneous polynomials by Lemma 2.1. Theorem 3.4 is stated in Zarhina (1959) and also referred to in Gelfand and Vilenkin (1964, p. 235), but the proof seems incomplete. Zarhina states that it is unknown whether $\Sigma^{(k)} \cap A_d^{(k)}$ is closed in $A_d^{(k)}$ and claims without proof that its closure is different from $(A_d^{(k)})_+$. Independently and almost simultaneously Theorems 3.2 and 3.4 were published by Berg *et al.* (1979) and Schmüdgen (1979). An explicit example of a function $\varphi \in \mathscr{P}(\mathbb{N}_0^2) \setminus \mathscr{H}(\mathbb{N}_0^2)$ has been given by Friedrich in a paper to appear in 1984 (or later). Only few papers contain sufficient conditions for a function $\varphi \in \mathscr{P}(\mathbb{N}_0^k)$, $k \geqq 2$, to be a moment function, cf. Devinatz (1957), Èskin (1960) and Nussbaum (1966). All these papers are based on results about the delicate problem of finding commuting self-adjoint extensions of symmetric operators which commute on a common dense domain in a Hilbert space. A survey of the operator theory approach to the moment problem is given by Fuglede (1983), cf. Exercise 3.15. The notion of $F$-moment sequences and the $F$-moment problem can be extended to $k$ dimensions, and the first solution of the problem is due to Haviland (1935, 1936). His result is given in Exercise 3.16. For results in case of concrete sets $F \subseteq \mathbb{R}^k$, see Exercise 3.17 and McGregor (1980). If $F$ is a compact convex set with nonvoid interior the problem is solved by Maserick (1977) and rediscovered by Cassier (1983). Semigroups of moments are considered in Berg (1984a), Buchwalter (1984) and Buchwalter and Cassier (1983a, b). In

Petersen (1982) it is proved that $\varphi \in \mathscr{H}(\mathbb{N}_0^k)$ is determinate provided each of the one-dimensional moment sequences $\varphi(n_1, 0, \ldots, 0), \ldots, \varphi(0, \ldots, 0, n_k)$ is determinate. The results in 3.6–3.13 were summarized in Berg and Christensen (1983b).

The two-sided moment problem has been studied by Jones *et al.* (1980, 1984). The first paper deals with the Stieltjes case where the representing measure is concentrated on $]0, \infty[$, and the authors use continued fractions as in Stieltjes' original work. In the second paper, where no restriction is required on the support of the measure, the authors study orthogonal Laurent polynomials and closely related Gaussian-type quadrature formulas to obtain the solution, thus following the method of Hamburger. Our approach can be called the Hahn–Banach method, first used by M. Riesz in connection with his solution of the Hamburger moment problem. This method was generalized to higher dimensions by Haviland in the papers cited above and later formulated in terms of adapted spaces by Choquet (1962). Theorem 4.3 is essentially in Wall (1931) although only extensions of $s: \mathbb{N}_0 \to \mathbb{R}$ to $\{k \in \mathbb{Z} \mid k \geqq -2n\}$ are discussed.

That the log-normal distribution is indeterminate for the classical Stieltjes moment problem was noticed by Stieltjes (1894) and Heyde (1963).

It seems difficult, but would be very interesting, to determine the structure of perfect semigroups. Theorems 5.4 and 5.5 show two possibilities of building perfect semigroups out of simpler ones. As another possibility let $(S_n)$ be a sequence of semigroups. Then the direct sum

$$S = \bigoplus_{n=1}^{\infty} S_n = \{s = (s_1, s_2, \ldots) \mid s_n \in S_n,\ s_n = 0 \text{ eventually}\}$$

is perfect if and only if each $S_n$ is perfect. We sketch the proof and remark first that $S^*$ is isomorphic to $\prod_{k=1}^{\infty} S_k^*$. This implies in particular that $S^*$ need not be locally compact for a perfect semigroup $S$. For $\varphi \in \mathscr{P}(S)$ and $n \in \mathbb{N}$ there is a measure $\mu_n \in E_+(\prod_{k=1}^{n} S_k^*)$ such that

$$\varphi(s_1, \ldots, s_n, 0, 0, \ldots) = \int \rho_1(s_1) \cdots \rho_n(s_n)\, d\mu_n(\rho_1, \ldots, \rho_n)$$

for $s_k \in S_k$, $k = 1, \ldots, n$. It suffices to prove the existence of a measure $\mu \in E_+(\prod_{k=1}^{\infty} S_k^*)$ such that $\mu^{p_n} = \mu_n$ for all $n$, where

$$p_n\colon \prod_{k=1}^{\infty} S_k^* \to \prod_{k=1}^{n} S_k^*$$

is the projection, and this is a consequence of a general result about projective systems, cf. Bourbaki (1965–1969, Chap. IX, p. 53).

The semigroup of dyadic numbers is perfect, and this may be used to prove that every countable 2-divisible semigroup with the identical involution is perfect, see Berg (1984b).

Continuous functions $f\colon ]a, b[ \mapsto \mathbb{R}$ such that $(x, y) \mapsto f(x + y)$ is a positive definite kernel on $]\frac{1}{2}a, \frac{1}{2}b[ \times ]\frac{1}{2}a, \frac{1}{2}b[$ were introduced by Bernstein who called them *exponentially convex functions*. Both Bernstein and Widder proved Theorem 5.12 in the special case of $H$ being one-dimensional, cf. Akhiezer (1965, p. 210). Akhiezer indicates how the result can be extended to $k$-dimensional intervals. See also Reeds (1979). Theorem 5.12 is also true if $H$ is an open convex set, cf. Devinatz (1955). His proof uses the spectral theorem for a family of commuting self-adjoint operators. A complete proof of Theorem 5.17 is given in Berg (1984b).

CHAPTER 7

# Hoeffding's Inequality and Multivariate Majorization

## §1. The Discrete Case

Many years ago Hoeffding (1956, Theorem 3) proved the following result: if $X_1, \ldots, X_n$ are independent Bernoulli random variables and if

$$\psi: \{0, 1, \ldots, n\} \to \mathbb{R}$$

is (strictly) concave, then

$$\mathbb{E}_{\bar{p}}\left[\psi\left(\sum_{i=1}^{n} X_i\right)\right] \leqq \mathbb{E}_{p_1, \ldots, p_n}\left[\psi\left(\sum_{i=1}^{n} X_i\right)\right], \tag{1}$$

where on the right-hand side it is assumed that $\mathbb{P}(X_i = 1) = p_i$ while on the left-hand side $\mathbb{P}(X_i = 1) = \bar{p} = (1/n) \sum_{i=1}^{n} p_i$ for all $i = 1, \ldots, n$.

Recently, Bickel and van Zwet (1980, Theorem 1.1) found a considerable extension of this result: let $\psi: \mathbb{R}^m \to \mathbb{R}$ and $k \in \mathbb{N}$ be given and let $X_1, \ldots, X_n$ be independent $m$-dimensional random vectors with distributions $\mu_1, \ldots, \mu_n$ and the extra condition that there is a finite subset $A \subseteq \mathbb{R}^m$ (depending on $\mu_1, \ldots, \mu_n$) with cardinality $k$ such that $\mu_j(A) = 1$ for all $j = 1, \ldots, n$. Then the following three conditions are equivalent:

(i) the inequality

$$\mathbb{E}_{\bar{\mu}}\left[\psi\left(\sum_{i=1}^{n} X_i\right)\right] \leqq \mathbb{E}_{\mu_1, \ldots, \mu_n}\left[\psi\left(\sum_{i=1}^{n} X_i\right)\right] \tag{2}$$

holds for all $n$ and all such $\mu_1, \ldots, \mu_n$;

(ii) the inequality (2) holds for $n = 2$ and all $\mu_1, \mu_2$ of the type described above;

(iii) the $k \times k$ matrix $(\psi(x_i + x_j))_{i, j=1, \ldots, k}$ is negative definite for every choice of $x_1, \ldots, x_k \in \mathbb{R}^m$.

The interpretation of (2) is analogous to that of (1): on the right-hand side $X_i$ has the distribution $\mu_i$ while on the left-hand side the random vectors $X_1, \ldots, X_n$ are identically distributed with the average distribution $\bar{\mu} = (1/n)\sum_{i=1}^n \mu_i$. Note that (2) reduces to (1) in case of $\{0, 1\}$-valued random variables.

The third condition in Bickel and van Zwet's result makes likely a certain kinship between negative definite functions and functions fulfilling inequalities of type (2) on more general semigroups than just euclidean spaces. This, in fact, is true and will lead us to interesting characterizations of negative definite functions as well as of completely negative definite functions on arbitrary abelian semigroups. This last-mentioned class of functions has not yet been defined, but appears naturally in this connection, and the lower bounded members of this class turn out to be nothing more than the completely alternating functions.

In this chapter we only consider abelian semigroups with the identical involution.

First let $S$ be an arbitrary nonempty set. If $(\Omega, \mathscr{A}, \mathbb{P})$ is a probability space then $X: \Omega \to S$ is called an *elementary random variable* if $\{X = s\} \in \mathscr{A}$ for all $s \in S$ and $|X(\Omega)| < \infty$, i.e. $X$ assumes only finitely many values. The distribution of an elementary random variable is, of course, a finite convex combination of one-point measures, or, as we have called it earlier, a molecular probability measure. If $S$ is an abelian semigroup and $X_1, \ldots, X_n$ are elementary $S$-valued random variables, then the sum $\sum_{i=1}^n X_i$ is of this type, too.

We shall make repeated use of the following straightforward identity:

**1.1. Lemma.** *Let $(a_{jk})$ be a symmetric real $n \times n$ matrix. Then for $c_1, \ldots, c_n, d_1, \ldots, d_n \in \mathbb{R}$ we have*

$$\sum_{j,k=1}^{n} a_{jk}\left(\frac{c_j + d_j}{2} \cdot \frac{c_k + d_k}{2} - c_j d_k\right) = \sum_{j,k=1}^{n} a_{jk}\frac{c_j - d_j}{2} \cdot \frac{c_k - d_k}{2}. \tag{3}$$

Proof. For $j \neq k$ the two factors of $a_{jk} = a_{kj}$ on the left-hand side sum up to

$$2 \cdot \frac{c_j + d_j}{2} \cdot \frac{c_k + d_k}{2} - c_j d_k - c_k d_j = 2 \cdot \frac{c_j - d_j}{2} \cdot \frac{c_k - d_k}{2}$$

and the factor of $a_{jj}$ is given by

$$\left(\frac{c_j + d_j}{2}\right)^2 - c_j d_j = \left(\frac{c_j - d_j}{2}\right)^2,$$

thereby proving the equality (3). □

The next result is now nearly obvious:

**1.2. Proposition.** *Let $S$ be a nonempty set and let $\psi: S \times S \to \mathbb{R}$ be symmetric. Then for independent elementary $S$-valued random variables $X$ and $Y$ the inequality*

$$\mathbb{E}_{\bar{\mu}}[\psi(X, Y)] \leqq \mathbb{E}_{\mu_1,\mu_2}[\psi(X, Y)] \tag{4}$$

*holds for all $\mu_1$ and $\mu_2$ if and only if $\psi$ is negative definite.*

(In (4) we define $\bar{\mu} = \frac{1}{2}(\mu_1 + \mu_2)$ and the interpretation is analogous to (1) and (2).)

PROOF. Let the two probability distributions $\mu_1$ and $\mu_2$ be concentrated on $\{s_1, \ldots, s_n\} \subseteq S$ and denote $\alpha_i = \mu_1(\{s_i\})$, $\beta_i = \mu_2(\{s_i\})$. Then by Lemma 1.1

$$\begin{aligned}
&\mathbb{E}_{\bar{\mu}}[\psi(X, Y)] - \mathbb{E}_{\mu_1,\mu_2}[\psi(X, Y)] \\
&= \sum_{i,j=1}^{n} \psi(s_i, s_j)\left(\frac{\alpha_i + \beta_i}{2}\frac{\alpha_j + \beta_j}{2} - \alpha_i\beta_j\right) \\
&= \sum_{i,j=1}^{n} \psi(s_i, s_j)\frac{\alpha_i - \beta_i}{2}\cdot\frac{\alpha_j - \beta_j}{2},
\end{aligned} \tag{5}$$

and this is clearly nonpositive if $\psi$ is negative definite.

Suppose on the other hand that (4) holds for all $\mu_1, \mu_2$. Let $s_1, \ldots, s_n \in S$ and $c_1, \ldots, c_n \in \mathbb{R}$ with $\sum c_i = 0$ be given. If not all $c_i$ are zero then $0 < \sum c_i^+ = \sum c_i^- =: c$, and putting

$$\mu_{1,2} := \sum_{i=1}^{n} \frac{c_i^\pm}{c} \varepsilon_{s_i},$$

the equality (5) shows

$$\sum_{i,j=1}^{n} c_i c_j \psi(s_i, s_j) \leqq 0,$$

which had to be shown. □

**1.3. Definition.** Let $S$ be an abelian semigroup. Then $\psi: S \to \mathbb{R}$ is said to fulfil *Hoeffding's inequality of order $n \geqq 2$* if for every sequence $X_1, \ldots, X_n$ of $n$ independent elementary $S$-valued random variables the inequality

$$\mathbb{E}_{\bar{\mu}}[\psi(X_1 + \cdots + X_n)] \leqq \mathbb{E}_{\mu_1,\ldots,\mu_n}[\psi(X_1 + \cdots + X_n)]$$

holds, where on the right-hand side $X_i$ has the distribution $\mu_i$, $1 \leqq i \leqq n$, while on the left-hand side all the $X_i$ have the same average distribution $\bar{\mu} := (1/n)\sum_{i=1}^{n} \mu_i$. The above inequality can also be expressed as

$$\int \psi \, d\bar{\mu}^{*n} \leqq \int \psi \, d(\mu_1 * \cdots * \mu_n).$$

Specializing in 1.2 the set $S$ to a semigroup gives

**1.4. Corollary.** *If $S$ is an abelian semigroup and $\psi: S \to \mathbb{R}$ a given function on $S$, then $\psi$ fulfils Hoeffding's inequality of order 2 if and only if $\psi$ is negative definite.*

A negative definite function on a semigroup need not however fulfil Hoeffding's inequalities of a higher order. As an example take $S = \mathbb{N}_0$ with usual addition. The function $\psi := 1_{\{1,3,5,\dots\}}$ is negative definite on $\mathbb{N}_0$ because $\psi(k) = \frac{1}{2}[1 - (-1)^k]$. If $X_1$, $X_2$, $X_3$ are independent Bernoulli random variables with parameters $p_1 = 0$, $p_2 = p_3 = \frac{3}{4}$, then $\bar{p} = \frac{1}{2}$ and

$$\mathbb{E}_{\bar{p}}[\psi(X_1 + X_2 + X_3)] = 3\bar{p}(1 - \bar{p})^2 + \bar{p}^3 = \tfrac{1}{2},$$

but

$$\mathbb{E}_{p_1, p_2, p_3}[\psi(X_1 + X_2 + X_3)] = \tfrac{3}{8}.$$

The next result will show that Hoeffding's inequality of order 3 actually implies a sharpened form of negative definiteness.

**1.5. Proposition.** *Let $\psi: S \to \mathbb{R}$ fulfil Hoeffding's inequality of order $n \geqq 3$. Then for all $a \in S$ the translate $E_{(n-2)a}\psi(s) = \psi(s + (n-2)a)$ is a negative definite function. In particular, $\psi \in \mathcal{N}(S)$.*

PROOF. Let $\mu, \nu$ be two molecular probability measures on $S$ and let $0 < p < 1$. We define $n$ new molecular probability measures $\mu_i$, $i = 1, \dots, n$ by

$$\mu_1 := p\varepsilon_a + (1 - p)\mu,$$

$$\mu_2 := p\varepsilon_a + (1 - p)\nu,$$

$$\mu_3 := \dots = \mu_n := p\varepsilon_a + (1 - p)\frac{\mu + \nu}{2}.$$

Then $\bar{\mu} = \mu_n$ and by easy calculation

$$\bar{\mu}^{*n} - \mu_1 * \mu_2 * \dots * \mu_n = \bar{\mu}^{*(n-2)} * (\bar{\mu}^{*2} - \mu_1 * \mu_2)$$

$$= \left[p\varepsilon_a + (1 - p)\frac{\mu + \nu}{2}\right]^{*(n-2)} * \left[(1 - p)\frac{\mu - \nu}{2}\right]^{*2},$$

hence, dividing out $(1 - p)^2$ and letting then $p$ tend to 1, we get

$$0 \geqq \int \psi \, d\left[\varepsilon_{(n-2)a} * \left(\frac{\mu - \nu}{2}\right)^{*2}\right] = \int E_{(n-2)a}\psi \, d\left[\left(\frac{\mu - \nu}{2}\right)^{*2}\right],$$

and therefore

$$\int E_{(n-2)a}\psi \, d\left[\left(\frac{\mu + \nu}{2}\right)^{*2}\right] \leqq \int E_{(n-2)a}\psi \, d(\mu * \nu).$$

By Corollary 1.4 $E_{(n-2)a}\psi$ is negative definite. □

The new property showing up in the above result will be given a name:

**1.6. Definition.** Let $S$ be an abelian semigroup. A function $\psi: S \to \mathbb{R}$ will be called *completely positive* (resp. *negative*) *definite* if $E_a\psi$ is positive (resp. negative) definite for all $a \in S$.

For $S = \mathbb{N}_0$ the completely positive definite functions are precisely the Stieltjes moment sequences, cf. Theorem 6.2.5. For $S = \mathbb{N}_0^k$, $k \geqq 2$, the completely positive definite functions are discussed in 6.3.7.

Our main result reads as follows:

**1.7. Theorem.** *Let S be an abelian semigroup and let $\psi$ be a real-valued function defined on S. Then the following three conditions are equivalent:*

(i) *$\psi$ fulfils Hoeffding's inequality of all orders;*
(ii) *$\psi$ fulfils Hoeffding's inequality of order* 3;
(iii) *$\psi$ is completely negative definite.*

PROOF. By Proposition 1.5 for $n = 3$ we know that (ii) implies (iii). Next let us assume that $\psi: S \to \mathbb{R}$ is completely negative definite. To see (i) it suffices by Corollary 1.4 to prove Hoeffding's inequality of order $n$ for $n \geqq 3$. We fix $n$ probability measures of finite support $\mu_1, \ldots, \mu_n$ on $S$ and may assume that they all are concentrated on $\{s_1, \ldots, s_k\} \subseteq S$. The function

$$\tilde{\psi}(s) := \int E_a\psi(s)\, d(\mu_3 * \cdots * \mu_n)(a)$$

is a mixture of negative definite functions, hence also of this type. Consequently, by Corollary 1.4, we have

$$\begin{aligned}\int \psi\, d(\mu_1 * \mu_2 * \cdots * \mu_n) &= \int \tilde{\psi}\, d(\mu_1 * \mu_2)\\ &\geqq \int \tilde{\psi}\, d\left[\left(\frac{\mu_1 + \mu_2}{2}\right)^{*2}\right]\\ &= \int \psi\, d\left(\frac{\mu_1 + \mu_2}{2} * \frac{\mu_1 + \mu_2}{2} * \mu_3 * \cdots * \mu_n\right).\end{aligned}$$

The proof will be finished by showing that suitable successive pairwise averaging of a given vector $x \in \mathbb{R}^n$ converges to the vector of averages $(\bar{x}, \bar{x}, \ldots, \bar{x}) \in \mathbb{R}^n$ where $\bar{x} = (1/n) \sum_{i=1}^n x_i$. To see this we define the doubly stochastic $n \times n$ matrix $T$ by

$$T = \begin{pmatrix} \frac{1}{2} & \frac{1}{2} & 0 & 0 & \cdots & 0\\ \frac{1}{2} & \frac{1}{2} & 0 & 0 & \cdots & 0\\ 0 & 0 & 1 & 0 & \cdots & 0\\ 0 & 0 & 0 & 1 & \cdots & 0\\ \vdots & \vdots & \vdots & \vdots & \ddots & \vdots\\ 0 & 0 & 0 & 0 & \cdots & 1 \end{pmatrix} \begin{pmatrix} 0 & 1 & 0 & 0 & \cdots & 0\\ 0 & 0 & 1 & 0 & \cdots & 0\\ 0 & 0 & 0 & 1 & \cdots & 0\\ \vdots & \vdots & \vdots & \vdots & \ddots & \vdots\\ 0 & 0 & 0 & 0 & \cdots & 1\\ 1 & 0 & 0 & 0 & \cdots & 0 \end{pmatrix}$$

$$= \begin{pmatrix} 0 & \frac{1}{2} & \frac{1}{2} & 0 & \cdots & 0\\ 0 & \frac{1}{2} & \frac{1}{2} & 0 & \cdots & 0\\ 0 & 0 & 0 & 1 & \cdots & 0\\ \vdots & \vdots & \vdots & \vdots & \ddots & \vdots\\ 0 & 0 & 0 & 0 & \cdots & 1\\ 1 & 0 & 0 & 0 & \cdots & 0 \end{pmatrix}$$

and observe that $T$ is the transition matrix of a finite irreducible and aperiodic Markov chain. A well-known limit theorem for Markov chains of this kind (cf., for example, Billingsley (1979, Theorem 8.7)) says that

$$T^m \to \begin{pmatrix} \pi_1 & \cdots & \pi_n \\ \pi_1 & \cdots & \pi_n \\ \vdots & \cdots & \vdots \\ \pi_1 & \cdots & \pi_n \end{pmatrix} \quad \text{as} \quad m \to \infty,$$

where $\pi_i \geqq 0$ for all $i$ and $\sum_{i=1}^n \pi_i = 1$. The limit matrix, however, must be doubly stochastic, too, and this is only possible if $\pi_1 = \pi_2 = \cdots = \pi_n = 1/n$. Therefore $T^m(x) \to (\bar{x}, \bar{x}, \ldots, \bar{x})$ for all $x \in \mathbb{R}^n$ as asserted.

Identifying a probability measure $\nu$ concentrated on $\{s_1, \ldots, s_k\}$ with the row vector $(\nu(\{s_1\}), \ldots, \nu(\{s_k\}))$ we now let $T$ operate on $n$-tuples $\mu = (\mu_1, \ldots, \mu_n)'$ of such probability measures by defining $T\mu$ as the $n \times k$ matrix

$$T\mu = \begin{pmatrix} (T\mu)_1 \\ \vdots \\ (T\mu)_n \end{pmatrix} = T\begin{pmatrix} \mu_1 \\ \vdots \\ \mu_n \end{pmatrix}.$$

The sequence $\{T^m\mu \mid m = 1, 2, \ldots\}$ converges to $\begin{pmatrix} \bar{\mu} \\ \vdots \\ \bar{\mu} \end{pmatrix}$ as $m \to \infty$, and as we have seen above

$$\mathbb{E}_\mu\left[\psi\left(\sum_{i=1}^n X_i\right)\right] \geqq \mathbb{E}_{T\mu}\left[\psi\left(\sum_{i=1}^n X_i\right)\right] \geqq \mathbb{E}_{T^2\mu}\left[\psi\left(\sum_{i=1}^n X_i\right)\right] \geqq \cdots,$$

so that the obvious continuity of $\mathbb{E}_\mu[\psi(\sum_{i=1}^n X_i)]$ as a function of $\mu$ finally shows

$$\mathbb{E}_\mu\left[\psi\left(\sum_{i=1}^n X_i\right)\right] \geqq \mathbb{E}_{\bar{\mu}}\left[\psi\left(\sum_{i=1}^n X_i\right)\right],$$

thereby finishing the proof. □

**1.8. Corollary.** *On a 2-divisible abelian semigroup negative definiteness is for each $n \geqq 2$ equivalent with Hoeffding's inequality of order $n$.*

Proof. It is sufficient to observe that a negative definite function $\psi$ on $S$ is automatically completely negative definite. To see this, let $a \in S$ be given, i.e. $a = b + b$ for some $b \in S$ by assumption. Then

$$\sum_{j,k=1}^n c_j c_k E_a \psi(s_j + s_k) = \sum_{j,k=1}^n c_j c_k \psi((b + s_j) + (b + s_k)) \leqq 0$$

if $c_1 + \cdots + c_n = 0$, so that all translates of $\psi$ are negative definite, too. □

**1.9. Remark.** Looking once more to the results proved so far, it will become obvious that, in fact, something more has been shown. We could introduce a *degree* $r$ by calling a kernel $\psi: S \times S \to \mathbb{R}$ negative definite of degree $r$ if all the matrices $(\psi(s_j, s_k))_{j,k \leqq m}$ are negative definite for $m = 1, 2, \ldots, r$. If $S$ is a semigroup then $\psi: S \to \mathbb{R}$ is called *negative definite of degree* $r$ if the kernel $\psi(s + t)$ has this property. Similarly, a degree could be added to Hoeffding's inequality of some given order $n$ by requiring all the measures $\mu_1, \ldots, \mu_n$ to be concentrated on one set of cardinality $r$.

The proof of Proposition 1.2 then shows that negative definiteness of degree $r$ is equivalent with the corresponding Hoeffding type inequality (4) of degree $r$. Proposition 1.5 tells us that Hoeffding's inequality of order 3 and degree $r + 1$ implies complete negative definiteness of degree $r$. Theorem 1.7 shows that a function which is completely negative definite of degree $r$ still fulfils Hoeffding's inequality of degree $r$ of all orders.

Applying these remarks for $r = 2$ to the additive semigroup $S = \mathbb{N}_0$ we conclude that a function $\psi: \mathbb{N}_0 \to \mathbb{R}$, such that all the $2 \times 2$ matrices

$$\begin{pmatrix} \psi(a + 2s) & \psi(a + s + t) \\ \psi(a + s + t) & \psi(a + 2t) \end{pmatrix}$$

with $a, s, t \in \mathbb{N}_0$ are negative definite, fulfils all Hoeffding inequalities of degree 2. But a $2 \times 2$ matrix $\begin{pmatrix} \alpha & \beta \\ \gamma & \delta \end{pmatrix}$ is negative definite if and only if $\gamma = \beta$ and $\alpha + \delta \leqq 2\beta$, hence complete negative definiteness of degree 2 is equivalent to $\psi$ being concave. We therefore get back Hoeffding's original result mentioned at the beginning of this chapter, even slightly sharpened, the measures $\mu_1, \ldots, \mu_n$ being allowed to be concentrated on any two-point set in $\mathbb{N}_0$, not just $\{0, 1\}$.

The proof of Corollary 1.8 makes obvious that on a 2-divisible semigroup negative definiteness of degree $r$ is equivalent to complete negative definiteness of degree $r$. Together with the preceding remarks this shows that the result of Bickel and van Zwet mentioned in the introduction generalizes from $\mathbb{R}^m$ to any 2-divisible abelian semigroup.

In Chapter 4 we have already met an interesting subclass of the negative definite functions, namely, the completely alternating functions. Their relation to completely negative definite functions will be clarified in the next theorem.

**1.10. Theorem.** *Let $S$ denote an abelian semigroup. Then a bounded function $\varphi: S \to \mathbb{R}$ is completely positive definite if and only if $\varphi$ is completely monotone, and a lower bounded function $\psi: S \to \mathbb{R}$ is completely negative definite if and only if $\psi$ is completely alternating.*

PROOF. If $\varphi: S \to \mathbb{R}$ is completely monotone then it has by Theorem 4.6.4 the integral representation

$$\varphi(s) = \int_{\hat{S}_+} \rho(s)\, d\mu(\rho),$$

$\mu$ being a Radon measure on the nonnegative bounded semicharacters. Then for any fixed $a \in S$ and $s_1, \ldots, s_n \in S$, $c_1, \ldots, c_n \in \mathbb{R}$ we get

$$\begin{aligned}\sum_{j,k=1}^{n} c_j c_k E_a \varphi(s_j + s_k) &= \sum_{j,k=1}^{n} c_j c_k \varphi(a + s_j + s_k) \\ &= \int_{\hat{S}_+} \rho(a) \left( \sum_{j=1}^{n} c_j \rho(s_j) \right)^2 d\mu(\rho) \geqq 0,\end{aligned}$$

showing that $E_a \varphi$ is positive definite.

Suppose on the other hand that $\varphi$ is bounded and completely positive definite. By Theorem 4.2.8 there are Radon measures $\mu_a$ on $\hat{S}$, $a \in S$, such that

$$E_a \varphi(s) = \varphi(a + s) = \int_{\hat{S}} \rho(s)\, d\mu_a(\rho), \qquad s \in S, \quad a \in S.$$

We put $\mu = \mu_0$ and have to show that $\mu$ is concentrated on $\hat{S}_+$. The open set $\hat{S} \setminus \hat{S}_+ = \bigcup_{a \in S} \mathcal{O}_a$ is the union of the open subsets $\mathcal{O}_a := \{\rho \in \hat{S} \mid \rho(a) < 0\}$, and $\mu$ being a Radon measure it is enough to show that $\mu(\mathcal{O}_a) = 0$ for all $a \in S$. We have

$$\varphi(a + s) = \int_{\hat{S}} \rho(s)\rho(a)\, d\mu(\rho) \qquad \text{for all} \quad s \in S,$$

and the unicity of the integral representation for positive definite functions (cf. Theorem 4.2.5) implies

$$\mu_a = \rho(a)\mu.$$

Therefore

$$0 \leqq \mu_a(\mathcal{O}_a) = \int_{\mathcal{O}_a} \rho(a)\, d\mu(\rho) \leqq 0$$

and we see $\mu(\mathcal{O}_a) = 0$.

The second statement concerning negative definite functions now follows easily: If $\psi$ is lower bounded then $\psi$ is completely negative definite if and only if $\exp(-t\psi)$ is completely positive definite and bounded for all $t > 0$, and by the first part of the proof this is true if and only if $\exp(-t\psi)$ is completely monotone which is equivalent with $\psi$ being completely alternating; see 3.2.2 and 4.6.10. □

A typical example of a completely negative definite function which is not bounded below is given on the additive semigroup $(\mathbb{R}, +)$ by $\psi(x) := -x^2$. If $c_1, \ldots, c_n \in \mathbb{R}$ and $\sum c_j = 0$ then

$$\sum_{j,k} c_j c_k [-(x_j + x_k)^2] = -2\left(\sum c_j x_j\right)^2 \leqq 0,$$

and $\mathbb{R}$ being 2-divisible $\psi$ is, in fact, completely negative definite. But then, of course, the restriction $\psi|\mathbb{Z}$ to the non-2-divisible semigroup $(\mathbb{Z}, +)$ is also completely negative definite. This however is no coincidence, as we shall now see.

We recall (cf. Definition 4.3.8) that a *quadratic form* $q$ on an abelian group $G$, considered as a semigroup with the involution $x^* = -x$, is a real-valued function $q$ on $G$ such that

$$q(x + y) + q(x - y) = 2q(x) + 2q(y) \tag{6}$$

holds for all $x, y \in G$. By Theorem 4.3.9 a nonnegative quadratic form is negative definite in the group sense on $G$, i.e.

$$\sum_{j,k} c_j c_k q(x_j - x_k) \leqq 0 \qquad \text{if } \sum c_j = 0,$$

so that in view of (6) the function $-q$ is negative definite in the semigroup sense on $G$. But we even have the following result:

**1.11. Proposition.** *If $q$ is a nonnegative quadratic form on the abelian group $G$, then $-q$ is completely negative definite in the semigroup sense, and hence fulfils all Hoeffding inequalities.*

Proof. In the proof of the already-mentioned Theorem 4.3.9 it was shown that

$$B(x, y) := q(x) + q(y) - q(x - y) = q(x + y) - q(x) - q(y)$$

is biadditive, i.e. $B(x_1 + x_2, y) = B(x_1, y) + B(x_2, y)$ and $B(x, y_1 + y_2) = B(x, y_1) + B(x, y_2)$ hold for all $x, x_1, \ldots \in G$.

We fix some $a \in G$ and consider the translate $E_a q(x) = q(a + x)$. If we put $h(x) := q(a + x) - q(x) - q(a)$ then

$$E_a q = q + h + q(a),$$

so $-E_a q$ is the sum of the negative definite function $-q$ and a term which is obviously negative definite, being a constant plus an additive function. □

**1.12. Exercise.** Let $S$ denote an abelian semigroup. Denote by $C_n$ the closed convex cone generated by molecular signed measures of the form $\bar{\mu}^{*n} - \mu_1 * \mu_2 * \cdots * \mu_n$, where $\mu_1, \ldots, \mu_n$ are molecular probability measures on $S$ and where $\bar{\mu} = (1/n) \sum_{i=1}^n \mu_i$. The closure is to be taken in $\mathrm{Mol}_{\mathbb{R}}(S)$ with respect to the finest locally convex topology on $\mathrm{Mol}_{\mathbb{R}}(S)$ which is just the topology induced by all real-valued functions on $S$ via the canonical duality. Let $\sigma \in \mathrm{Mol}_{\mathbb{R}}(S)$. Show that:

(i) if $\int \psi \, d\sigma \leqq 0$ for all negative definite functions $\psi: S \to \mathbb{R}$ then $\sigma \in C_2$;
(ii) if $\int \psi \, d\sigma \leqq 0$ for all completely negative definite $\psi: S \to \mathbb{R}$ then $\sigma \in C_3$.

**1.13. Exercise.** Show directly, by using the integral representation, that each completely alternating function fulfils all Hoeffding inequalities.

**1.14. Exercise.** If $h: S \to \mathbb{R}$ is additive then $-\exp(h)$ is completely negative definite.

**1.15. Exercise.** The function $\psi: S \to \mathbb{R}$ is completely negative definite if and only if for all $t > 0$ the function $-\exp(-t\psi)$ has this property.

**1.16. Exercise.** Let $(S, +)$ denote an abelian semigroup and let $\varphi: S \to \mathbb{R}$ be bounded. Then the *reverse Hoeffding inequality*

$$E_{\bar{\mu}}[\varphi(X_1 + \cdots + X_n)] \geqq E_{\mu_1, \ldots, \mu_n}[\varphi(X_1 + \cdots + X_n)]$$

holds for all $n$ if and only if it holds for $n = 3$, and if and only if $\varphi$ has the form $\varphi = c + \varphi'$ for some $c \in \mathbb{R}$ and a completely monotone function $\varphi'$.

**1.17. Exercise.** Let $p_1, p_2, \cdots \in [0, 1]$ be given such that $p = \sum p_i < \infty$. Then the infinite convolution $*_{i=1}^{\infty} B(1, p_i)$ of the Bernoulli distributions $B(1, p_i) = (1 - p_i)\varepsilon_0 + p_i \varepsilon_1$ exists, and for any concave function $\psi: \mathbb{N}_0 \to \mathbb{R}_+$ we have

$$\int \psi \, d\pi_p \leqq \int \psi d\left(\mathop{*}_{i=1}^{\infty} B(1, p_i)\right),$$

where $\pi_p$ is the Poisson distribution with parameter $p$. *Hint*: Use Exercise 5.3.16.

## §2. Extension to Nondiscrete Semigroups

We know from the preceding section that negative definite and completely negative definite functions can be characterized by Hoeffding's inequality involving only probability measures of finite support. For applications, however, one would certainly also like to have Hoeffding's inequality in a nondiscrete situation. We begin with a counterpart to Proposition 1.2.

**2.1. Theorem.** *Let $S$ be a Hausdorff space and let $\psi: S \times S \to \mathbb{R}$ be a continuous negative definite kernel. Then for all Radon probability measures $\mu_1, \mu_2$ on $S$ we have*

$$\int \psi \, d(\bar{\mu} \otimes \bar{\mu}) \leqq \int \psi \, d(\mu_1 \otimes \mu_2), \qquad \bar{\mu} := \tfrac{1}{2}(\mu_1 + \mu_2),$$

*as soon as $\psi$ is integrable with respect to $\bar{\mu} \otimes \bar{\mu}$ or $\psi \geqq 0$.*

PROOF. In a first step we assume that both $\mu_1$ and $\mu_2$ have compact support, i.e. $\mu_1(K) = \mu_2(K) = 1$ for some compact subset $K \subseteq S$. By Proposition 2.3.5 there are nets $(\mu_{1,\alpha})$ and $(\mu_{2,\alpha})$ of molecular probability measures on $K$ converging weakly to $\mu_1$ (resp. $\mu_2$) (it is no restriction to assume that both nets have the same index set). Then $\bar{\mu}_\alpha := \frac{1}{2}(\mu_{1,\alpha} + \mu_{2,\alpha}) \to \bar{\mu}$ and by Theorem 2.3.3 we also have

$$\mu_{1,\alpha} \otimes \mu_{2,\alpha} \to \mu_1 \otimes \mu_2, \qquad \bar{\mu}_\alpha \otimes \bar{\mu}_\alpha \to \bar{\mu} \otimes \bar{\mu}$$

in the weak topology. On $K \times K$ the kernel $\psi$ is a bounded continuous function, so that the portmanteau theorem (2.3.1) and Proposition 1.2 lead to the conclusion

$$\int \psi \, d(\bar{\mu} \otimes \bar{\mu}) = \lim_\alpha \int \psi \, d(\bar{\mu}_\alpha \otimes \bar{\mu}_\alpha)$$

$$\leqq \lim_\alpha \int \psi \, d(\mu_{1,\alpha} \otimes \mu_{2,\alpha}) = \int \psi \, d(\mu_1 \otimes \mu_2). \tag{1}$$

In the second step we make use of the conditional probabilities

$$\mu_{1,K}(B) := \frac{\mu_1(B \cap K)}{\mu_1(K)}, \qquad \mu_{2,K}(B) := \frac{\mu_2(B \cap K)}{\mu_2(K)}, \qquad B \in \mathscr{B}(S),$$

which certainly are defined for large enough $K \in \mathscr{K}(S)$ and which immediately are seen to be again Radon measures.

Putting $\overline{\mu_K} := \frac{1}{2}(\mu_{1,K} + \mu_{2,K})$ we have by (1) the inequality

$$\int \psi \, d(\overline{\mu_K} \otimes \overline{\mu_K}) \leqq \int \psi(\mu_{1,K} \otimes \mu_{2,K}) \tag{2}$$

for the above compact subsets $K \subseteq S$, and the integrability of $\psi$ with respect to $\bar{\mu} \otimes \bar{\mu}$ gives, by Exercise 2.3.9,

$$\int \psi \, d(\overline{\mu_K} \otimes \overline{\mu_K}) = \tfrac{1}{4}\Big[\int \psi \, d(\mu_{1,K} \otimes \mu_{1,K}) + 2\int \psi \, d(\mu_{1,K} \otimes \mu_{2,K})$$

$$+ \int \psi \, d(\mu_{2,K} \otimes \mu_{2,K})\Big]$$

$$= \tfrac{1}{4}\Big[\frac{1}{(\mu_1(K))^2}\int_{K \times K} \psi \, d(\mu_1 \otimes \mu_1)$$

$$+ \frac{2}{\mu_1(K)\mu_2(K)}\int_{K \times K} \psi \, d(\mu_1 \otimes \mu_2)$$

$$+ \frac{1}{(\mu_2(K))^2}\int_{K \times K} \psi \, d(\mu_2 \otimes \mu_2)\Big]$$

$$\underset{K}{\to} \tfrac{1}{4}\Big[\int \psi \, d(\mu_1 \otimes \mu_1) + 2\int \psi \, d(\mu_1 \otimes \mu_2) + \int \psi \, d(\mu_2 \otimes \mu_2)\Big]$$

$$= \int \psi \, d(\bar{\mu} \otimes \bar{\mu})$$

as well as

$$\int \psi \, d(\mu_{1,K} \otimes \mu_{2,K}) = \frac{1}{\mu_1(K)\mu_2(K)} \int_{K \times K} \psi \, d(\mu_1 \otimes \mu_2) \to \int \psi \, d(\mu_1 \otimes \mu_2).$$

The same conclusion holds by Proposition 2.1.7 if $\psi \geqq 0$. Therefore inequality (2) extends to the limit and we get

$$\int \psi \, d(\bar{\mu} \otimes \bar{\mu}) \leqq \int \psi \, d(\mu_1 \otimes \mu_2)$$

as asserted. □

Let now $(S, +)$ be a Hausdorff topological abelian semigroup with a neutral element. For two Radon probability measures $\mu, \nu$ on $S$ the convolution $\mu * \nu$ is then a well-defined new Radon measure and furthermore $(M_+^1(S), *)$ is again a Hausdorff abelian topological semigroup, see Corollary 2.3.4. Theorem 2.1 has the following obvious consequence:

**2.2. Corollary.** *Let $\psi$ be a continuous negative definite function on the abelian Hausdorff topological semigroup $S$. Then for all $\mu_1, \mu_2 \in M_+^1(S)$ we have*

$$\int \psi \, d(\bar{\mu} * \bar{\mu}) \leqq \int \psi \, d(\mu_1 * \mu_2), \qquad \bar{\mu} := \tfrac{1}{2}(\mu_1 + \mu_2)$$

*as soon as $\psi$ is integrable with respect to $\bar{\mu} * \bar{\mu}$ or $\psi \geqq 0$.*

**2.3. Example.** Consider the semigroup $S = (\mathbb{N}_0, +)$. We have already mentioned that the indicator function $\psi = 1_{\{1,3,5,\ldots\}}$ is negative definite on $\mathbb{N}_0$. If $X, Y$ are independent $\mathbb{N}_0$-valued random variables with varying distributions $\mu$ (resp. $\nu$), but with a fixed given average distribution $\kappa = \frac{1}{2}(\mu + \nu)$, then

$$\mathbb{P}_{\mu,\nu}(X + Y \text{ is odd}) = \mathbb{E}_{\mu,\nu}[\psi(X + Y)]$$

is minimized for $\mu = \nu = \kappa$.

**2.4. Example.** For $\alpha \in ]0, 2]$ the function $x \mapsto |x|^\alpha$ is negative definite in the group sense on $\mathbb{R}$ (this follows, for example, from 3.1.10 and 3.2.10). If $X, Y$ are independent real-valued random variables with distributions $\mu$ (resp. $\nu$), then under the restriction of a given average distribution $\kappa = \frac{1}{2}(\mu + \nu)$ the expectation

$$\mathbb{E}_{\mu,\nu}[|X - Y|^\alpha]$$

is minimized for $\mu = \nu = \kappa$.

Theorem 1.7, the main result of §1, has the following nondiscrete extension:

**2.5. Theorem.** *Let $\psi$ be a completely negative definite continuous function defined on the abelian Hausdorff topological semigroup $S$. Then for all Radon* *probability measures $\mu_1, \ldots, \mu_n$ on $S$ we have*

$$\int \psi \, d(\bar{\mu}^{*n}) \leqq \int \psi \, d(\mu_1 * \cdots * \mu_n), \qquad \bar{\mu} := \frac{1}{n} \sum_{i=1}^{n} \mu_i \tag{3}$$

*as soon as $\psi$ is integrable with respect to $\bar{\mu}^{*n}$, and without this restriction if $\psi$ is nonnegative.*

PROOF. Of course, $\psi$ is integrable with respect to $\mu_1 * \cdots * \mu_n$, too, in case $\int |\psi| \, d(\bar{\mu}^{*n}) < \infty$. As in the proof of Theorem 2.1 we first assume that $\mu_1(K) = \mu_2(K) = \cdots = \mu_n(K) = 1$ for some compact set $K \subseteq S$. Then both $\operatorname{supp}(\mu_1 * \cdots * \mu_n)$ and $\operatorname{supp}(\bar{\mu}^{*n})$ are contained in the compact set $K_n := K + \cdots + K$ ($n$ summands). Approximating $\mu_1, \ldots, \mu_n$ by molecular measures, the validity of (3) for these measures (Theorem 1.7) extends to the limit.

Now we use again the conditional probability measures

$$\mu_{i,K}(B) := \frac{\mu_i(B \cap K)}{\mu_i(K)}, \qquad B \in \mathscr{B}(S), \quad 1 \leqq i \leqq n, \quad K \in \mathscr{K}(S)$$

being well defined for large enough $K$, and have to go to the limit in the inequality

$$\int \psi \, d((\overline{\mu_K})^{*n}) \leqq \int \psi \, d(\mu_{1,K} * \cdots * \mu_{n,K}), \tag{4}$$

where $\overline{\mu_K} := (1/n) \sum_{i=1}^{n} \mu_{i,K}$.

Evaluating the left-hand side we get

$$\int \psi \, d((\overline{\mu_K})^{*n}) = \frac{1}{n^n} \sum_{i_1, \ldots, i_n = 1}^{n} \frac{1}{\prod_{j=1}^{n} \mu_{i_j}(K)} \int_{K_n} \psi \, d(\mu_{i_1} * \cdots * \mu_{i_n})$$

and this converges, if either $\int |\psi| \, d(\bar{\mu}^{*n}) < \infty$ or $\psi \geqq 0$, to $\int \psi \, d(\bar{\mu}^{*n})$. For the right-hand side of (4) we obtain

$$\int \psi \, d(\mu_{1,K} * \cdots * \mu_{n,K}) = \frac{1}{\prod_{i=1}^{n} \mu_i(K)} \int_{K_n} \psi \, d(\mu_1 * \cdots * \mu_n)$$

$$\xrightarrow[K]{} \int \psi \, d(\mu_1 * \cdots * \mu_n),$$

thereby finishing our proof. □

**2.6. Example.** On the additive semigroup $(\mathbb{R}, +)$ the function $\psi(x) = -x^2$ is completely negative definite, as we have seen above. Theorem 2.5 now implies that for independent real-valued random variables $X_1, \ldots, X_n$ with distributions $\mu_1, \ldots, \mu_n$ (resp. the same average distribution $\bar{\mu} = (1/n) \sum_{i=1}^{n} \mu_i$)

$$\mathbb{E}_{\bar{\mu}}(S_n^2) \geqq \mathbb{E}_{\mu_1, \ldots, \mu_n}(S_n^2), \qquad S_n = \sum_{i=1}^{n} X_i.$$

If the first moments of $\mu_1, \ldots, \mu_n$ exist, then subtracting $[\mathbb{E}_{\bar{\mu}}(S_n)]^2 = [\mathbb{E}_{\mu_1, \ldots, \mu_n}(S_n)]^2$ in the above inequality we get

$$\operatorname{var}_{\bar{\mu}}(S_n) \geqq \operatorname{var}_{\mu_1, \ldots, \mu_n}(S_n),$$

i.e. the variance of the sum gets maximal, if all the $X_i$ have the same distribution. On the other hand from

$$\sum (X_i - \bar{X})^2 = \sum X_i^2 - \frac{S_n^2}{n}, \qquad \bar{X} = \frac{S_n}{n},$$

and $\mathbb{E}_{\bar{\mu}}(\sum X_i^2) = \mathbb{E}_{\mu_1, \ldots, \mu_n}(\sum X_i^2)$ we see that (in case the second moments are finite)

$$\mathbb{E}_{\bar{\mu}}\left[\frac{1}{n-1} \sum (X_i - \bar{X})^2\right] \leqq \mathbb{E}_{\mu_1, \ldots, \mu_n}\left[\frac{1}{n-1} \sum (X_i - \bar{X})^2\right],$$

i.e. the mean empirical variance gets minimal in the i.i.d.-case.

**2.7. Exercise.** Let $X$, $Y$ be independent $\mathbb{Z}$-valued random variables with distribution $\mu$ (resp. $\nu$) and a fixed average distribution $\kappa = \frac{1}{2}(\mu + \nu)$. Show that

$$\mathbb{P}_{\mu, \nu}(X - Y \text{ is odd})$$

gets minimal for $\mu = \nu$.

**2.8. Exercise.** Let $X_1, \ldots, X_n$ be independent real-valued random variables with distributions $\mu_1, \ldots, \mu_n$. Show that for any (continuous) increasing function $\psi\colon \mathbb{R} \to \mathbb{R}$ the inequality

$$\mathbb{E}_{\bar{\mu}}\left[\psi\left(\max_{1 \leqq j \leqq n} X_j\right)\right] \leqq \mathbb{E}_{\mu_1, \ldots, \mu_n}\left[\psi\left(\max_{1 \leqq j \leqq n} X_j\right)\right]$$

holds as soon as the left-hand side is defined. *Hint*: Assume first that $\psi$ is bounded below and adjoin the neutral element $-\infty$ to the semigroup $(\mathbb{R}, \max)$.

**2.9. Exercise.** Let $\psi$ be a Borel measurable completely alternating function defined on a Hausdorff topological abelian semigroup $S$, and assume that $(\rho, s) \mapsto \rho(s)$ is Borel measurable on $\hat{S} \times S$. Then

$$\int \psi \, d\bar{\mu}^{*n} \leqq \int \psi \, d(\mu_1 * \cdots * \mu_n)$$

holds for all Radon probabilities $\mu_1, \ldots, \mu_n$ on $S$. *Hint*: Use Fubini's theorem.

## §3. Completely Negative Definite Functions and Schur-Monotonicity

An important and widely used ordering on $\mathbb{R}^n$ is given by the so-called *majorization*. A vector $x \in \mathbb{R}^n$ is majorized by $y \in \mathbb{R}^n$ if—intuitively—its components are "less spread out" than those of $y$; this means precisely

$$\sum_{i=1}^{k} x_{[i]} \leqq \sum_{i=1}^{k} y_{[i]}, \qquad \text{for} \quad k = 1, \ldots, n-1$$

and

$$\sum_{i=1}^{n} x_{[i]} = \sum_{i=1}^{n} y_{[i]},$$

where

$$x_{[1]} \geqq x_{[2]} \geqq \cdots \geqq x_{[n]}$$

are the components of $x$ in decreasing order; we write

$$x \prec y$$

in this case. An elementary example is given by

$$\left(\frac{1}{n}, \ldots, \frac{1}{n}\right) \prec \left(\frac{1}{n-1}, \ldots, \frac{1}{n-1}, 0\right) \prec \cdots \prec (\tfrac{1}{2}, \tfrac{1}{2}, 0, \ldots, 0)$$
$$\prec (1, 0, \ldots, 0).$$

Many interesting characterizations of majorization have been found. Of fundamental importance are the following four equivalent statements (where $\Omega_n$ denotes the set of all $n \times n$ doubly stochastic matrices, cf. Exercise 2.5.15, and where $y_\pi$ for $y \in \mathbb{R}^n$ and a permutation $\pi$ of $\{1, \ldots, n\}$ denotes the vector $(y_{\pi(1)}, \ldots, y_{\pi(n)})$):

(i) $x \prec y$;
(ii) $x' = Py'$ for some $P \in \Omega_n$;
(iii) $x \in \operatorname{conv}\{y_\pi \mid \pi \text{ a permutation of } \{1, \ldots, n\}\}$;
(iv) $\sum_{i=1}^{n} f(x_i) \leqq \sum_{i=1}^{n} f(y_i)$ for each convex function $f: \mathbb{R} \to \mathbb{R}$.

For a proof of these results we refer to the monograph of Marshall and Olkin (1979), which furthermore contains a wealth of material concerning majorization and a bibliography on this subject of approximately 450 items.

The real-valued functions preserving the order of majorization are called Schur-convex, i.e. $f: A \to \mathbb{R}$ (where $A \subseteq \mathbb{R}^n$) is *Schur-convex* if $x, y \in A$ and $x \prec y$ implies $f(x) \leqq f(y)$. If $f$ instead reverses the order it is called *Schur-concave*. For example, $f(x) = \sum_{i=1}^{n} (x_i - \bar{x})^2$ is Schur-convex on $\mathbb{R}^n$ ($\bar{x} := \sum x_i/n$), and the well-known entropy function $H(p_1, \ldots, p_n) = -\sum p_i \log p_i$ is Schur-concave on $[0, 1]^n$. Another example is given in the following

**3.1. Lemma.** *On* $\mathbb{R}_+^n$ *the function* $f(x) = \prod_{i=1}^n x_i$ *is Schur-concave.*

PROOF. Let $x, y \in \mathbb{R}_+^n$ be given with $x \prec y$. Since the log-function is concave on $\mathbb{R}_+$ (zero included in the obvious sense), the equivalence of (i) and (iv) above implies

$$\prod_{i=1}^n x_i = e^{\sum \log x_i} \geqq e^{\sum \log y_i} = \prod_{i=1}^n y_i. \qquad \square$$

Among the four equivalent statements (i)–(iv) property (ii) immediately leads to a multivariate generalization of the majorization order:

**3.2. Definition.** Let $A$ be a convex subset of some real linear space $E$. For two vectors $x = (x_1, \ldots, x_n)$ and $y = (y_1, \ldots, y_n)$ whose components $x_i$ and $y_i$ belong to $A$, we say that $x$ is *majorized* by $y$ and write $x \prec y$ if $x' = Py'$ for some $P = (p_{ij}) \in \Omega_n$ in the sense that

$$x_i = \sum_{j=1}^n p_{ij} y_j$$

holds for $i = 1, \ldots, n$. A real-valued function $f$ on $A^n$ is *Schur-convex* if $x, y \in A^n$ and $x \prec y$ implies $f(x) \leqq f(y)$; $f$ is called *Schur-concave* if $-f$ is Schur-convex.

In the following $A$ will be a convex subset of the family of all probability measures on a given measurable space, for example, the set of all molecular probabilities on an abstract set $S$.

If $\mu_1, \ldots, \mu_n$ are probability measures and $\bar{\mu} = (1/n) \sum_{i=1}^n \mu_i$, then

$$\begin{pmatrix} \bar{\mu} \\ \vdots \\ \bar{\mu} \end{pmatrix} = J_n \cdot \begin{pmatrix} \mu_1 \\ \vdots \\ \mu_n \end{pmatrix},$$

where $J_n$ is the doubly stochastic matrix with equal entries $1/n$, hence $(\bar{\mu}, \ldots, \bar{\mu})$ is majorized by $(\mu_1, \ldots, \mu_n)$. It seems natural to examine if the Hoeffding-type inequalities proved so far are only special cases of more general results stating that

$$\mathbb{E}_{\mu_1, \ldots, \mu_n}\left[\psi\left(\sum_{i=1}^n X_i\right)\right]$$

is under appropriate assumptions on $\psi$ a Schur-convex function of $(\mu_1, \ldots, \mu_n)$. For $n = 2$ this turns out to be true in full generality.

**3.3. Proposition.** *Let* $\psi: S \times S \to \mathbb{R}$ *be a negative definite kernel and let* $\nu_1, \nu_2, \mu_1, \mu_2$ *be molecular probability measures on S. If* $(\nu_1, \nu_2) \prec (\mu_1, \mu_2)$ *then*

$$\mathbb{E}_{\nu_1, \nu_2}[\psi(X, Y)] \leqq \mathbb{E}_{\mu_1, \mu_2}[\psi(X, Y)].$$

(The interpretation is analogous to that of (4) in §1.)

Proof. For some $P \in \Omega_2$ we have $\nu = P\mu$ where $\nu = (\nu_1, \nu_2)'$ and $\mu = (\mu_1, \mu_2)'$, and $P$ has the form

$$P = \begin{pmatrix} \alpha & 1-\alpha \\ 1-\alpha & \alpha \end{pmatrix}$$

for some $\alpha \in [0, 1]$. Let $\mu_1$ and $\mu_2$ both be concentrated on $\{s_1, \ldots, s_n\} \subseteq S$. Then $\nu_1$ and $\nu_2$ are also concentrated there and

$$\begin{aligned} &\mathbb{E}_\nu[\psi(X, Y)] - \mathbb{E}_\mu[\psi(X, Y)] \\ &\quad = \sum_{j,k=1}^{n} \psi(s_j, s_k)[\nu_1(\{s_j\})\nu_2(\{s_k\}) - \mu_1(\{s_j\})\mu_2(\{s_k\})]. \end{aligned}$$

For $j \neq k$ we find

$$\begin{aligned} &[\nu_1(\{s_j\})\nu_2(\{s_k\}) - \mu_1(\{s_j\})\mu_2(\{s_k\})] + [\{\nu_1(\{s_k\})\nu_2(\{s_j\}) \\ &\quad - \mu_1(\{s_k\})\mu_2(\{s_j\})] \\ &\quad = 2\alpha(1-\alpha)[\mu_1(\{s_j\}) - \mu_2(\{s_j\})] \cdot [\mu_1(\{s_k\}) - \mu_2(\{s_k\})], \end{aligned}$$

furthermore

$$\nu_1(\{s_j\})\nu_2(\{s_j\}) - \mu_1(\{s_j\})\mu_2(\{s_j\}) = \alpha(1-\alpha)[\mu_1(\{s_j\}) - \mu_2(\{s_j\})]^2,$$

so that

$$\mathbb{E}_\nu[\psi(X, Y)] - \mathbb{E}_\mu[\psi(X, Y)] = \alpha(1-\alpha) \sum_{j,k=1}^{n} c_j c_k \psi(s_j, s_k),$$

with $c_j = \mu_1(\{s_j\}) - \mu_2(\{s_j\})$, $j = 1, \ldots, n$. Since $\sum c_j = 0$ the expression above is indeed $\leqq 0$. □

**3.4. Example.** For independent integer-valued random variables $X$, $Y$ with distributions $\mu_1, \mu_2$ the probability

$$\mathbb{P}_{\mu_1,\mu_2}(X - Y \text{ is odd})$$

is Schur-convex in $\mu_1, \mu_2$; in case $X \geqq 0$, $Y \geqq 0$ so is

$$\mathbb{P}_{\mu_1,\mu_2}(X + Y \text{ is odd}).$$

**3.5. Example.** Let $X$, $Y$ be independent with values in a separable Hilbert space. Then

$$\mathbb{E}_{\mu_1,\mu_2}(\langle X, Y\rangle)$$

is Schur-concave in $\mu_1, \mu_2$.

**3.6. Definition.** Let $S$ be an abelian semigroup. A function $\psi: S \to \mathbb{R}$ is called *Schur-increasing of order n*, $n = 1, 2, \ldots,$ if

$$\mathbb{E}_{\mu_1,\ldots,\mu_n}\left[\psi\left(\sum_{i=1}^{n} X_i\right)\right]$$

is a Schur-convex function of $(\mu_1, \ldots, \mu_n) \in \mathrm{Mol}^1_+(S)^n$, i.e. if $\nu = (\nu_1, \ldots, \nu_n)$ and $\mu = (\mu_1, \ldots, \mu_n)$ in $\mathrm{Mol}^1_+(S)^n$ are such that $\nu \prec \mu$ then

$$\int \psi \, d(\nu_1 * \cdots * \nu_n) \leqq \int \psi \, d(\mu_1 * \cdots * \mu_n).$$

Moreover, $\psi$ is called *Schur-increasing*, if it is Schur-increasing of order $n$ for all $n = 1, 2, \ldots$. Finally, $\psi$ is called *Schur-decreasing* (resp. *of order n*), if $-\psi$ is Schur-increasing (resp. of order $n$).

Any constant function is both Schur-increasing and Schur-decreasing, and it is not difficult to see that the functions which are both Schur-increasing and Schur-decreasing are precisely the functions $\psi(s) = c + h(s)$ where $c \in \mathbb{R}$ and $h: S \to \mathbb{R}$ is additive, cf. Exercise 3.16.

The set of Schur-increasing functions of order $n$ is a closed convex cone in $\mathbb{R}^S$ which decreases with $n$. If $\psi$ is Schur-increasing of order $n$ then $\psi$ satisfies, of course, Hoeffding's inequality of order $n$, and for $n = 2$ each of these properties is equivalent with $\psi \in \mathcal{N}(S)$ by Proposition 3.3 and Corollary 1.4. By Theorem 1.7 it is natural to examine if a completely negative definite function is Schur-increasing. Closely related is the question if every completely positive definite function is Schur-decreasing. These questions will be answered in the negative below, cf. 3.13, but the answer is positive under suitable boundedness conditions on the function as the following result shows.

**3.7. Theorem.** *Let S be an abelian semigroup. Then:*

(i) *every bounded completely positive definite function is Schur-decreasing;*
(ii) *every lower bounded completely negative definite function is Schur-increasing.*

PROOF. A bounded completely positive definite function $\varphi$ is completely monotone by Theorem 1.10 and has therefore by Theorem 4.6.4 a representation

$$\varphi(s) = \int_{\hat{S}_+} \rho(s) \, d\kappa(\rho),$$

where $\kappa \in M_+(\hat{S}_+)$. It therefore suffices to prove that every $\rho \in \hat{S}_+$ is Schur-decreasing. We prove that every $\rho \in S^*_+$ has this property. Let $\nu = (\nu_1, \ldots, \nu_n)$ and $\mu = (\mu_1, \ldots, \mu_n) \in \mathrm{Mol}^1_+(S)^n$ satisfy

$$\nu_i = \sum_{j=1}^n p_{ij} \mu_j, \qquad i = 1, \ldots, n,$$

where $(p_{ij}) \in \Omega_n$. Putting $b_i = \int \rho \, d\nu_i$, $a_j = \int \rho \, d\mu_j$, $b = (b_1, \ldots, b_n)$ and $a = (a_1, \ldots, a_n)$ we have

$$b_i = \sum_{j=1}^n p_{ij} a_j$$

hence $b \prec a$, so by Lemma 3.1

$$\int \rho \, d(\nu_1 * \cdots * \nu_n) = \prod_i b_i \geqq \prod_i a_i = \int \rho \, d(\mu_1 * \cdots * \mu_n).$$

If $\psi$ is lower bounded and completely negative definite then $\exp(-t\psi)$ is bounded and completely positive definite hence Schur-decreasing by (i) for $t > 0$. Finally,

$$\psi = \lim_{t \to 0} \frac{1}{t}(1 - e^{-t\psi})$$

is Schur-increasing as pointwise limit of Schur-increasing functions. □

The above theorem leaves open the question if, for example, $\psi(x) = -x^2$ is Schur-increasing on $(\mathbb{R}, +)$. This in fact is true and could even be proved directly, but we will see below that a much more general result can be shown. First of all we consider the case where the underlying semigroup is perfect. We need the following:

**3.8. Lemma.** *Let $S$ be a perfect semigroup and let $\varphi: S \to \mathbb{R}$ be completely positive definite. Then the Radon measure representing $\varphi$ is concentrated on $S_+^*$.*

PROOF. Let $\mu \in M_+(S^*)$ be the Radon measure such that

$$\varphi(s) = \int \rho(s) \, d\mu(\rho), \qquad s \in S.$$

By assumption there are uniquely determined measures $\mu_a \in M_+(S^*)$, $a \in S$, such that

$$E_a \varphi(s) = \varphi(a + s) = \int \rho(s) \, d\mu_a(\rho), \qquad s \in S$$

implying

$$\mu_a = \rho(a)\mu.$$

The sets $T_a := \{\rho \in S^* \mid \rho(a) < 0\}$ are open in $S^*$ and

$$\mu_a(T_a) = \int_{T_a} \rho(a) \, d\mu(\rho)$$

shows that $\mu(T_a) = 0$; therefore

$$\mu(S^* \backslash S_+^*) = \mu\left(\bigcup_{a \in S} T_a\right) = 0$$

since $\mu$ is a Radon measure. □

**3.9. Theorem.** *For an abelian semigroup S the following conditions are equivalent:*

(i) *Every completely positive definite function is Schur-decreasing.*
(ii) *Every completely negative definite function is Schur-increasing.*

*If S is perfect or if* $S = \mathbb{N}_0$ *then* (i) *and* (ii) *are verified.*

PROOF. "(i) ⇒ (ii)" If $\psi$ is completely negative definite then $\exp(-t\psi)$ is completely positive definite, hence Schur-decreasing for $t > 0$. Therefore

$$\psi = \lim_{t \to 0} \frac{1}{t}(1 - e^{-t\psi})$$

is Schur-increasing.

"(ii) ⇒ (i)" If $\varphi$ is completely positive definite then $-\varphi$ is completely negative definite and hence Schur-increasing.

Every $\rho \in S_+^*$ is Schur-decreasing as shown in the proof of Theorem 3.7, and therefore every moment function $\varphi \in \mathscr{H}(S)$ with representing measure $\mu$ supported by $S_+^*$ is Schur-decreasing. This shows that (i) holds for $S = \mathbb{N}_0$ because a completely positive definite function on $\mathbb{N}_0$ is a Stieltjes moment sequence, cf. 6.2.5. Furthermore (i) holds for a perfect semigroup by Lemma 3.8. □

**3.10. Corollary.** *Let S be a Hausdorff topological abelian semigroup containing a dense perfect semigroup. Then every continuous completely positive* (*resp. negative*) *definite function* $\varphi\colon S \to \mathbb{R}$ *is Schur-decreasing* (*resp. increasing*).

PROOF. Let two vectors $\mu = (\mu_1, \ldots, \mu_n)$ and $\nu = (\nu_1, \ldots, \nu_n)$ of molecular probability measures on $S$ be given such that $\nu \prec \mu$. There is a finite set $A \subseteq S$ on which all $\mu_i$ and hence all $\nu_i$ are concentrated. Approximating each $s \in A$ by a suitable net of points from the given dense perfect subsemigroup $P$ we can find nets $\mu^{(i)} = (\mu_1^{(i)}, \ldots, \mu_n^{(i)})$ and $\nu^{(i)} = (\nu_1^{(i)}, \ldots, \nu_n^{(i)})$ from $\mathrm{Mol}_+^1(P)^n$ such that $\nu^{(i)} \prec \mu^{(i)}$ and

$$\int \varphi \, d(\nu_1 * \cdots * \nu_n) = \lim_i \int \varphi \, d(\nu_1^{(i)} * \cdots * \nu_n^{(i)})$$

$$\geq \lim_i \int \varphi \, d(\mu_1^{(i)} * \cdots * \mu_n^{(i)}) = \int \varphi \, d(\mu_1 * \cdots * \mu_n),$$

where we have used that $\varphi | P$ is completely positive definite and hence Schur-decreasing. The case of $\varphi$ being completely negative definite is treated similarly. □

This corollary applies, in particular, to the important semigroups $(\mathbb{R}_+^k, +)$ and $(\mathbb{R}^k, +)$. Since they are 2-divisible, we conclude that continuous negative definite functions are the same as continuous Schur-increasing functions on these semigroups.

In Proposition 1.11 we saw that a nonnegative quadratic form $q$ on an abelian group $G$ fulfils the reverse Hoeffding inequality of each order. As we shall see now $q$ is even Schur-decreasing. It should be noticed that $G$ is considered as a semigroup with the identical involution, so there is no contradiction in the fact that $q$ is negative definite in the group sense and $-q$ is completely negative definite in the semigroup sense.

**3.11. Proposition.** *A nonnegative quadratic form $q$ on an abelian group $G$ is Schur-decreasing.*

PROOF. We consider first the case where $G = \mathbb{Z}^k$ for some $k \geqq 1$. In the proof of Theorem 4.3.9 it was shown that

$$B(s, t) := q(s) + q(t) - q(s - t), \qquad s, t \in \mathbb{Z}^k$$

is biadditive and, furthermore, that $B$ is a positive definite kernel. Denoting $e_1 = (1, 0, \ldots, 0), \ldots, e_k = (0, \ldots, 0, 1)$ we see, in particular, that the matrix $(B(e_i, e_j))_{i,j=1,\ldots,k}$ is positive definite. Noting $q(0) = 0$ this implies

$$q(s) = \tfrac{1}{2}B(s, s) = \tfrac{1}{2} \sum_{i,j=1}^{k} s_i s_j B(e_i, e_j)$$

and $q$ is therefore the restriction of a continuous nonnegative quadratic form on $\mathbb{R}^k$. The result follows now from 3.10 and 1.11 since $\mathbb{R}^k$ contains the dense perfect semigroup $\mathbb{Q}^k$.

If $G$ is arbitrary, let $\mu_1, \ldots, \mu_n$ be all concentrated on the finite subset $A = \{s_1, \ldots, s_k\} \subseteq S$ implying $\nu_i(A) = 1$ for all $i = 1, \ldots, n$ if $\nu \prec \mu$. We define a canonical group homomorphism $h: \mathbb{Z}^k \to G$ by $h(z_1, \ldots, z_k) := \sum_{j=1}^{k} z_j s_j$. Then $q \circ h$ is a nonnegative quadratic form on $\mathbb{Z}^k$. Let $\tilde{\mu}$ and $\tilde{\nu}$ on $\mathbb{Z}^k$ be defined by

$$\tilde{\mu}_i := \sum_{j=1}^{k} \mu_i(\{s_j\})\varepsilon_{e_j}, \qquad \tilde{\nu}_i := \sum_{j=1}^{k} \nu_i(\{s_j\})\varepsilon_{e_j}.$$

Then $\tilde{\mu}_i^h = \mu_i$, $\tilde{\nu}_i^h = \nu_i$ for $i = 1, \ldots, n$ and $\tilde{\nu} \prec \tilde{\mu}$, implying

$$\begin{aligned}
\int q \, d(\nu_1 * \cdots * \nu_n) &= \int q \, d(\tilde{\nu}_1^h * \cdots * \tilde{\nu}_n^h) \\
&= \int q \, d[(\tilde{\nu}_1 * \cdots * \tilde{\nu}_n)^h] \\
&= \int q \circ h \, d(\tilde{\nu}_1 * \cdots * \tilde{\nu}_n) \\
&\geqq \int q \circ h \, d(\tilde{\mu}_1 * \cdots * \tilde{\mu}_n) \\
&= \int q \, d(\mu_1 * \cdots * \mu_n),
\end{aligned}$$

where we have used the result of Exercise 2.3.12. □

**3.12.** As in the case of Hoeffding's inequality we can extend the results proved so far in this section to nondiscrete distributions, assuming, of course, continuity of the positive and negative definite functions (kernels) to be integrated, and assuming furthermore that the probabilities involved are Radon measures. To prove these extensions one uses the same ideas as in §2; cf., in particular, the proofs of Theorems 2.1 and 2.5.

We will now show the existence of completely negative definite functions on $\mathbb{N}_0^2$ which are not Schur-increasing. By Theorem 3.9 this is equivalent with the existence of completely positive definite functions on $\mathbb{N}_0^2$ which are not Schur-decreasing. We prove even more:

**3.13. Proposition.** *There exist completely positive definite functions on* $\mathbb{N}_0^2$ *which are not Schur-decreasing of order* 3.

PROOF. Let $B$ denote the set of polynomials

$$p(x, y) = \sum c_{jk} x^j y^k$$

in two variables where $c_{jk} \geqq 0$ and $p(1, 1) = \sum c_{jk} = 1$. In accordance with Definition 3.2 we consider the ordering $\prec$ on $B^3$ and define the set

$$\tilde{B} = \{q_1 q_2 q_3 - p_1 p_2 p_3 \mid p = (p_1, p_2, p_3), q = (q_1, q_2, q_3) \in B^3, q \prec p\}.$$

For a function $\varphi\colon \mathbb{N}_0^2 \to \mathbb{R}$ and a polynomial $p \in A^{(2)}$ given as above we define

$$\langle \varphi, p \rangle = \sum \varphi((j, k)) c_{jk},$$

and the vector spaces $\mathbb{R}^{\mathbb{N}_0^2}$ and $A^{(2)}$ form a dual pair under $\langle \cdot, \cdot \rangle$. With $\mu \in \mathrm{Mol}_+^1(\mathbb{N}_0^2)$ is associated a polynomial $p \in B$ by defining the coefficient of $x^j y^k$ as $c_{jk} = \mu(\{(j, k)\})$. The mapping $\mu \mapsto p$ is clearly an affine isomorphism of $\mathrm{Mol}_+^1(\mathbb{N}_0^2)$ onto $B$ and convolution of measures corresponds to products of polynomials. Therefore $\varphi\colon \mathbb{N}_0^2 \to \mathbb{R}$ is Schur-decreasing of order 3 if and only if

$$\langle \varphi, r \rangle \geqq 0 \qquad \text{for all} \quad r \in \tilde{B}.$$

In the proof of Theorem 6.3.7 we introduced the cone $C^{(2)}$ with the property that

$$\langle \varphi, r \rangle \geqq 0 \qquad \text{for all } r \in C^{(2)}$$

if and only if $\varphi$ is completely positive definite. Since $C^{(2)}$ is a closed convex cone by Lemma 6.3.9 it suffices to prove that $\tilde{B} \setminus C^{(2)} \neq \emptyset$. In fact, having found a polynomial $p_0 \in \tilde{B} \setminus C^{(2)}$, there exists a function $\varphi\colon \mathbb{N}_0^2 \to \mathbb{R}$ by the separation theorem (1.2.3) such that

$$\langle \varphi, p_0 \rangle < 0 \quad \text{and} \quad \langle \varphi, r \rangle \geqq 0 \qquad \text{for all} \quad r \in C^{(2)}.$$

This function $\varphi$ is completely positive definite but not Schur-decreasing of order 3.

The proof will be finished by the following lemma:

**3.14. Lemma.** *Let $p = (x, y, 1) \in B^3$, let $P_\alpha \in \Omega_3$ be the matrix*

$$P_\alpha = \begin{pmatrix} 0 & \alpha & 1-\alpha \\ 1-\alpha & 0 & \alpha \\ \alpha & 1-\alpha & 0 \end{pmatrix}, \qquad \alpha \in [0, 1],$$

*and let $q = (q_1, q_2, q_3)$ be defined by $q' = P_\alpha p'$. Then $r_\alpha := q_1 q_2 q_3 - p_1 p_2 p_3$ is given by*

$$r_\alpha(x, y) = \alpha(1-\alpha)[\alpha(x + y^2 + x^2 y) + (1-\alpha)(y + x^2 + xy^2) - 3xy],$$

*and $r_\alpha \in \tilde{B} \backslash C^{(2)}$ if and only if $\alpha \in [0, 1] \backslash \{0, \frac{1}{2}, 1\}$.*

Proof. The formula for $r_\alpha$ is easily verified and we see that $r_0 = r_1 = 0$ and that

$$\begin{aligned} 8r_{1/2} &= x + y + x^2 + y^2 - 6xy + x^2 y + xy^2 \\ &= (x-y)^2 + x(1-y)^2 + y(1-x)^2, \end{aligned}$$

which shows that $r_{1/2} \in C^{(2)}$.

We show next that $\tilde{r}_\alpha := \{\alpha(1-\alpha)\}^{-1} r_\alpha \in C^{(2)}$ implies $\alpha = \frac{1}{2}$.

If $\tilde{r}_\alpha = a + xb + yc + xyd$ with $a, b, c, d \in \Sigma^{(2)}$ we see that $a$, $b$ and $c$ are of even degree $\leqq 2$ and $d$ is a nonnegative constant because $\tilde{r}_\alpha$ is of degree 3. By assumption

$$a = \sum a_i^2, \qquad b = \sum b_i^2, \qquad c = \sum c_i^2,$$

where $a_i, b_i, c_i \in A_1^{(2)}$. We have

$$\tilde{r}_\alpha(0, 0) = 0 = \sum a_i(0, 0)^2,$$

$$\tilde{r}_\alpha(x, 0) = \alpha x + (1-\alpha)x^2 = \sum a_i(x, 0)^2 + x \sum b_i(x, 0)^2,$$

$$\tilde{r}_\alpha(0, y) = (1-\alpha)y + \alpha y^2 = \sum a_i(0, y)^2 + y \sum c_i(0, y)^2,$$

which shows that $a_i(0, 0) = 0$, $b_i(x, 0) = \beta_i$, $c_i(0, y) = \gamma_i$ for certain constants $\beta_i, \gamma_i \in \mathbb{R}$. This means that we can write

$$a_i(x, y) = \delta_i x + \varepsilon_i y,$$

$$b_i(x, y) = \beta_i + \beta_i' y,$$

$$c_i(x, y) = \gamma_i + \gamma_i' x,$$

for certain constants $\delta_i, \varepsilon_i, \beta_i', \gamma_i' \in \mathbb{R}$.

If we insert these expressions and compare coefficients we find

$$\alpha = \sum \beta_i^2 = \sum \varepsilon_i^2 = \sum \gamma_i'^2,$$

$$1 - \alpha = \sum \gamma_i^2 = \sum \delta_i^2 = \sum \beta_i'^2,$$

$$-3 = 2\sum \delta_i \varepsilon_i + 2\sum \beta_i \beta_i' + 2\sum \gamma_i \gamma_i' + d.$$

We apply the Cauchy–Schwarz inequality to the sums in the last formula and get

$$\tfrac{1}{2}(d+3) \leqq 3\sqrt{\alpha(1-\alpha)},$$

and since $d \geqq 0$ this implies $\sqrt{\alpha(1-\alpha)} \geqq \frac{1}{2}$ which is equivalent to $(\alpha - \frac{1}{2})^2 \leqq 0$, hence $\alpha = \frac{1}{2}$. □

**3.15. Remarks.** (i) If $r \in \tilde{B}$ then $r(x, y) \geqq 0$ for $(x, y) \in \mathbb{R}_+^2$ and $r(1, 1) = 0$. The positivity assertion follows from Lemma 3.1. The existence of a polynomial $p_0 \in \tilde{B} \setminus C^{(2)}$ also implies that there exists a completely positive definite function on $\mathbb{N}_0^2$ which is not a Stieltjes moment function. This assertion is weaker than the result of Theorem 6.3.7.

(ii) The result of 3.13 is best possible in the sense that every positive definite function and hence *a fortiori* every completely positive definite function on an abelian semigroup is Schur-decreasing of order 2.

**3.16. Exercise.** Show that the sum of a constant function and an additive function is both Schur-increasing and Schur-decreasing. Show conversely that if $\varphi: S \to \mathbb{R}$ is both Schur-increasing and Schur-decreasing of order 2, then $\varphi$ is sum of a constant and an additive function.

**3.17. Exercise.** Show that if $\varphi: S \to \mathbb{R}$ is Schur-increasing of order $n + 1$ then $E_a\varphi$ is Schur-increasing of order $n$ for all $a \in S$. Show that if $\varphi$ is Schur-increasing then so is $E_a\varphi$ for all $a \in S$.

**3.18. Exercise.** Show that

$$\left(\frac{x_1 + \cdots + x_k}{k}\right)^k - x_1 x_2 \cdots x_k \in C^{(k)},$$

where $C^{(k)} \subseteq A^{(k)}$ is the cone introduced in the proof of Theorem 6.3.7. *Hint*: If the polynomial in question is not in $C^{(k)}$, then there exists a completely positive definite function on $\mathbb{N}_0^k$ which does not verify the reverse Hoeffding inequality of order $k$.

**3.19. Exercise.** Let $X_1, \ldots, X_n$ be independent integrable real-valued random variables with distributions $\mu_1, \ldots, \mu_n$. Then

$$\operatorname{var}_{\mu_1, \ldots, \mu_n}\left(\sum_{i=1}^n X_i\right)$$

is a Schur-concave function of $\mu_1, \ldots, \mu_n$, whereas

$$\mathbb{E}_{\mu_1, \ldots, \mu_n}\left[\frac{1}{n-1}\sum_{i=1}^n (X_i - \bar{X})^2\right]$$

is Schur-convex in $\mu_1, \ldots, \mu_n$ (if $\mathbb{E}(X_i^2) < \infty$ for all $i$).

**3.20. Exercise.** Show that the semigroup $(\mathbb{Z}, +)$ has the two equivalent properties of Theorem 3.9. *Hint*: See Exercise 6.4.7.

**3.21. Exercise.** A doubly stochastic matrix $P \in \Omega_n$ is called *elementary* if at most two off-diagonal entries of $P$ are nonzero. Let $\Gamma_n \subseteq \Omega_n$ be the set of all products of finitely many elementary doubly stochastic $n \times n$ matrices. If $\psi: S \to \mathbb{R}$ is completely negative definite and $\nu, \mu \in (\mathrm{Mol}_+^1(S))^n$ are such that $\nu' = P\mu'$ for some $P \in \overline{\Gamma_n}$, the closure of $\Gamma_n$, then

$$\int \psi \, d(\nu_1 * \cdots * \nu_n) \leqq \int \psi \, d(\mu_1 * \cdots * \mu_n).$$

**3.22. Exercise.** Let $B^{(k)}$ denote the set of polynomials $p(x) = \sum c_\alpha x^\alpha$ in $k$ variables with $c_\alpha \geqq 0$ and $p(\mathbf{1}) = \sum c_\alpha = 1$, where $\mathbf{1} = (1, \ldots, 1) \in \mathbb{R}^k$. Let further

$$B_n^{(k)} = \{q_1 \cdots q_n - p_1 \cdots p_n \mid p = (p_1, \ldots, p_n), q = (q_1, \ldots, q_n) \in (B^{(k)})^n, q \prec p\}$$

for $n = 2, 3, \ldots$. Show that

(i) $B_2^{(k)} \subseteq B_3^{(k)} \subseteq \cdots, B_n^{(k)} \subseteq \{p \in A_+^{(k)}([0, \infty[^k) \mid p(\mathbf{1}) = 0\}$;

(ii)
$$C(B_2^{(k)}) = \{p \in \Sigma^{(k)} \mid p(\mathbf{1}) = 0\},$$
$$C(B_3^{(k)}) \supseteq \{p \in C^{(k)} \mid p(\mathbf{1}) = 0\},$$

where the smallest convex cone in $A^{(k)}$ containing a subset $M \subseteq A^{(k)}$ is denoted $C(M)$, and $C^{(k)}$ is defined in 6.3.7;

(iii) $\varphi: \mathbb{N}_0^k \to \mathbb{R}$ is Schur-decreasing of order $n$ if and only if

$$\langle \varphi, p \rangle \geqq 0 \qquad \text{for all} \quad p \in B_n^{(k)}.$$

## Notes and Remarks

Most of the results in §§1 and 2 were published in Christensen and Ressel (1981) and in Ressel (1982a). The nondiscrete extension of Hoeffding's inequality is shown in these two papers to hold even for $\tau$-smooth probability measures on arbitrary abelian topological semigroups (the Hausdorff property is not essential). Theorem 1.7 shows that Hoeffding's inequality of order 3 plays a particular role; it is unclear if the number 3 can be replaced by some larger $k \in \mathbb{N}$. Exercise 1.17 is taken from Karlin and Novikoff (1963).

Our Definition 3.2 of multivariate majorization coincides for $A \subseteq \mathbb{R}^m$ with that given in Marshall and Olkin (1979, Ch. 15, A.2). On page 14 of this monograph it is stated that the classical Schur-convex functions should rather be called Schur-increasing, since they preserve the Schur majorization order $x \prec y$. This is the reason for the terminology introduced in 3.6. Sharpening the majorization order on $(\mathrm{Mol}_+^1(S))^n$ by requiring the doubly stochastic matrix in the definition of that ordering to belong to $\Gamma_n$ (cf. Exercise 3.21), we get what Marshall and Olkin (1979, Ch. 15, D.1) called *chain majorization*. Lemma 3.14 together with Exercise 3.21 imply, in particular, that $\Gamma_3$ is not dense in $\Omega_3$. More precisely the matrices $P_\alpha$ defined in Lemma 3.14 belong to $\Omega_3 \setminus \overline{\Gamma_3}$ for $\alpha \in ]0, \frac{1}{2}[ \cup ]\frac{1}{2}, 1[$. But even $P_{1/2} \notin \overline{\Gamma_3}$ and this follows from the fact that the set of doubly stochastic $n \times n$ matrices $P = (p_{ij})$ such that $|\det P| \leqq \prod_{i=1}^n p_{i,\pi(i)}$ for at least one permutation $\pi$ of $\{1, \ldots, n\}$ is a closed subsemigroup of $\Omega_n$ containing $\Gamma_n$, and that (for $n = 3$) the matrix $P_{1/2}$ does not have this property (Fuglede and Johansen, private communication). The existence of completely positive definite functions which are not Schur-decreasing of order 3 was announced in Berg and Christensen (1983b).

The result of Exercise 3.18 has been shown by explicit formulas by Motzkin (1967), but for $k \geqq 3$ such formulas are complicated.

After the completion of this manuscript the following result has been obtained: On a 2-divisible semigroup every negative definite function is Schur-increasing. See Berg (1984b).

# CHAPTER 8
# Positive and Negative Definite Functions on Abelian Semigroups Without Zero

The integral representation theorems for positive and negative definite functions obtained so far were all proved under the assumption that the semigroup contained a neutral element, i.e. a "zero" with respect to the additively written semigroup operation. We also saw that some boundedness conditions were necessary for the functions under consideration in order to prove the main representation results.

In this chapter we shall see that on abelian semigroups without zero, too, some interesting integral representation theorems for positive and negative definite functions can be proved, as long as unboundedness (resp. unboundedness below) is only allowed "near the (nonexisting) zero element", a condition to be made precise below. All semigroups considered in this chapter are assumed to carry the identical involution.

## §1. Quasibounded Positive and Negative Definite Functions

Let $(H, +)$ denote an abelian semigroup. We now have in mind a semigroup without zero, but formally this will not be required. An element $h \in H$ is called *maximal* if it is not of the form $h_1 + h_2$ with $h_1, h_2 \in H$, cf. Choquet (1954, p. 262). Let $H' := H + H$ be the subsemigroup of nonmaximal elements of $H$. Note that $H' = H$ whenever $H$ is 2-divisible, but, for example, $H' = H$ also for the non-2-divisible semigroup $H = (]-1, 1[, \cdot)$. Put $S := H \cup \{0\}$ with some element $0 \notin H$. Let $0 + h = h + 0 := h$ for any $h \in H$, and $0 + 0 := 0$; then $S$ is an abelian semigroup with neutral element 0, and $\hat{S}$ contains the absorbing element $1_{\{0\}}$, cf. 4.1.3.

If we use the original definition of positive and negative definite functions also for semigroups without zero, then clearly the values at maximal elements of $H$ do not enter into the defining property. A *positive* (resp. *negative*) *definite function* on $H$ is therefore a function $\varphi\colon H' \to \mathbb{R}$ such that the kernel $(h, k) \mapsto \varphi(h + k)$ is positive (resp. negative) definite on $H \times H$. The set of positive (resp. negative) definite functions on $H$ is denoted $\mathscr{P}(H)$ (resp. $\mathscr{N}(H)$). If $H$ already contains a neutral element, then $H' = H$ and the notion agrees with that already given in Chapter 4.

Let us agree in calling a function $\varphi\colon H' \to \mathbb{R}$ *quasibounded* (resp. *quasibounded below*) if

$$\sup_{s\in S}|\varphi(h' + s)| < \infty \qquad \text{for all} \quad h' \in H'$$

(resp.

$$\inf_{s\in S}\varphi(h' + s) > -\infty \qquad \text{for all} \quad h' \in H').$$

Furthermore we introduce $\mathscr{P}^q(H)$ to be the set of all quasibounded positive definite functions on $H$. Some simple properties of $\mathscr{P}^q(H)$ are stated below.

**1.1. Lemma.** (i) *The set $\mathscr{P}^q(H)$ is a closed convex cone in $\mathbb{R}^{H'}$, stable under multiplication.*

(ii) *For $\varphi \in \mathscr{P}^q(H)$ and $s \in S$ the two functions $h' \mapsto \varphi(h') \pm \varphi(s + h')$ belong also to $\mathscr{P}^q(H)$.*

(iii) *If $\varphi \in \mathscr{P}^q(H)$ and $h, k \in H$, then*

$$s \mapsto \varphi(s + 2h) + \varphi(s + 2k) - 2\varphi(s + h + k)$$

*belongs to $\mathscr{P}^b(S)$.*

Proof. (i) If $\varphi \in \mathscr{P}^q(H)$ then $s \mapsto \varphi(s + 2h)$ is bounded and positive definite on $S$, hence by Proposition 4.1.12 $\sup_{s\in S}|\varphi(s + 2h)| = \varphi(2h)$ and furthermore

$$|\varphi(h + k + s)| \leqq [\varphi(2h)\varphi(2k + 2s)]^{1/2},$$

showing immediately the closedness of $\mathscr{P}^q(H)$. Stability under multiplication is obvious.

(ii) If $h_1, \ldots, h_n \in H$ and $c_1, \ldots, c_n \in \mathbb{R}$ then $\Phi\colon S \to \mathbb{R}$ defined by

$$\Phi(s) := \sum_{i,j=1}^{n} c_i c_j \varphi(s + h_i + h_j)$$

is easily seen to belong to $\mathscr{P}^b(S)$, cf. the proof of Theorem 4.1.14, hence, again by 4.1.12, $|\Phi(s)| \leqq \Phi(0)$ for all $s \in S$, i.e.

$$\sum_{i,j=1}^{n} c_i c_j[\varphi(h_i + h_j) \pm \varphi(s + h_i + h_j)] \geqq 0.$$

(iii) Take $\Phi$ as defined above for $n = 2$, $h_1 = h$, $h_2 = k$, $c_1 = 1$ and $c_2 = -1$. □

Let $\varphi \in \mathscr{P}^q(H)$ and put $\varphi_{h,k}(s) := \varphi(s + h + k)$ for $s \in S$ and $h, k \in H$. We mentioned already that $\varphi_h := \varphi_{h,h} \in \mathscr{P}^b(S)$, hence by Theorem 4.2.8 there are uniquely determined Radon measures $\mu_h \in M_+(\hat{S})$ such that

$$\varphi_h(s) = \int_{\hat{S}} \rho(s)\, d\mu_h(\rho) \qquad \text{for all} \quad s \in S.$$

Furthermore, in view of the above lemma, there are measures $\nu_{h,k} \in M_+(\hat{S})$ with

$$\varphi(s + 2h) - 2\varphi(s + h + k) + \varphi(s + 2k) = \int_{\hat{S}} \rho(s)\, d\nu_{h,k}(\rho)$$

for all $s \in S$ and $h, k \in H$. Put $\mu_{h,k} := \frac{1}{2}(\mu_h + \mu_k - \nu_{h,k})$, then $\mu_{h,k} \in M(\hat{S})$, $\mu_{h,h} = \mu_h \in M_+(\hat{S})$ and

$$\varphi(s + h + k) = \int_{\hat{S}} \rho(s)\, d\mu_{h,k}(\rho) \qquad \text{for} \quad s \in S,\ h, k \in H.$$

Our aim is to get an integral representation for quasibounded positive definite functions on $H$, but in order to avoid pathologies we have to restrict ourselves to the subcone

$$\mathscr{P}^q_0(H) := \{\varphi \in \mathscr{P}^q(H) \,|\, \mu_{h,k}(\{1_{\{0\}}\}) = 0 \text{ for all } h, k \in H\}.$$

Notice that $\varphi \in \mathscr{P}^q(H)$ belongs to $\mathscr{P}^q_0(H)$ if and only if

$$\mu_h(\{1_{\{0\}}\}) = \nu_{h,k}(\{1_{\{0\}}\}) = 0 \qquad \text{for all} \quad h, k \in H,$$

and from this remark it follows that $\mathscr{P}^q_0(H)$ is an extreme subset of $\mathscr{P}^q(H)$.

The example $H = ([1, \infty[, +)$ and $\varphi = 1_{\{2\}}$ shows the possibility of $\mathscr{P}^q_0(H) \subsetneqq \mathscr{P}^q(H)$. For semigroups without maximal elements, however, examples of this kind are not possible, as the next lemma shows.

**1.2. Lemma.** *Let $\varphi$ belong to $\mathscr{P}^q(H)$. Then we have:*

(i) *If $2H + S = H'$ and $\mu_h(\{1_{\{0\}}\}) = 0$ for all $h \in H$, it follows $\varphi \in \mathscr{P}^q_0(H)$.*
(ii) *If $H = H'$, then $\mathscr{P}^q_0(H) = \mathscr{P}^q(H)$.*

PROOF. (i) Let $h, k \in H$; we can find $h_0 \in H$ and $a \in S$ such that $h + k = 2h_0 + a$. For any $s \in S$ we have

$$\int_{\hat{S}} \rho(s)\, d\mu_{h,k}(\rho) = \varphi(s + h + k) = \varphi(s + 2h_0 + a)$$

$$= \varphi_{h_0}(s + a) = \int_{\hat{S}} \rho(s)\rho(a)\, d\mu_{h_0}(\rho),$$

implying $\mu_{h,k} = \rho(a)\mu_{h_0}$ by unicity of the integral representation. Hence

$$\mu_{h,k}(\{1_{\{0\}}\}) = 1_{\{0\}}(a) \cdot \mu_{h_0}(\{1_{\{0\}}\}) = 0.$$

(ii) For given $h, k \in H$ we find $h_1, h_2 \in H$ such that $h = h_1 + h_2$. From

$$\int_{\hat{S}} \rho(s)\, d\mu_{h,k}(\rho) = \varphi(s + h + k) = \varphi_{h_1,h_2}(s + k)$$

$$= \int_{\hat{S}} \rho(s)\rho(k)\, d\mu_{h_1,h_2}(\rho)$$

for all $s \in S$ we deduce $\mu_{h,k} = \rho(k)\mu_{h_1,h_2}$, and therefore

$$\mu_{h,k}(\{1_{\{0\}}\}) = 0. \qquad \square$$

**1.3. Remarks.** (1) If $H$ contains already a neutral element, then it is easily seen that

$$\hat{S}\backslash\{1_{\{0\}}\} \to \hat{H}, \qquad \rho \mapsto \rho|H$$

is a topological semigroup isomorphism, showing, in particular, that $1_{\{0\}}$ in this case is an isolated point of $\hat{S}$.

(2) The set $H'' := 2H + S$ is always a subsemigroup of $H'$. For $H = (\mathbb{N}, +)$ we have $H'' = H' \subsetneqq H$. That $H'' \subsetneqq H'$ is possible is shown by the semigroup $H = (\mathbb{N}\backslash\{1\}, +)$.

**1.4. Theorem.** *Let $\varphi \in \mathscr{P}^q_0(H)$ be given. Then there exists a uniquely determined Radon measure $\mu$ on the locally compact space $\check{H} := \hat{S}\backslash\{1_{\{0\}}\}$ such that*

$$\varphi(h') = \int_{\check{H}} \rho(h')\, d\mu(\rho) \qquad \textit{for all} \quad h' \in H'.$$

PROOF. For any $h_1, h_2, k_1, k_2 \in H$ we have

$$\int_{\hat{S}} \rho(s + h_1 + h_2)\, d\mu_{k_1,k_2}(\rho) = \varphi(s + h_1 + h_2 + k_1 + k_2)$$

$$= \int_{\hat{S}} \rho(s + k_1 + k_2)\, d\mu_{h_1,h_2}(\rho)$$

for all $s \in S$, giving us

$$\rho(h_1 + h_2)\mu_{k_1,k_2} = \rho(k_1 + k_2)\mu_{h_1,h_2}.$$

Let $\mathcal{O}_h = \{\rho \in \hat{S} \mid \rho(h) \neq 0\}$; then $\mathcal{O}_h$ is open and $\bigcup_{h\in H} \mathcal{O}_h = \check{H}$. Setting

$$\mu = \frac{1}{\rho^2(h)}\, \mu_h \qquad \text{on } \mathcal{O}_h$$

we obtain by Theorem 2.1.18 a well-defined Radon measure $\mu$ on $\check{H}$. We claim that

$$\rho(h_1 + h_2)\mu = \mu_{h_1,h_2} \qquad \text{on } \check{H}.$$

In fact, on $\mathcal{O}_k$ we have

$$\rho(h_1 + h_2)\mu = \frac{\rho(h_1 + h_2)}{\rho(k + k)}\, \mu_k = \mu_{h_1,h_2}.$$

(This also shows that $\mu_{h_1,h_2}$ is concentrated on $\mathcal{O}_{h_1} \cap \mathcal{O}_{h_2}$.) Therefore, we get for $h' = h_1 + h_2 \in H'$

$$\int_{\check{H}} \rho(h')\, d\mu(\rho) = \mu_{h_1,h_2}(\check{H}) = \mu_{h_1,h_2}(\hat{S}) = \varphi(h'),$$

where we used that by hypothesis $\varphi \in \mathscr{P}_0^q(H)$.

The unicity of $\mu$ is easily seen in the following way: if $\nu$ is another Radon measure on $\check{H}$ representing $\varphi$ then

$$\varphi(s + 2h) = \int_{\hat{S}} \rho(s)\rho^2(h)\, d\mu(\rho) = \int_{\hat{S}} \rho(s)\rho^2(h)\, d\nu(\rho),$$

implying $\mu|\mathcal{O}_h = \nu|\mathcal{O}_h$ for each $h \in H$ and therefore $\mu = \nu$ by 2.1.18. □

**1.5. Example.** Let us consider the open half-line $H = ]0, \infty[$ with usual addition. We have $S = [0, \infty[$, $\hat{S} \cong [0, \infty]$ (cf. 4.4.1) where $\infty$ corresponds to the semicharacter $s \mapsto e^{-\infty \cdot s}$, i.e. to $1_{\{0\}}$. Hence the locally compact space $\check{H}$ can be identified with $[0, \infty[$. In view of Lemma 1.2 and Theorem 1.4 each quasibounded positive definite function $\varphi\colon ]0, \infty[ \to \mathbb{R}$ has the representation

$$\varphi(s) = \int_{[0,\infty[} e^{-\lambda s}\, d\mu(\lambda)$$

for some uniquely determined Radon measure $\mu$ on $\mathbb{R}_+$. This shows that quasibounded positive definite functions on $]0, \infty[$ are exactly the completely monotone functions in the classical sense, cf. Corollary 4.6.14. A well-known example is the function $1/s$, for which the measure $\mu$ in the above representation is just Lebesgue measure on $\mathbb{R}_+$.

**1.6. Example.** Let $H$ be the additive semigroup of natural numbers, i.e. $H = (\mathbb{N}, +)$. Then $S = \mathbb{N}_0$, $\hat{S} = [-1, 1]$ and $\check{H}$ can be identified with $[-1, 0[ \cup ]0, 1]$. The function $n \mapsto 1/n$ is positive definite and even bounded on $\mathbb{N}$. Its integral representation is simply

$$\frac{1}{n} = \int_0^1 t^{n-1}\, dt = \int_{]0,1[} t^n \cdot \frac{dt}{t}.$$

The function $n \mapsto (-1)^n \cdot 1/n$ is positive definite, too, and has the representing measure $|t|^{-1} \cdot 1_{[-1,0[}(t)\, dt$.

A natural question arises at this point: Under what conditions can a given quasibounded positive definite function on $H$ be extended to a bounded positive definite function on $S$? An answer is given by

**1.7. Proposition.** *Let $\varphi$ belong to $\mathscr{P}_0^q(H)$ and let $\mu$ be its desintegrating measure given by Theorem 1.4. Then $\varphi$ can be extended to a function $\Phi \in \mathscr{P}^b(S)$ if and only if $\mu(\check{H}) < \infty$.*

PROOF. If $\mu(\check{H})$ is finite then $\Phi(s) := \int_{\check{H}} \rho(s)\, d\mu(\rho)$ defines a positive definite extension.

On the other hand, suppose that $\Phi(s) = \int_{\hat{S}} \rho(s)\, d\nu(\rho)$ with $\nu \in M_+(\hat{S})$ is an extension of $\varphi$. Then by the unicity statement in Theorem 1.4 we get $\mu = \nu$ on $\check{H}$, in particular, $\mu(\check{H}) < \infty$. □

From this result we see, for example, that the positive definite function $n \mapsto 1/n$ on $\mathbb{N}$, although being bounded, cannot be extended to a positive definite function on $\mathbb{N}_0$.

We introduce now $\mathcal{N}^q(H)$, the set of all negative definite functions $\psi: H' \to \mathbb{R}$ which are quasibounded below. If $\psi \in \mathcal{N}^q(H)$ and $h \in H$ then $s \mapsto \psi(2h + s)$ is negative definite and bounded below, hence $\psi(2h + s) \geqq \psi(2h)$ by Corollary 4.3.2. Some simple properties of $\mathcal{N}^q(H)$ are contained in the following lemma:

**1.8. Lemma.** (i) *$\mathcal{N}^q(H)$ is a closed convex cone in $\mathbb{R}^{H'}$.*
(ii) $-\mathcal{P}^q(H) \subseteq \mathcal{N}^q(H)$.
(iii) *$\psi \in \mathcal{N}^q(H)$ if and only if $\exp(-t\psi) \in \mathcal{P}^q(H)$ for all $t > 0$.*
(iv) *Put $\Delta_a\psi(h') := \psi(a + h') - \psi(h')$ for $h' \in H'$ and $a \in S$. Then if $\psi \in \mathcal{N}^q(H)$ we get $\Delta_a\psi \in \mathcal{P}^q(H)$ for all $a \in S$.*

PROOF. (ii) is a trivial property. (iii) follows from Theorem 3.2.2. (iii) together with Lemma 1.1 implies (i). Concerning (iv) we observe that if $\psi \in -\mathcal{P}^q(H)$ then $\Delta_a\psi \in \mathcal{P}^q(H)$ by Lemma 1.1. Using the closedness of $\mathcal{P}^q(H)$ and the fact that

$$\psi = \lim_{t \to 0} \frac{1}{t}[1 - \exp(-t\psi)],$$

we get $\Delta_a\psi \in \mathcal{P}^q(H)$ for every $\psi \in \mathcal{N}^q(H)$. □

Let us consider for a moment the semigroup $S$ with neutral element. If $\psi \in \mathcal{N}(S)$ is bounded below and

$$\psi(s) = c + \alpha(s) + \int_{\hat{S}\setminus\{1\}} (1 - \rho(s))\, d\mu(\rho)$$

is the uniquely determined representation given by Theorem 4.3.20, where $c \in \mathbb{R}$, $\alpha: S \to \mathbb{R}_+$ is additive and $\mu \in M_+(\hat{S}\setminus\{1\})$, and if $\Delta_a\psi(s) = \psi(s + a) - \psi(s) = \hat{\sigma}_a(s)$, then

$$\alpha(a) = \sigma_a(\{1\}) \tag{1}$$

and

$$(1 - \rho(a))\mu = \sigma_a \qquad \text{on } \hat{S}\setminus\{1\},$$

implying $\mu(\{1_{\{0\}}\}) = \sigma_a(\{1_{\{0\}}\})$ for all $a \in S\setminus\{0\} = H$. Motivated by this result and taking into account the last lemma we define

$$\mathcal{N}_0^q(H) := \{\psi \in \mathcal{N}^q(H) \mid \Delta_a\psi \in \mathcal{P}_0^q(H) \text{ for all } a \in H\}.$$

It is easily seen that $\mathcal{N}^q_0(H)$ is an extreme subcone of $\mathcal{N}^q(H)$, and from Lemma 1.2 it follows immediately that the two cones coincide in case $H$ is without maximal elements.

**1.9. Theorem.** *Let $\psi \in \mathcal{N}^q_0(H)$ be given. Then there exist an additive function $\alpha: S \to \mathbb{R}_+$ and a Radon measure $\mu$ on the locally compact space $\check{H}\setminus\{1\}$ such that for $h', k' \in H'$*

$$\psi(h') - \psi(k') = \int_{\check{H}\setminus\{1\}} (\rho(k') - \rho(h'))\, d\mu(\rho) + \alpha(h') - \alpha(k').$$

*The pair $(\alpha, \mu)$ is uniquely determined and given in the following way:*

$$\alpha(s) = \lim_{n\to\infty} \frac{\psi(2h + ns)}{n} \qquad \text{independent of} \quad h \in H,$$

$$\mu = (1 - \rho(a))^{-1}\sigma_a \qquad \text{on the open set} \quad \mathcal{O}_a = \{\rho \in \check{H} \mid \rho(a) < 1\},$$

*where $\sigma_a$ is the Radon measure on $\check{H}$ satisfying*

$$\Delta_a\psi(h') = \int_{\check{H}} \rho(h')\, d\sigma_a(\rho), \qquad h' \in H'.$$

PROOF. From $\Delta_a\psi \in \mathcal{P}^q(H)$ we get, using Lemma 1.8, that $-\Delta_a\psi \in \mathcal{N}^q(H)$ and hence $\Delta_b(-\Delta_a\psi) \in \mathcal{P}^q(H)$. We have

$$\begin{aligned}
[\Delta_b(-\Delta_a\psi)](h') &= -\Delta_a\psi(h' + b) + \Delta_a\psi(h') \\
&= -\psi(a + b + h') + \psi(h' + b) + \psi(h' + a) - \psi(h') \\
&= \int_{\check{H}} \rho(h')\, d\sigma_a(\rho) - \int_{\check{H}} \rho(h' + b)\, d\sigma_a(\rho) \\
&= \int_{\check{H}} \rho(h')[1 - \rho(b)]\, d\sigma_a(\rho) = \int_{\check{H}} \rho(h')[1 - \rho(a)]\, d\sigma_b(\rho)
\end{aligned}$$

for all $h' \in H'$, and the unicity statement in Theorem 1.4 tells us that

$$[1 - \rho(b)]\sigma_a = [1 - \rho(a)]\sigma_b \qquad \text{on } \check{H}.$$

Taking into account that the open sets $\mathcal{O}_a$, $a \in H$, cover $\check{H}\setminus\{1\}$, and invoking again Theorem 2.1.18 we conclude that

$$\mu := [1 - \rho(a)]^{-1}\sigma_a \qquad \text{on } \mathcal{O}_a$$

is a well-defined Radon measure on $\check{H}\setminus\{1\}$ satisfying $[1 - \rho(a)]\mu = \sigma_a$ on $\check{H}\setminus\{1\}$. Hence

$$\Delta_a\psi(2h) = \int_{\check{H}\setminus\{1\}} \rho^2(h)[1 - \rho(a)]\, d\mu(\rho) + \sigma_a(\{1\})$$

for all $h \in H$ and $a \in S$.

Now we observe that $\Delta_s\psi(2h) = \psi(2h+s) - \psi(2h) =: \psi_h(s)$ is non-negative and negative definite as a function of $s$; hence $\psi_h$ has the representation

$$\psi_h(s) = \alpha_h(s) + \int_{\hat{S}\setminus\{1\}} [1 - \rho(s)]\, d\mu_h(\rho)$$

with an additive function $\alpha_h\colon S \to \mathbb{R}_+$ and $\mu_h \in M_+(\hat{S}\setminus\{1\})$. We get

$$\begin{aligned}\psi(2h+s) - \psi(2h) &= \int_{\check{H}\setminus\{1\}} \rho^2(h)[1 - \rho(s)]\, d\mu(\rho) + \sigma_s(\{1\}) \\ &= \alpha_h(s) + \int_{\hat{S}\setminus\{1\}} [1 - \rho(s)]\, d\mu_h(\rho)\end{aligned}$$

for all $h \in H$, $s \in S$. From relation (1) we learn that $\alpha_h(s) = \tau_{s,h}(\{1\})$ where $\hat{\tau}_{s,h} = \Delta_s\psi_h$. Put $h' = 2h + a$, then

$$\begin{aligned}\Delta_s\psi(h') &= \psi(2h+a+s) - \psi(2h+a) = \Delta_s\psi_h(a) \\ &= \int_{\check{H}} \rho(h')\, d\sigma_s(\rho) = \int_{\check{H}} \rho(a)\rho^2(h)\, d\sigma_s(\rho),\end{aligned}$$

and therefore

$$\tau_{s,h} = \rho^2(h)\sigma_s$$

implying $\alpha_h(s) = \tau_{s,h}(\{1\}) = \sigma_s(\{1\})$, independent of $h$ and additive in $s \in S$. Put $\alpha(s) := \sigma_s(\{1\})$, then indeed

$$\begin{aligned}\psi(h'+s) - \psi(h') = \Delta_s\psi(h') &= \int_{\check{H}} \rho(h')\, d\sigma_s(\rho) \\ &= \int_{\check{H}\setminus\{1\}} \rho(h')\, d\sigma_s(\rho) + \sigma_s(\{1\}) \\ &= \int_{\check{H}\setminus\{1\}} \rho(h')[1 - \rho(s)]\, d\mu(\rho) + \alpha(s)\end{aligned}$$

for all $h' \in H'$ and $s \in S$. Computing in this way $\psi(h'+k') - \psi(h')$ and $\psi(h'+k') - \psi(k')$ for $h', k' \in H'$ gives us, finally,

$$\psi(h') - \psi(k') = \int_{\check{H}\setminus\{1\}} [\rho(k') - \rho(h')]\, d\mu(\rho) + \alpha(h') - \alpha(k')$$

as asserted.

From Theorem 4.3.20 we know that

$$\alpha(s) = \alpha_h(s) = \lim_{n\to\infty} \frac{1}{n}\psi_h(ns) = \lim_{n\to\infty} \frac{1}{n}\psi(2h+ns)$$

independent of $h \in H$. Hence $\alpha$ is uniquely determined and so then is $\mu$, because any representing measure must coincide on $\mathcal{O}_a$ with $[1 - \rho(a)]^{-1}\sigma_a$ by immediate calculation, using the unicity statement in Theorem 1.4. □

The integral representations of 1.4 and 1.9 lead naturally to the question about a converse statement, to be answered now.

**1.10. Proposition.** *Put* $\chi_{h'}(\rho) := \rho(h')$ *for* $h' \in H'$ *and* $\rho \in \check{H}$.

(i) *If* $\mu$ *is a Radon measure on* $\check{H}$ *such that* $\chi_{h'} \in \mathscr{L}^1(\mu)$ *for each* $h' \in H'$, *then* $\varphi: H' \to \mathbb{R}$ *defined by* $\varphi(h') := \int_{\check{H}} \chi_{h'}\, d\mu$ *belongs to* $\mathscr{P}^q_0(H)$.

(ii) *If* $\mu$ *is a Radon measure on* $\check{H}\setminus\{1\}$ *such that* $\chi_{h'} - \chi_{k'} \in \mathscr{L}^1(\mu)$ *for all* $h', k' \in H'$, *and if* $\alpha: S \to \mathbb{R}_+$ *is additive, then* $\psi: H' \to \mathbb{R}$ *defined by*

$$\psi(h') := \int_{\check{H}\setminus\{1\}} (\chi_{k'} - \chi_{h'})\, d\mu + \alpha(h'),$$

$k' \in H'$ *arbitrary but fixed, belongs to* $\mathscr{N}^q_0(H)$.

Proof. (i) Of course $\varphi$ is positive definite and for each $h' \in H'$

$$\sup_{s\in S}|\varphi(h' + s)| \leqq \sup_{s\in S} \int |\rho(h' + s)|\, d\mu(\rho) \leqq \int |\rho(h')|\, d\mu(\rho) < \infty$$

showing $\varphi$ to be quasibounded. The measures $\mu_{h,k}$ occurring in the definition of $\mathscr{P}^q_0(H)$ are given by

$$\mu_{h,k} = \rho(h)\rho(k)\mu$$

so that $\mu_{h,k}(\{1_{\{0\}}\}) = 0$ for all $h, k \in H$ and therefore $\varphi \in \mathscr{P}^q_0(H)$.

(ii) Let $h_1, \ldots, h_n \in H$ and $c_1, \ldots, c_n \in \mathbb{R}$ be given such that $\sum c_i = 0$. Then

$$\begin{aligned}\sum_{i,j=1}^{n} c_i c_j \psi(h_i + h_j) &= \int \sum_{i,j} c_i c_j[\rho(k') - \rho(h_i + h_j)]\, d\mu(\rho)\\ &\quad + \sum_{i,j} c_i c_j[\alpha(h_i) + \alpha(h_j)]\\ &= -\int \left[\sum_{i=1}^{n} c_i \rho(h_i)\right]^2 d\mu(\rho) \leqq 0.\end{aligned}$$

Furthermore, for $h' = h_1 + h_2 \in H'$

$$\begin{aligned}\psi(h' + s) &\geqq \int [\rho(k') - \rho(h' + s)]\, d\mu(\rho)\\ &\geqq \int [\rho(k') - \sqrt{\rho(2h_1)}\cdot\sqrt{\rho(2h_2 + 2s)}]\, d\mu(\rho)\\ &\geqq \int [\rho(k') - \tfrac{1}{2}\rho(2h_1) - \tfrac{1}{2}\rho(2h_2 + 2s)]\, d\mu(\rho)\\ &\geqq \tfrac{1}{2}\int [\rho(k') - \rho(2h_1)]\, d\mu(\rho) + \tfrac{1}{2}\int [\rho(k') - \rho(2h_2)]\, d\mu(\rho)\\ &> -\infty\end{aligned}$$

for all $s \in S$; hence $\psi$ is quasibounded below. Finally, for $a \in S$ and $h' \in H'$

$$\Delta_a \psi(h') = \int_{\check{H}\setminus\{1\}} [1 - \rho(a)]\rho(h')\, d\mu(\rho) + \alpha(a)$$

$$= \int_{\check{H}} \rho(h')\, d\sigma_a(\rho),$$

where $\sigma_a(\rho) = [1 - \rho(a)]\mu$ on $\check{H}\setminus\{1\}$ and $\sigma_a(\{1\}) = \alpha(a)$, implying $\Delta_a \psi \in \mathscr{P}_0^q(H)$ by (i), hence indeed $\psi \in \mathscr{N}_0^q(H)$ as asserted. □

Looking once more at the proof of 1.9, we see that only the property $\{\Delta_a \psi \mid a \in H\} \subseteq \mathscr{P}_0^q(H)$ has been used. Together with the above proposition this means, that any function $\psi: H' \to \mathbb{R}$ such that $\Delta_a \psi \in \mathscr{P}_0^q(H)$ for all $a \in H$ belongs already to $\mathscr{N}_0^q(H)$, a partial converse to Lemma 1.8(iv).

**1.11. Example.** Let again $H$ be the additive semigroup $]0, \infty[$. In 1.5 we saw that $\check{H}$ can be identified with $[0, \infty[$, where the point 0 corresponds to the semicharacter 1, hence $\check{H}\setminus\{1\} \cong ]0, \infty[$. We have $\mathscr{N}_0^q(H) = \mathscr{N}^q(H)$ because $H$ is 2-divisible. By Theorem 1.9 any negative definite function $\psi: ]0, \infty[ \to \mathbb{R}$, quasibounded below, has the representation

$$\psi(s) - \psi(t) = \int_{]0,\infty[} (e^{-\lambda t} - e^{-\lambda s})\, d\mu(\lambda) + \alpha(s - t)$$

for all $s, t \in ]0, \infty[$, where $\alpha \in \mathbb{R}_+$ and $\mu \in M_+(]0, \infty[)$. From this and Proposition 1.10 it follows that the functions in $\mathscr{N}^q(]0, \infty[)$ can be characterized as

$$\psi(s) = c + \alpha s + \int_0^\infty (e^{-\lambda} - e^{-\lambda s})\, d\mu(\lambda), \qquad s > 0, \tag{2}$$

where $c \in \mathbb{R}$, $\alpha \geqq 0$ and $\mu \in M_+(]0, \infty[)$ is such that

$$\int_0^1 \lambda\, d\mu(\lambda) < \infty \qquad \text{and} \qquad \int_1^\infty e^{-s\lambda}\, d\mu(\lambda) < \infty, \qquad \text{for all} \quad s > 0.$$

From (2) we see that $\psi$ is $C^\infty$ and that

$$\psi'(s) = \alpha + \int_0^\infty \lambda e^{-\lambda s}\, d\mu(\lambda)$$

is completely monotone. Conversely, if $\psi: ]0, \infty[ \to \mathbb{R}$ is $C^\infty$ and $\psi'$ is completely monotone, we have by Corollary 4.6.14 that $\psi' = \mathscr{L}\sigma$ for some $\sigma \in M_+([0, \infty[)$ and therefore

$$\psi'(s) = \alpha + \int_0^\infty \lambda e^{-\lambda s}\, d\mu(\lambda),$$

with $\alpha = \sigma(\{0\})$ and $\mu = \lambda^{-1}(\sigma \mid ]0, \infty[)$, hence $\psi$ has the form (2) and belongs to $\mathscr{N}^q(]0, \infty[)$.

An important example is given by the log-function whose integral representation reads

$$\log s - \log t = \int_0^\infty (e^{-\lambda t} - e^{-\lambda s}) \frac{d\lambda}{\lambda}, \qquad s, t > 0,$$

or, taking $t = 1$,

$$\log s = \int_0^\infty (e^{-\lambda} - e^{-\lambda s}) \frac{d\lambda}{\lambda}, \qquad s > 0.$$

**1.12. Example.** For the additive semigroup $\mathbb{N}$ we found in 1.6 that $\check{\mathbb{N}} \cong [-1, 1]\setminus\{0\}$, hence $\check{\mathbb{N}}\setminus\{1\} \cong [-1, 1]\setminus\{0, 1\}$. A measure $\mu$ on $\check{\mathbb{N}}\setminus\{1\}$ for which Proposition 1.10(ii) can be applied, is, for example, given by

$$\mu := \frac{dt}{|t|(1-t)}.$$

The corresponding negative definite function $\psi\colon \mathbb{N}' = \{2, 3, 4, \ldots\} \to \mathbb{R}$ has the values (choosing $k' = 2$)

$$\psi(2) = 0, \qquad \psi(3) = \psi(4) = 1, \qquad \psi(5) = \psi(6) = 1 + \tfrac{1}{2}, \ldots,$$

$$\psi(2n+1) = \psi(2n+2) = 1 + \tfrac{1}{2} + \tfrac{1}{3} + \cdots + \frac{1}{n}.$$

The log-function restricted to $\mathbb{N}'$ is of course also negative definite there. It has the integral representation

$$\log n - \log m = \int_{]0,1[} (t^m - t^n) \frac{dt}{-t \log t}.$$

From the description of $\mathscr{P}^q(]0, \infty[)$ and $\mathscr{N}^q(]0, \infty[)$ given in the Examples 1.5 and 1.11 we have the following result:

**1.13. Theorem.** *If $\varphi \in \mathscr{P}^q(]0, \infty[)$ then $\psi\colon ]0, \infty[ \to \mathbb{R}$ defined by $\psi(s) := \int_1^s \varphi(t)\, dt$ belongs to $\mathscr{N}^q(]0, \infty[)$. Conversely, the derivative of any function in $\mathscr{N}^q(]0, \infty[)$ belongs to $\mathscr{P}^q(]0, \infty[)$.*

**1.14. Exercise.** Let $H$ be $\mathbb{N}^k$ with componentwise addition. Show that $\check{H}$ can be identified with $([-1, 0[ \cup ]0, 1])^k$. Show that both $n = (n_1, \ldots, n_k) \mapsto (n_1 + \cdots + n_k)^{-1}$ and $n \mapsto (-1)^{n_1 + \cdots + n_k}(n_1 + \cdots + n_k)^{-1}$ are positive definite (and quasibounded) and try to find their integral representations.

**1.15. Exercise.** Let $H$ be the real line with maximum operation. Then $\mathscr{P}^q(H)$ is the set of all nonnegative decreasing functions, and $\mathscr{N}^q(H)$ consists of all increasing functions on $\mathbb{R}$. The same is true if we replace the line by an arbitrary nonempty subset of $\mathbb{R}$.

**1.16. Exercise.** If $H$ is the additive semigroup $]0, \infty[^n$, then $\check{H} = \mathbb{R}_+^n$ (by natural identification). Show that all first partial derivatives of a function in $\mathcal{N}^q(]0, \infty[^n)$ belong to $\mathcal{P}^q(]0, \infty[^n)$.

**1.17. Exercise.** Show that $\|x\|^{-1}$ is not positive definite on $]0, \infty[^n$ for $n > 1$, $\|\cdot\|$ denoting the euclidean norm.

**1.18. Exercise.** Let $\psi$ be a strictly positive, negative definite function on the abelian semigroup $H$. Then if $f \in \mathcal{P}^q(]0, \infty[)$ we have $f \circ \psi \in \mathcal{P}^q(H)$, and if $g \in \mathcal{N}^q(]0, \infty[)$, then $g \circ \psi \in \mathcal{N}^q(H)$.

**1.19. Exercise.** Use 1.18 to show that $x \mapsto \log(\sum_{i=1}^n x_i)$ is negative definite on $]0, \infty[^n$ and that $x \mapsto (\sum_{i=1}^n x_i)^{-t}$ is positive definite for all $t > 0$.

**1.20. Exercise.** Let $K$ be a compact Hausdorff space and let $H$ be the set of all strictly positive continuous functions on $K$ with pointwise addition. Show that $\check{H}$ can be identified with $M_+(K)$, the set of Radon measures on $K$.

**1.21. Exercise.** Let $\psi \in \mathcal{N}_0^q(H)$ have the representation given in Theorem 1.9. Then $\psi$ can be extended to a negative definite function on $S$, bounded below, if and only if $\int_{\check{H}\setminus\{1\}} [1 - \rho(h)]\, d\mu(\rho) < \infty$ for all $h \in H$.

**1.22. Exercise.** Let $S$ be a perfect semigroup with zero element 0 and identical involution, and assume that $H := S\setminus\{0\}$ fulfils $H = H + H$. Then $H$ is a *perfect semigroup without zero* in the sense that for every $\varphi \in \mathcal{P}(H)$ there is a unique Radon measure $\mu$ on $S^*\setminus\{1_{\{0\}}\}$ such that

$$\varphi(h) = \int \rho(h) d\mu(\rho), \qquad h \in H.$$

## §2. Completely Monotone and Completely Alternating Functions

Again let $(H, +)$ denote an arbitrary abelian semigroup and put, as in the preceding section, $S := H \cup \{0\}$, $\check{H} := \hat{S}\setminus\{1_{\{0\}}\}$. In the definition of completely monotone (resp. alternating) functions on $H$ no change is necessary compared with the definition for functions on semigroups with neutral element (cf. 4.6.1): we again require

$$\nabla_{h_1} \cdots \nabla_{h_n} \varphi(h) \geqq 0 \qquad (\text{resp. } \leqq 0)$$

for all $n \in \mathbb{N}$, $(h_1, \ldots, h_n) \in H^n$ and all $h \in H$. If one of the $h_j$'s is 0, this expression is still well defined and equal to zero, so that the condition is not changed by allowing $(h_1, \ldots, h_n) \in S^n$. Of course a completely monotone function is also required to be nonnegative.

We denote by $\mathscr{M}(H)$ (resp. $\mathscr{A}(H)$) the set of all completely monotone (resp. alternating) functions on $H$. Note that in contrast with the positive and negative definite functions which in a natural way are only defined on the subsemigroup $H'$ of nonmaximal elements of $H$, the completely monotone (resp. alternating) functions are defined on all of $H$.

For $f: H \to \mathbb{R}$ and $h \in H$ let $f_h: S \to \mathbb{R}$ be defined by $f_h(s) := f(h+s)$, and for fixed $s \in S$ define $\Delta_s f: H \to \mathbb{R}$ by $\Delta_s f(h) := f(h+s) - f(h)$. Let us first state some simple properties of $\mathscr{M}(H)$ and $\mathscr{A}(H)$.

**2.1. Lemma.** (i) *Both $\mathscr{M}(H)$ and $\mathscr{A}(H)$ are closed convex cones in $\mathbb{R}^H$.*

(ii) $-\mathscr{M}(H) \subseteq \mathscr{A}(H)$.

(iii) *A function $\varphi: H \to \mathbb{R}$ belongs to $\mathscr{M}(H)$ (resp. $\mathscr{A}(H)$) if and only if $\varphi_h \in \mathscr{M}(S)$ (resp. $\varphi_h \in \mathscr{A}(S)$) for all $h \in H$.*

(iv) *A function $\psi: H \to \mathbb{R}$ belongs to $\mathscr{A}(H)$ if and only if for every $a \in S$ the function $\Delta_a \psi$ belongs to $\mathscr{M}(H)$.*

(v) *$\varphi_1, \varphi_2 \in \mathscr{M}(H)$ implies $\varphi_1 \cdot \varphi_2 \in \mathscr{M}(H)$.*

(vi) *$\psi \in \mathscr{A}(H)$ if and only if $e^{-t\psi} \in \mathscr{M}(H)$ for all $t > 0$.*

Proof. (i) and (ii) need no proof. (iii) follows from the identity

$$\nabla_{a_1} \cdots \nabla_{a_n} \varphi_h(s) = \nabla_{a_1} \cdots \nabla_{a_n} \varphi(h+s)$$

valid for all $h \in H$, $s \in S$, $n \in \mathbb{N}$, $(a_1, \ldots, a_n) \in S^n$.

We get (iv) from

$$\nabla_{a_1} \cdots \nabla_{a_n}(\Delta_a \psi) = -\nabla_{a_1} \cdots \nabla_{a_n} \nabla_a \psi,$$

which holds for all $a \in S$, $n \in \mathbb{N}$ and $(a_1, \ldots, a_n) \in S^n$.

Using (iii) and the fact that $\mathscr{M}(S)$ is closed under pointwise multiplication we get property (v) observing also that $(\varphi_1 \cdot \varphi_2)_h = (\varphi_1)_h \cdot (\varphi_2)_h$. In the same way (vi) follows from Proposition 4.6.10. □

Let $\varphi$ be a completely monotone function on $H$. Then the above lemma tells us that for each $h \in H$ the translate $\varphi_h$ belongs to $\mathscr{M}(S)$ implying by Theorem 4.6.5 the representation

$$\varphi(h+s) = \int_{\hat{S}_+} \rho(s)\, d\mu_h(\rho)$$

with uniquely determined Radon measures $\mu_h$ on $\hat{S}_+$. Again we have to require that $\mu_h(\{1_{\{0\}}\}) = 0$ for all $h \in H$ in order to get an integral representation on $H$. We put

$$\mathscr{M}_0(H) := \{\varphi \in \mathscr{M}(H) \mid \mu_h(\{1_{\{0\}}\}) = 0 \text{ for all } h \in H\},$$

which is an extreme subcone of $\mathscr{M}(H)$.

**2.2. Lemma.** *If $H = H'$, in particular, if $H$ is 2-divisible or if $H$ already contains a neutral element, then $\mathscr{M}(H) = \mathscr{M}_0(H)$.*

PROOF. Let $\varphi \in \mathscr{M}(H)$ and let $h \in H$ have the form $h = h_1 + h_2$ with $h_1, h_2 \in H$. From

$$\varphi(h + s) = \int_{\hat{S}_+} \rho(s)\, d\mu_h(\rho) = \int_{\hat{S}_+} \rho(s)\rho(h_2)\, d\mu_{h_1}(\rho)$$

for all $s \in S$ we conclude

$$\mu_h = \rho(h_2)\mu_{h_1},$$

whence

$$\mu_h(\{1_{\{0\}}\}) = 1_{\{0\}}(h_2)\mu_{h_1}(\{1_{\{0\}}\}) = 0. \qquad \square$$

**2.3. Theorem.** *For every $\varphi \in \mathscr{M}_0(H)$ there exists a uniquely determined Radon measure $\mu \in M_+(\check{H}_+)$ such that*

$$\varphi(h) = \int_{\check{H}_+} \rho(h)\, d\mu(\rho) \qquad \textit{for all} \quad h \in H.$$

PROOF. It is easy to see that

$$\rho(h)\mu_k = \rho(k)\mu_h$$

for all $h, k \in H$; hence the Radon measure $\mu$ defined by

$$\mu := \frac{1}{\rho(h)}\mu_h \qquad \text{on} \quad \mathcal{O}_h := \{\rho \in \hat{S}_+ \mid \rho(h) > 0\}$$

is well defined on $\check{H}_+$ and satisfies

$$\rho(h)\mu = \mu_h$$

on $\check{H}_+$. Furthermore,

$$\varphi(h + s) = \int_{\check{H}_+} \rho(s)\, d\mu_h(\rho) = \int_{\check{H}_+} \rho(h + s)\, d\mu(\rho)$$

gives for $s = 0$

$$\varphi(h) = \int_{\check{H}_+} \rho(h)\, d\mu(\rho)$$

for all $h \in H$. This proves existence of $\mu$ and the uniqueness follows in the usual way. $\square$

By Lemma 2.1 we know that $\psi \in \mathscr{A}(H)$ if and only if $\Delta_a\psi \in \mathscr{M}(H)$ for each $a \in S$. We therefore introduce.

$$\mathscr{A}_0(H) := \{\psi \in \mathscr{A}(H) \mid \Delta_a\psi \in \mathscr{M}_0(H) \text{ for all } a \in S\}$$

and note that $\mathscr{A}(H) = \mathscr{A}_0(H)$ if $H = H'$ by Lemma 2.2.

**2.4. Theorem.** *Let $\psi \in \mathscr{A}_0(H)$ be given. Then there exist an additive function $\alpha: S \to \mathbb{R}_+$ and a Radon measure $\mu \in M_+(\check{H}_+ \setminus \{1\})$ such that for all $h, k \in H$*

$$\psi(h) - \psi(k) = \int_{\check{H}_+ \setminus \{1\}} [\rho(k) - \rho(h)]\, d\mu(\rho) + \alpha(h) - \alpha(k).$$

*Both $\alpha$ and $\mu$ are uniquely determined.*

We do not give the proof because it is very similar to that of Theorem 1.9. Proposition 1.10 has also its obvious counterpart.

From Chapter 4, §6 we know that for an abelian semigroup $S$ with neutral element we always have $\mathscr{M}(S) \subseteq \mathscr{P}^b(S)$ and $\mathscr{A}(S) \subseteq \mathscr{N}^l(S)$ with equality in case $S$ is 2-divisible. Similar relations are, of course, also to be expected on semigroups without zero. We state them in the following theorem.

**2.5. Theorem.** *The restriction mappings $\mathscr{M}_0(H) \to \mathscr{P}_0^q(H)$ (resp. $\mathscr{A}_0(H) \to \mathscr{N}_0^q(H)$) given by $f \mapsto f \mid H'$ are injective. If $H' = H$ then $\mathscr{M}_0(H)$ is an extreme subcone of $\mathscr{P}_0^q(H)$ and $\mathscr{A}_0(H)$ is extreme in $\mathscr{N}_0^q(H)$. For 2-divisible $H$ we have $\mathscr{M}_0(H) = \mathscr{P}_0^q(H)$ and $\mathscr{A}_0(H) = \mathscr{N}_0^q(H)$.*

PROOF. Let $\varphi \in \mathscr{M}_0(H)$ be given. Then $\varphi$ has an integral representation $\varphi(h) = \int \rho(h)\, d\mu(\rho)$ as stated in Theorem 2.3, from which it is immediately seen that $\varphi | H'$ is positive definite and, in fact, belongs to $\mathscr{P}_0^q(H)$. If $\varphi_1, \varphi_2 \in \mathscr{M}_0(H)$ have the same restrictions to $H'$, then their representing measures coincide on each $\mathscr{O}_{2h} = \{\rho \in \hat{S}_+ \mid \rho(2h) > 0\}$ and are therefore equal on $\check{H}_+$.

If $H = H'$, then $\mathscr{M}_0(H) \subseteq \mathscr{P}_0^q(H)$ and $\mathscr{M}_0(H)$ consists exactly of those functions in $\mathscr{P}_0^q(H)$ whose representing measure is concentrated on $\check{H}_+$, showing $\mathscr{M}_0(H)$ to be extreme in $\mathscr{P}_0^q(H)$. In the 2-divisible case, of course, $\check{H}_+ = \check{H}$ implying $\mathscr{M}_0(H) = \mathscr{P}_0^q(H)$.

The corresponding statements concerning negative definite and completely alternating functions are proved analogously. □

**2.6. Example.** If $H = (\mathbb{N}, +)$ then $\check{H} = [-1, 0[ \cup ]0, 1]$, see 1.6. Hence the positive part $\check{H}_+$ is given by the interval $]0, 1]$. The function $n \mapsto 1/n$ is completely monotone, its representing measure $1_{]0,1]}(t) \cdot dt/t$ being concentrated on $\check{H}_+$. Likewise on $\mathbb{N}^k$ the function

$$n = (n_1, \ldots, n_k) \mapsto (n_1 + \cdots + n_k)^{-1}$$

is completely monotone. Since the character $1 \in \hat{\mathbb{N}}_0$ corresponds to $1 \in [-1, 1]$, we have $\check{\mathbb{N}}_+ \setminus \{1\} \cong ]0, 1[$, and we see that the log-function already treated in Example 1.12 is indeed completely alternating with representing measure $1_{]0,1[}(t) \cdot dt/(-t \log t)$.

**2.7.** Let $(H, +)$ be an idempotent abelian semigroup without neutral element, i.e. $h + h = h$ for all $h \in H$. Then $H = H'$ and defining the ordering $\leqq$ on $H$ by $h \leqq k$ if $h + k = k$ as in 4.4.16 we see that every positive definite function $\varphi: H \to \mathbb{R}$ is nonnegative, decreasing and quasibounded, hence by Lemma 1.2, Lemma 2.2 and Theorem 2.5 that $\mathscr{P}(H) = \mathscr{P}^q(H) = \mathscr{P}^q_0(H) = \mathscr{M}(H) = \mathscr{M}_0(H)$. Similarly, every negative definite function $\psi: H \to \mathbb{R}$ is increasing and quasibounded below and $\mathscr{N}(H) = \mathscr{N}^q(H) = \mathscr{N}^q_0(H) = \mathscr{A}(H) = \mathscr{A}_0(H)$.

If $\rho \in \check{H}$ then $J_\rho := \{h \in H \mid \rho(h) = 1\}$ is a subsemigroup of $H$ which is hereditary on the left and $\rho \mid H = 1_{J_\rho}$, cf. 4.4.16, and it is easily seen that $\rho \mapsto J_\rho$ is a bijection of $\check{H}$ onto the set $\mathscr{S}$ of (nonempty) subsemigroups of $H$ which are hereditary on the left. The bijection becomes a homeomorphism if $\mathscr{S}$ is equipped with the coarsest topology in which the sets

$$\tilde{h} := \{J \in \mathscr{S} \mid h \in J\}, \qquad h \in H$$

are clopen (i.e. open and closed). Each set $\tilde{h}$ is compact in $\mathscr{S}$.

With the above notation we can specialize the Theorems 2.3 and 2.4 to give the following extension of Proposition 4.4.17:

**2.8. Proposition.** *Let $(H, +)$ be an idempotent semigroup. For every $\varphi \in \mathscr{P}(H)$ there is a unique $\mu \in M_+(\mathscr{S})$ such that*

$$\varphi(h) = \mu(\tilde{h}), \qquad h \in H,$$

*and for $\psi \in \mathscr{N}(H)$ there is a unique $\mu \in M_+(\mathscr{S} \setminus \{H\})$ such that*

$$\psi(h) - \psi(k) = \mu(\tilde{k} \setminus \tilde{h}) - \mu(\tilde{h} \setminus \tilde{k}), \qquad h, k \in H.$$

**2.9. Example.** Let $X$ be a locally compact space and let $H = (\mathscr{K}, \cap)$ be the semigroup of compact subsets of $X$, the semigroup operation being intersection $\cap$. The empty set is an absorbing element, and $H$ has a neutral element if and only if $X$ is compact. The semigroup $H$ is idempotent and the ordering of 2.7 is given by $K \leqq L \Leftrightarrow K \supseteq L$. The space $\mathscr{S}$ consists of the nonempty families $J \subseteq \mathscr{K}$ with the following properties

$$K, L \in J \Rightarrow K \cap L \in J,$$

$$K \subseteq L,\ K \in J \Rightarrow L \in J,$$

and the topology on $\mathscr{S}$ is the coarsest in which the sets

$$\tilde{K} = \{J \in \mathscr{S} \mid K \in J\}, \qquad K \in \mathscr{K}$$

are clopen.

In continuation of the results of 4.6.15–4.6.19 we will study the *completely monotone capacities* on $\mathscr{K}$, i.e. the functions $\varphi \in \mathscr{M}(H)$ which are continuous on the right in the sense of 4.6.15. We show below that such functions admit an integral representation over the set of semicharacters on $\mathscr{K}$ which are continuous on the right.

We first show that the set of semicharacters on $\mathscr{K}$ which are continuous on the right can be identified with $\mathscr{K}$.

Let $\mathscr{S}_r$ denote the set of $J \in \mathscr{S}$ for which $1_J$ is continuous on the right.

**2.10. Lemma.** *For* $L \in \mathscr{K}$ *let* $J_L = \{K \in \mathscr{K} \mid L \subseteq K\}$. *Then* $L \mapsto J_L$ *is a bijection of* $\mathscr{K}$ *onto* $\mathscr{S}_r$, *and the inverse mapping is* $J \mapsto \bigcap_{K \in J} K$.

PROOF. It is obvious that $J_L \in \mathscr{S}_r$ and that $\bigcap_{K \in J_L} K = L$. Conversely, if $J \in \mathscr{S}_r$ we consider the set $L \in \mathscr{K}$ defined by

$$L = \bigcap_{K \in J} K$$

and have clearly $J_L \supseteq J$. Assume $L \notin J$. Since $1_J$ is continuous on the right there is an open neighbourhood $G$ of $L$ such that $\mathscr{V}(L, G) \cap J = \emptyset$ where $\mathscr{V}(L, G) = \{K \in \mathscr{K} \mid L \subseteq K \subseteq G\}$. We have

$$\bigcap_{K \in J} K \backslash G = L \backslash G = \emptyset,$$

in particular, for $K_0 \in J$

$$\bigcap_{K \in J} (K_0 \cap K) \backslash G = \emptyset.$$

By the finite intersection property for the compact space $K_0 \backslash G$ we can find $K_1, \ldots, K_n \in J$ such that

$$(K_0 \cap K_1 \cap \cdots \cap K_n) \backslash G = \emptyset,$$

hence $K \subseteq G$ for $K := K_0 \cap K_1 \cap \cdots \cap K_n \in J$, so we get the contradiction $K \in \mathscr{V}(L, G) \cap J$. We therefore have $L \in J$, hence $J_L = J$. □

We now introduce the so-called *myope topology* on $\mathscr{K}$. Let $\mathscr{F}$ (resp. $\mathscr{G}$) denote the set of closed (resp. open) subsets of $X$ and define for a subset $B \subseteq X$

$$\mathscr{K}^B = \{K \in \mathscr{K} \mid K \cap B = \emptyset\},$$

$$\mathscr{K}_B = \{K \in \mathscr{K} \mid K \cap B \neq \emptyset\} = \mathscr{K} \backslash \mathscr{K}^B.$$

The myope topology on $\mathscr{K}$ is the coarsest topology in which the sets $\mathscr{K}^F$, $F \in \mathscr{F}$ and $\mathscr{K}_G$, $G \in \mathscr{G}$ are open. The sets $\mathscr{K}^F \cap \mathscr{K}_{G_1} \cap \cdots \cap \mathscr{K}_{G_n}$ with $F \in \mathscr{F}, G_1, \ldots, G_n \in \mathscr{G}$ form a base for the topology which is easily seen to be Hausdorff. It is well known that $\mathscr{K}$ is locally compact, cf. Matheron (1975), and a proof of this is included in Remark 2.12 below.

**2.11. Proposition.** *The mapping* $b: \mathscr{S} \to \mathscr{K}$ *defined by*

$$b(J) = \bigcap_{K \in J} K, \qquad J \in \mathscr{S}$$

*is continuous and proper and maps* $\mathscr{S}_r$ *bijectively onto* $\mathscr{K}$.

PROOF. By Lemma 2.10 we already know that $b$ is a bijection of $\mathscr{S}_r$ onto $\mathscr{K}$. For the continuity of $b$ it suffices to prove that $b^{-1}(\mathscr{K}^F)$ and $b^{-1}(\mathscr{K}_G)$ are open in $\mathscr{S}$ for $F \in \mathscr{F}$ and $G \in \mathscr{G}$. We first prove the following biimplication for $J \in \mathscr{S}$ and $G \in \mathscr{G}$:

$$b(J) \subseteq G \Leftrightarrow \exists L \in J : L \subseteq G. \tag{1}$$

Here "$\Leftarrow$" is clear, and if $b(J) \subseteq G$ then

$$\bigcap_{K \in J} K \backslash G = \emptyset,$$

and as in the proof of Lemma 2.10 we see that there is $L \in J$ such that $L \backslash G = \emptyset$. By (1) it follows that

$$b^{-1}(\mathscr{K}^F) = \{J \in \mathscr{S} \mid b(J) \subseteq F^c\} = \bigcup_{\substack{L \in \mathscr{K} \\ L \subseteq F^c}} \tilde{L}$$

which shows that $b^{-1}(\mathscr{K}^F)$ is open in $\mathscr{S}$ when $F \in \mathscr{F}$. For $G \in \mathscr{G}$ and $J_0 \in b^{-1}(\mathscr{K}_G)$ we choose open relatively compact sets $\omega_1, \omega_2 \subseteq X$ such that $b(J_0) \subseteq \omega_1 \subseteq \bar{\omega}_1 \subseteq \omega_2$ and a point $x_0 \in b(J_0) \cap G$. We next choose an open neighbourhood $V$ of $x_0$ such that $\bar{V} \subseteq G$ and define

$$\mathscr{O} = \{J \in \mathscr{S} \mid \bar{\omega}_1 \in J, \bar{\omega}_2 \backslash V \notin J\},$$

which is an open subset of $\mathscr{S}$. We have $J_0 \in \mathscr{O}$ because $b(J_0) \subseteq \omega_1$ implies by (1) that $\bar{\omega}_1 \in J_0$. Furthermore, $\bar{\omega}_2 \backslash V \in J_0$ is impossible since $x_0 \in V \cap b(J_0)$. Finally, we claim that $\mathscr{O} \subseteq b^{-1}(\mathscr{K}_G)$ which shows that $b^{-1}(\mathscr{K}_G)$ is open in $\mathscr{S}$. In fact, for $J \in \mathscr{O}$, we have $b(J) \subseteq \bar{\omega}_1 \subseteq \omega_2$ and $b(J) \nsubseteq \omega_2 \backslash \bar{V}$, hence $b(J) \cap \bar{V} \neq \emptyset$ and *a fortiori* $J \in b^{-1}(\mathscr{K}_G)$. To see that $b(J) \nsubseteq \omega_2 \backslash \bar{V}$ we remark that the assumption $b(J) \subseteq \omega_2 \backslash \bar{V}$ leads by (1) to $\bar{\omega}_2 \backslash V \in J$.

We show finally that $b$ is proper, i.e. that $b^{-1}(\mathscr{V})$ is compact for each compact set $\mathscr{V} \subseteq \mathscr{K}$. Let $\mathscr{G}_0$ denote the family of all open relatively compact subsets of $X$. Then $(\mathscr{K}^{G^c})_{G \in \mathscr{G}_0}$ is an open covering of $\mathscr{K}$ so if $\mathscr{V} \subseteq \mathscr{K}$ is compact there are $G_1, \ldots, G_n \in \mathscr{G}_0$ such that $\mathscr{V}$ is covered by the sets $\mathscr{K}^{G_i^c}$, $i = 1, \ldots, n$. It is therefore sufficient to show that $b^{-1}(\mathscr{K}^{G^c})$ is relatively compact for $G \in \mathscr{G}_0$, but this is clear since

$$b^{-1}(\mathscr{K}^{G^c}) = \{J \in \mathscr{S} \mid b(J) \subseteq G\} \subseteq \tilde{\bar{G}},$$

and $\tilde{K}$ is compact for $K \in \mathscr{K}$, cf. 2.7. □

**2.12. Remark.** The above proof shows that a compact subset $\mathscr{V}$ of $\mathscr{K}$ is contained in a set of the form

$$\mathscr{K}^{K^c} = \{L \in \mathscr{K} \mid L \subseteq K\}$$

for a suitable $K \in \mathscr{K}$. That $\mathscr{K}^{K^c}$ is compact in $\mathscr{K}$ follows from Proposition 2.11 since

$$b(\tilde{K}) = \mathscr{K}^{K^c} \qquad \text{for} \quad K \in \mathscr{K}.$$

This shows in particular that $\mathscr{K}$ is locally compact. In fact, if $K_0 \in \mathscr{K}^F$ where $F \in \mathscr{F}$, then $\mathscr{K}^{K^c}$ is a compact neighbourhood of $K_0$ contained in $\mathscr{K}^F$ provided $K$ is a compact neighbourhood of $K_0$ disjoint with $F$. And if $K_0 \in \mathscr{K}_G$ where $G \in \mathscr{G}$, then $\mathscr{K}^{K^c} \cap \mathscr{K}_H$ is a compact neighbourhood of $K_0$ contained in $\mathscr{K}_G$ provided $K$ is a compact neighbourhood of $K_0$ and $H$ is a closed subset of $G$ such that $\mathring{H} \cap K_0 \neq \emptyset$.

**2.13. Theorem.** *For every $\varphi \in \mathscr{M}(\mathscr{K}, \cap)$ which is continuous on the right there exists a unique $\mu \in M_+(\mathscr{K})$ such that*

$$\varphi(K) = \mu(\mathscr{K}^{K^c}) \qquad \text{for} \quad K \in \mathscr{K}, \tag{2}$$

*and for every $\mu \in M_+(\mathscr{K})$ this formula defines a function $\varphi \in \mathscr{M}(\mathscr{K}, \cap)$ which is continuous on the right.*

PROOF. Suppose first that $\varphi \in \mathscr{M}(\mathscr{K}, \cap)$ is continuous on the right. By Proposition 2.8 there exists $\nu \in M_+(\mathscr{S})$ such that $\varphi(K) = \nu(\tilde{K})$ for $K \in \mathscr{K}$. As in the proof of Theorem 4.6.18 we consider the image measure $\mu = \nu^b$, which is a Radon measure on $\mathscr{K}$ because $b$ is continuous and proper. For fixed $K \in \mathscr{K}$ we define a decreasing net $(\tilde{L})_{L \in A}$ where $A = \{L \in \mathscr{K} \mid K \subseteq \mathring{L}\}$ and the ordering of $A$ is defined by $L_1 \leqq L_2 \Leftrightarrow L_1 \supseteq L_2$, and we have

$$\bigcap_{L \in A} \tilde{L} = \{J \in \mathscr{S} \mid b(J) \subseteq K\}$$

as an easy consequence of (1). Using that $\varphi$ is decreasing with respect to $\leqq$ and continuous on the right we find by Lemma 2.1.3 that

$$\varphi(K) = \inf\{\varphi(L) \mid L \in A\} = \inf\{\nu(\tilde{L}) \mid L \in A\}$$

$$= \nu\left(\bigcap_{L \in A} \tilde{L}\right) = \nu(\{J \in \mathscr{S} \mid b(J) \subseteq K\}) = \mu(\mathscr{K}^{K^c}).$$

Conversely, if $\mu \in M_+(\mathscr{K})$ then (2) defines a nonnegative real-valued function since $\mathscr{K}^{K^c}$ is compact in $\mathscr{K}$. By induction it is easy to establish that

$$\nabla_{K_1} \nabla_{K_2} \cdots \nabla_{K_n} \varphi(K) = \mu(\mathscr{K}^{K^c} \cap \mathscr{K}_{K_1^c} \cap \cdots \cap \mathscr{K}_{K_n^c}),$$

which shows that $\varphi$ is completely monotone on $(\mathscr{K}, \cap)$. For fixed $K \in \mathscr{K}$ and $A$ as above $(\mathscr{K}^{L^c})_{L \in A}$ is a decreasing net of compact subsets of $\mathscr{K}$ with intersection $\mathscr{K}^{K^c}$. By Lemma 2.1.3 follows that

$$\varphi(K) = \inf\{\varphi(L) \mid L \in A\},$$

which shows that $\varphi$ is continuous on the right. The proof of uniqueness of the representing measure is similar to the proof in Theorem 4.6.18 and is left as an exercise to the reader. □

**2.14. Exercise.** Give an example of a function $\varphi \in \mathscr{P}_0^q(\mathbb{N})$ whose representing measure is concentrated on $\check{\mathbb{N}}_+$, but such that $\varphi$ cannot be extended to a function in $\mathscr{M}_0(\mathbb{N})$.

**2.15. Exercise.** Let $\varphi: H \to \mathbb{R}$ be a function such that for all $h \in H$ the translate $s \mapsto \varphi(h + s)$ is positive definite and bounded on $S$ (resp. negative definite and bounded below on $S$). Then $\varphi$ is completely monotone (resp. alternating) on $H$. Show that the converse statement holds also.

**2.16. Exercise.** Let $\psi \in \mathscr{A}_0(H)$ be strictly positive. Then $f \circ \psi \in \mathscr{M}_0(H)$ for $f \in \mathscr{P}^q(]0, \infty[)$ and $g \circ \psi \in \mathscr{A}_0(H)$ for $g \in \mathscr{N}^q(]0, \infty[)$.

**2.17. Exercise.** Let $X$ be a locally compact space and $(\mathcal{K}, \cap)$ the semigroup of compact subsets of $X$. Then the mapping $(a, \mu) \mapsto \psi$, where

$$\psi(K) = a - \mu(\tilde{K} \setminus \{\mathcal{K}\}) \quad \text{for } K \in \mathcal{K},$$

is an affine bijection of $\mathbb{R} \times M_+(\mathcal{S} \setminus \{\mathcal{K}\})$ onto $\mathcal{A}(\mathcal{K}, \cap)$. Furthermore $(a, \mu) \mapsto \psi$, where

$$\psi(K) = a - \mu(\mathcal{K}^{K^c} \setminus \{\varnothing\}) \quad \text{for } K \in \mathcal{K},$$

is an affine bijection of $\mathbb{R} \times M_+(\mathcal{K} \setminus \{\varnothing\})$ onto the set of $\psi \in \mathcal{A}(\mathcal{K}, \cap)$ which are continuous on the right.

## Notes and Remarks

Many of the results in this chapter were published in Ressel (1979). Choquet (1954, §47) introduced completely monotone and alternating functions on arbitrary abelian semigroups, but his main interest was directed towards the special case where $S = (\mathcal{K}(X), \cap)$ is the semigroup of all compact subsets of some given locally compact space $X$ with respect to intersection. Theorem 2.13 is due to Choquet (1954, §50) and is also formulated in Talagrand (1976). The results of 2.9–2.13 are similar to the results of 4.6.15–4.6.19. Our method of proof seems to be new.

Some of the results of §1 (for example Theorem 1.4) have obvious extensions to semigroups with involutions other than the identical one which we assumed throughout this chapter in order to make the exposition more readable.

It is also possible to weaken the condition of quasi-boundedness and still have unique integral representation; see Stochel (1983) for some results in this direction.

# References


AKHIEZER, N. I. (1965). *The Classical Moment Problem.* Edinburgh: Oliver and Boyd.

ALFSEN, E. M. (1971). *Compact Convex Sets and Boundary Integrals.* Berlin–Heidelberg–New York: Springer-Verlag.

ANDENAES, P. R. (1970). Hahn–Banach extensions which are maximal on a given cone. *Math. Ann.* **188**, 90–96.

ARONSZAJN, N. (1944). La théorie générale de noyaux réproduisants et ses applications. *Proc. Cambridge Philos. Soc.* **39**, 133–153.

ARONSZAJN, N. (1950). Theory of reproducing kernels. *Trans. Amer. Math. Soc.* **68**, 337–404.

ARVESON, W. B. (1969). Subalgebras of C*-algebras. *Acta Math.* **123**, 141–224.

ATZMON, A. (1975). A moment problem for positive measures on the unit disc. *Pacific J. Math.* **59**, 317–325.

BANACH, S. (1932). *Théorie des opérations linéaires.* Monografie Matematyczne 1. Warszawa.

BATT, J. (1973). Die Verallgemeinerungen des Darstellungssatzes von F. Riesz und ihre Anwendungen. *Jahresber. Deutsch. Math.-Verein.* **74**, 147–181.

BAUER, H. (1978). *Wahrscheinlichkeitstheorie und Grundzüge der Masstheorie*, 3. Auflage. Berlin–New York: de Gruyter.

BERG, C. (1980). Quelques remarques sur le cône de Stieltjes. In *Séminaire de Théorie du Potentiel Paris,* No. 5. Lecture Notes in Mathematics 814, pp. 70–79. Berlin–Heidelberg–New York: Springer-Verlag.

BERG, C. (1984a). Semi-groupes de moments. *Math. Scand.* **54**.

BERG, C. (1984b). Fonctions définies négatives et majoration de Schur. In *Colloque de Théorie du Potentiel Jacques Deny—Orsay 1983.* Lecture Notes in Mathematics. Berlin–Heidelberg–New York: Springer-Verlag.

BERG, C. and J. P. R. CHRISTENSEN (1981). Density questions in the classical theory of moments. *Ann. Inst. Fourier (Grenoble)* **31** (3), 99–114.

BERG, C. and J. P. R. CHRISTENSEN (1983a). Exposants critiques dans le problème des moments. *C. R. Acad. Sci. Paris,* Série I, **296**, 661–663.

BERG, C. and J. P. R. CHRISTENSEN (1983b). Suites complètement définies positives, majoration de Schur et le problème des moments de Stieltjes en dimension k. *C. R. Acad. Sci. Paris,* Série I, **297**, 45–48.

BERG, C. and G. FORST (1975). *Potential Theory on Locally Compact Abelian Groups*. Berlin–Heidelberg–New York: Springer-Verlag.

BERG, C. and G. FORST (1979). Infinitely divisible probability measures and potential kernels. In *Probability Measures on Groups*. Lecture Notes in Mathematics 706, pp. 22–35. Berlin–Heidelberg–New York: Springer-Verlag.

BERG, C. and P. H. MASERICK (1982). Polynomially positive definite sequences. *Math. Ann.* **259**, 487–495.

BERG, C. and P. H. MASERICK (1984). Exponentially bounded positive definite functions. *Illinois J. Math.* **28**, 162–179.

BERG, C. and P. RESSEL (1978). Une forme abstraite du théorème de Schoenberg. *Arch. Math. (Basel)* **30**, 55–61.

BERG, C., J. P. R. CHRISTENSEN and C. U. JENSEN (1979). A remark on the multidimensional moment problem. *Math. Ann.* **243**, 163–169.

BERG, C., J. P. R. CHRISTENSEN and P. RESSEL (1976). Positive definite functions on abelian semigroups. *Math. Ann.* **223**, 253–274.

BICKEL, P. J. and W. R. VAN ZWET (1980). On a theorem of Hoeffding. In *Asymptotic Theory of Statistical Tests and Estimation* (Ed. by I. M. Chakravarti), pp. 307–324. New York: Academic Press.

BILLINGSLEY, P. (1968). *Convergence of Probability Measures*. New York: Wiley.

BILLINGSLEY, P. (1979). *Probability and Measure*. New York: Wiley.

BOCHNER, S. (1955). *Harmonic Analysis and the Theory of Probability*. Berkeley and Los Angeles: University of California Press.

BOURBAKI, N. (1965–1969). *Éléments de Mathématique*. Livre VI, Intégration, 2ᵉ éd., Ch. 1–9. Paris: Hermann.

BRELOT, M. (1971). *On Topologies and Boundaries in Potential Theory*. Lecture Notes in Mathematics 175. Berlin–Heidelberg–New York: Springer-Verlag.

BRETAGNOLLE, J., D. DACUNHA-CASTELLE and J. L. KRIVINE (1966). Lois stables et espaces $L^p$. *Ann. Inst. H. Poincaré*, Sect. B, II, **3**, 231–259.

BUCHWALTER, H. (1984). Semi-groupes de moments sur un convexe compact de $\mathbb{R}^p$. *Math. Scand*, **54**.

BUCHWALTER, H. and G. CASSIER (1983a). Semi-groupes dans le problème des moments. *C. R. Acad. Sci. Paris*, Série I, **296**, 389–391.

BUCHWALTER, H. and G. CASSIER (1983b). Semi-groupes dans le problème des moments. *J. Funct. Anal.* **52**, 129–145.

CASSIER, G. (1983). Le problème des moments pour un convexe compact de $\mathbb{R}^n$. *C. R. Acad. Sci. Paris*, Série I, **296**, 195–197.

CHOQUET, G. (1954). Theory of capacities. *Ann. Inst. Fourier (Grenoble)* **5**, 131–295.

CHOQUET, G. (1962). Le problème des moments. *Séminaire d'initiation à l'analyse*, 1ère année. Inst. Henri Poincaré, Paris.

CHOQUET, G. (1969). *Lectures on Analysis*, Vol. 1. Reading, Massachusetts: Benjamin.

CHRISTENSEN, J. P. R. and P. RESSEL (1978). Functions operating on positive definite matrices and a theorem of Schoenberg. *Trans. Amer. Math. Soc.* **243**, 89–95.

CHRISTENSEN, J. P. R. and P. RESSEL (1981). A probabilistic characterisation of negative definite and completely alternating functions. *Z. Wahrsch. verw. Gebiete* **57**, 407–417.

CHRISTENSEN, J. P. R. and P. RESSEL (1982). Positive definite kernels on the complex Hilbert sphere. *Math. Z.* **180**, 193–201.

CHRISTENSEN, J. P. R. and P. RESSEL (1983). Norm-dependent positive definite functions on B-spaces. In *Probability in Banach Spaces* IV. Lecture Notes in Mathematics 990, pp. 47–53. Berlin–Heidelberg–New York: Springer-Verlag.

CLIFFORD, A. H. and G. B. PRESTON (1961). *The Algebraic Theory of Semigroups*, Vols. I & II. Mathematical Surveys 7. Providence: American Mathematical Society.

COURRÈGE, P. (1964). Générateur infinitésimal d'un semi-groupe de convolution sur $\mathbb{R}^n$, et formule de Lévy–Khinchine. *Bull. Sci. Math.* **88** (2), 3–30.

DAVIS, P. J. (1963). *Interpolation and Approximation.* New York–Toronto–London: Blaisdell.

DELLACHERIE, C. and P.-A. MEYER (1978). *Probabilities and Potential.* Amsterdam–New York–Oxford: North-Holland.

DEVINATZ, A. (1955). The representation of functions as Laplace–Stieltjes integrals. *Duke Math. J.* **22**, 185–192.

DEVINATZ, A. (1957). Two parameter moment problems. *Duke Math. J.* **24**, 481–498.

DIEUDONNÉ, J. (1981). *History of Functional Analysis.* Amsterdam–New York–Oxford: North-Holland.

DONOGHUE, Jr., W. F. (1969). *Distributions and Fourier Transforms.* New York–London: Academic Press.

DONOGHUE, Jr., W. F. (1974). *Monotone Matrix Functions and Analytic Continuation.* Berlin–Heidelberg–New York: Springer-Verlag.

DOUGLAS, R. G. (1964). On extremal measures and subspace density. *Michigan Math. J.* **11**, 243–246.

DUNKL, C. F. and D. E. RAMIREZ (1975). *Representations of Commutative Semitopological Semigroups.* Lecture Notes in Mathematics 435. Berlin–Heidelberg–New York: Springer-Verlag.

DVORETZKY, A. (1961). Some results on convex bodies and Banach spaces. *Proceedings of the International Symposium on Linear Spaces*, Jerusalem, pp. 123–160.

EINHORN, S. J. (1969). Functions positive definite on C[0, 1]. *Proc. Amer. Math. Soc.* **22**, 702–703.

ÈSKIN, G. I. (1960). A sufficient condition for the solvability of the moment problem in several dimensions. *Soviet Math. Dokl.* **1**, 895–898. (Transl. from *Dokl. Akad. Nauk SSSR* **133**, 540–543.)

EVANS, D. E. and J. T. LEWIS (1977). *Dilations of Irreversible Evolutions in Algebraic Quantum Theory.* Communications of the Dublin Institute for Advanced Studies, Series A, No. 24, Dublin.

FARAUT, J. and K. HARZALLAH (1974). Distances hilbertiennes invariantes sur un espace homogène. *Ann. Inst. Fourier* (*Grenoble*) **24** (3), 171–217.

FIGIEL, T. (1976). A short proof of Dvoretzky's theorem on almost spherical sections of convex bodies. *Compositio Math.* **33**, 297–301.

FIL'ŠTINSKIĬ, V. A. (1964). The power moment problem on the entire axis with a finite number of empty intervals in the spectrum. *Zap. Meh.-Mat. Fak. i Ḥarkov. Mat. Obšč.* **30** (4), 186–200 (in Russian).

FINE, N. J. and P. H. MASERICK (1970). On the simplex of completely monotonic functions on a commutative semigroup. *Canad. J. Math.* **22**, 317–326.

FITZGERALD, C. H. and R. A. HORN (1977). On fractional Hadamard powers of positive definite matrices. *J. Math. Anal. Appl.* **61**, 633–642.

FORST, G. (1976). The Lévy–Hinčin representation of negative definite functions. *Z. Wahrsch. verw. Gebiete* **34**, 313–318.

FRIEDRICH, J. (1984). A note on the two-dimensional moment problem. *Math. Nachr.* To appear.

FUGLEDE, B. (1983). The multidimensional moment problem. *Expo. Math.* **1**, 47–65.

GELFAND, I. M. and N. Ya. VILENKIN (1964). *Generalized Functions*, Vol. 4. New York–London: Academic Press.

GRAHAM, C. C. (1977). Functional calculus and positive definite functions. *Trans. Amer. Math. Soc.* **231**, 215–231.

GRAHAM, C. C. and O. C. MCGEHEE (1979). *Essays in Commutative Harmonic Analysis.* Berlin–Heidelberg–New York: Springer-Verlag.

HALL, P. (1981). A comedy of errors: the canonical form for a stable characteristic function. *Bull. London. Math. Soc.* **13**, 23–27.

HARN VAN, K. and F. W. STEUTEL (1980). Opgave 92. *Statist. Neerlandica* **34** (4), 224–225.

HARZALLAH, K. (1969). Fonctions opérant sur les fonctions définies-négatives. *Ann. Inst. Fourier (Grenoble)* **19** (2), 527–532.

HAVILAND, E. K. (1935). On the momentum problem for distributions in more than one dimension. *Amer. J. Math.* **57**, 562–568.

HAVILAND, E. K. (1936). On the momentum problem for distribution functions in more than one dimension, II. *Amer. J. Math.* **58**, 164–168.

HELMS, L. L. (1969). *Introduction to Potential Theory*. New York: Wiley.

HERZ, C. S. (1963a). Fonctions opérant sur les fonctions définies-positives. *Ann. Inst. Fourier (Grenoble)* **13**, 161–180.

HERZ, C. S. (1963b). Une ébauche d'une théorie générale des fonctions définies-négatives. *Séminaire Brelot-Choquet-Deny (Théorie du Potentiel)*, 7e année. Inst. Henri Poincaré, Paris.

HEWITT, E. and H. S. ZUCKERMANN (1956). The $l_1$-algebra of a commutative semigroup. *Trans. Amer. Math. Soc.* **83**, 70–97.

HEYDE, C. C. (1963). On a property of the lognormal distribution. *J. Roy. Statist. Soc.*, Ser. B, **25**, 392–393.

HEYER, H. (1977). *Probability Measures on Locally Compact Groups*. Berlin–Heidelberg–New York: Springer-Verlag.

HILBERT, D. (1888). Über die Darstellung definiter Formen als Summe von Formenquadraten. *Math. Ann.* **32**, 342–350. (In *Ges. Abh.*, Vol. 2, pp. 154–161. Berlin: Springer-Verlag, 1933).

HILBERT, D. (1904). Grundzüge einer allgemeinen Theorie der linearen Integralgleichungen. *Nachr. Göttinger Akad. Wiss. Math. Phys. Klasse*, 49–91.

HILDEBRANDT, T. H. and I. J. SCHOENBERG (1933). On linear functional operators and the moment problem for a finite interval in one or several dimensions. *Ann. of Math.* **34**, 317–328.

HILLE, E. (1972). Introduction to general theory of reproducing kernels. *Rocky Mountain J. Math.* **2**, 321–368.

HIRSCH, F. (1972). Familles résolvantes, générateurs, cogénérateurs, potentiels. *Ann. Inst. Fourier (Grenoble)* **22** (1), 89–210.

HOEFFDING, W. (1956). On the distribution of the number of successes in independent trials. *Ann. Math. Stat.* **27**, 713–721.

HOFFMANN-JØRGENSEN, J. and P. RESSEL (1977). On completely monotone functions on $C_+(X)$. *Math. Scand.* **40**, 79–93.

HOFMANN, K. H. (1976). Topological semigroups. History, theory, applications. *Jahresber. Deutsch. Math.-Verein.* **78**, 9–59.

HOFMANN, K. H. and P. S. MOSTERT (1966). *Elements of Compact Semigroups*. Columbus: Merrill.

HORN, R. A. (1969a). The theory of infinitely divisible matrices and kernels. *Trans. Amer. Math. Soc.* **136**, 269–286.

HORN, R. A. (1969b). Infinitely divisible positive definite sequences. *Trans. Amer. Math. Soc.* **136**, 287–303.

JOHANSEN, S. (1967). *Anvendelser af extremalpunktsmetoder i sandsynlighedsregningen*. Institut for Matematisk Statistik, København Universitet.

JONES, W. B., W. J. THRON and H. WAADELAND (1980). A strong Stieltjes moment problem. *Trans. Amer. Math. Soc.* **261**, 503–528.

JONES, W. B., O. NJÅSTAD and W. J. THRON (1984). Orthogonal Laurent polynomials and the strong Hamburger moment problem. *J. Math. Anal. Appl.* **98**, 528–554.

KAHANE, J.-P. (1979). Sur les fonctions de type positif et de type négatif. Seminar on Harmonic Analysis 1978–79, pp. 21–37. *Publ. Math. Orsay* **79**, No. 7, Univ. Paris XI, Orsay.

KAHANE, J.-P. (1981). Hélices et quasi-hélices. In *Mathematical Analysis and Applications* (Ed. by L. Nachbin). Essays dedicated to Laurent Schwartz on the occasion of this 65th birthday. Part B. Adv. Math. Suppl. Studies 7B. New York–London: Academic Press.

KARLIN, S. and A. NOVIKOFF (1963). Generalized convex inequalities. *Pacific J. Math.* **13**, 1251–1279.

KINGMAN, J. F. C. (1972). *Regenerative Phenomena.* London–New York: Wiley.

KISYŃSKI, J. (1968). On the generation of tight measures. *Studia Math.* **30**, 141–151.

KOLMOGOROV, A. N. (1941). Stationary sequences in Hilbert space. *Bull. Math. Univ. Moscow* **2**, 1–40.

KONHEIM, A. G. and B. WEISS (1965). Functions which operate on characteristic functions. *Pacific J. Math.* **15**, 1279–1293.

KUELBS, J. (1973). Positive definite symmetric functions on linear spaces. *J. Math. Anal. Appl.* **42**, 413–426.

LINDAHL, R. J. and P. H. MASERICK (1971). Positive-definite functions on involution semigroups. *Duke Math. J.* **38**, 771–782.

LOÈVE, M. (1963). *Probability Theory*, 3rd ed. Princeton: van Nostrand.

LUKACS, E. (1960). *Characteristic Functions.* London: Griffin.

MARCZEWSKI, E. and C. RYLL-NARDZEWSKI (1953). Remarks on the compactness and non-direct products of measures. *Fund. Math.* **40**, 165–170.

MARSHALL, A. W. and I. OLKIN (1979). *Inequalities: Theory of Majorization and its Applications.* New York–London: Academic Press.

MASERICK, P. H. (1975). BV-functions, positive definite functions and moment problems. *Trans. Amer. Math. Soc.* **214**, 137–152.

MASERICK, P. H. (1977). Moments of measures on convex bodies. *Pacific J. Math.* **68**, 135–152.

MASERICK, P. H. (1978). A Lévy–Khinchin formula for semigroups with involutions. *Math. Ann.* **236**, 209–216.

MASERICK, P. H. (1981). Disintegration with respect to $L_p$-density functions and singular measures. *Z. Wahrsch. verw. Gebiete* **57**, 311–326.

MASERICK, P. H. and F. H. SZAFRANIEC (1984). Equivalent definitions of positive-definiteness. *Pacific J. Math.* **110**, 315–324.

MATHERON, G. (1975). *Random Sets and Integral Geometry.* New York: Wiley.

MAZUR, S. and S. ULAM (1932). Sur les transformations isométriques d'espaces vectoriels normés. *C. R. Acad. Sci. Paris* **194**, 946–948.

McGREGOR, J. L. (1980). Solvability criteria for certain n-dimensional moment problems. *J. Approx. Theory* **30**, 315–333.

MERCER, J. (1909). Functions of positive and negative type, and their connection with the theory of integral equations. *Philos. Trans. Royal Soc. London*, Ser. A, **209**, 415–446.

MICCHELLI, C. A. (1978). A characterization of M. W. Wilson's criterion for non-negative expansions of orthogonal polynomials. *Proc. Amer. Math. Soc.* **71**, 69–72.

MICCHELLI, C. A. and R. A. WILLOUGHBY (1979). On functions which preserve the class of Stieltjes matrices. *Linear Algebra Appl.* **23**, 141–156.

MLAK, W. (1978). Dilations of Hilbert space operators (general theory). *Dissertationes Math.* **153**, 1–61.

MORANDO, P. (1969). Mesures aléatoires. In *Séminaire de Probabilité* III. Lecture Notes in Mathematics 88, pp. 190–229. Berlin–Heidelberg–New York: Springer-Verlag.

MOTZKIN, T. S. (1967). The arithmetic-geometric inequality. In *Inequalities* (Ed. by O. Shisha). New York–London: Academic Press.

NEUMANN, M. (1983). A maximality theorem concerning extreme points. *J. Math. Anal. Appl.* **96**, 148–152.

NUSSBAUM, A. E. (1966). Quasi-analytic vectors. *Ark. Mat.* **6**, 179–191.

OXTOBY, J. C. (1971). *Measure and Category.* New York–Heidelberg–Berlin: Springer-Verlag.

PAALMAN-DE MIRANDA, A. B. (1964). *Topological Semigroups.* Mathematical Centre Tracts 11. Amsterdam: Mathematisch Centrum.

PACHL, J. (1974). Free uniform measures. *Comment. Math. Univ. Carolin.* **15**, 541–553.

PARTHASARATHY, K. R. (1967). *Probability Measures on Metric Spaces.* New York–London: Academic Press.

PARTHASARATHY, K. R. and K. SCHMIDT (1972). *Positive Definite Kernels, Continuous Tensor Products, and Central Limit Theorems of Probability Theory.* Lecture Notes in Mathematics 272. Berlin–Heidelberg–New York: Springer-Verlag.

PARTHASARATHY, K. R., R. R. RAO and S. R. S. VARADHAN (1963). Probability distributions on locally compact abelian groups. *Illinois J. Math.* **7**, 337–369.

PARZEN, E. (1971). *Time Series Analysis Papers.* San Francisco: Holden Day.

PETERSEN, L. C. (1982). On the relation between the multidimensional moment problem and the one-dimensional moment problem. *Math. Scand.* **51**, 361–366.

PHELPS, R. R. (1966). *Lectures on Choquet's Theorem.* Princeton: van Nostrand.

POLLARD, D. and F. TOPSØE (1975). A unified approach to Riesz type representation theorems. *Studia Math.* **54**, 173–190.

REDÉI, L. (1965). *The Theory of Finitely Generated Commutative Semigroups.* New York: Pergamon Press.

REEDS, J. (1979). Widder theorem characterization of Laplace transforms of positive measures on $\mathbb{R}^k$. Report from the Department of Statistics, University of California, Berkeley.

RESSEL, P. (1974). Laplace-Transformation nichtnegativer und vektorwertiger Masse. *Manuscripta Math.* **13**, 143–152.

RESSEL, P. (1976). A short proof of Schoenberg's theorem. *Proc. Amer. Math. Soc.* **57**, 66–68.

RESSEL, P. (1977). Some continuity and measurability results on spaces of measures. *Math. Scand.* **40**, 69–78.

RESSEL, P. (1979). Positive definite functions on abelian semigroups without zero. In *Studies in Analysis* (Ed. by G.-C. Rota). Adv. Math. Suppl. Studies 4, pp. 291–310. New York–London: Academic Press.

RESSEL, P. (1982a). A general Hoeffding type inequality. *Z. Wahrsch. verw. Gebiete* **61**, 223–235.

RESSEL, P. (1982b). A topological version of Slutsky's theorem. *Proc. Amer. Math. Soc.* **85**, 272–274.

RESSEL, P. (1983). De Finetti-type theorems: an analytical approach. Technical report Nr. 209, Department of Statistics, Stanford University.

ROBERTSON, A. P. and W. J. ROBERTSON (1964). *Topological Vector Spaces.* Cambridge: The University Press.

ROBINSON, R. M. (1969). Some definite polynomials which are not sums of squares of real polynomials. *Notices Amer. Math. Soc.* **16**, 554.

ROBINSON, R. M. (1973). Some definite polynomials which are not sums of squares of real polynomials. In *Selected Questions of Algebra and Logic*, pp. 264–282. (A collection dedicated to the memory of A. I. Mal'cev.) Izdat. "Nauka" Sibirsk. Otdel., Novosibirsk (in Russian).

RUDIN, W. (1959). Positive definite sequences and absolutely monotonic functions. *Duke Math. J.* **26**, 617–622.

RUDIN, W. (1962). *Fourier Analysis on Groups.* New York: Interscience Publishers.

RUDIN, W. (1973). *Functional Analysis.* New York: McGraw-Hill.

SCHAEFER, H. H. (1971). *Topological Vector Spaces.* New York–Heidelberg–Berlin: Springer-Verlag.

SCHEMPP, W. (1977). On functions of positive type on commutative monoids. *Math. Z.* **156**, 115–121.

SCHMÜDGEN, K. (1979). An example of a positive polynomial which is not a sum of squares of polynomials. A positive, but not strongly positive functional. *Math. Nachr.* **88**, 385–390.

SCHOENBERG, I. J. (1938a). Metric spaces and completely monotone functions. *Ann. of Math.* **39**, 811–841.

SCHOENBERG, I. J. (1938b). Metric spaces and positive definite functions. *Trans. Amer. Math. Soc.* **44**, 522–536.

SCHOENBERG, I. J. (1942). Positive definite functions on spheres. *Duke Math. J.* **9**, 96–108.

SCHUR, I. (1911). Bemerkungen zur Theorie der beschränkten Bilinearformen mit unendlich vielen Veränderlichen. *J. Reine Angew. Math.* **140**, 1–29.

SCHWARTZ, L. (1973). *Radon Measures on Arbitrary Topological Spaces and Cylindrical Measures.* London: Oxford University Press.

SHOHAT, J. A. and J. D. TAMARKIN (1943). *The Problem of Moments.* Math. Surveys 1. Providence: American Mathematical Society.

SIBONY, D. (1967–1968). *Cônes de fonctions et potentiels.* Cours de 3ème cycle à la Faculté des Sciences de Paris.

STEWART, J. (1976). Positive definite functions and generalizations, a historical survey. *Rocky Mountain J. Math.* **6**, 409–434.

STIELTJES, T. J. (1894). Recherches sur les fractions continues. *Ann. Fac. Sci. Toulouse* **8**, 1–122, and **9**, 1–47. (In *Oeuvres Complètes*, Vol. II, pp. 402–567. Groningen: Noordhoff, 1918.)

STOCHEL, J. (1983). The Bochner type theorem for $*$-definite kernels on abelian $*$-semigroups without neutral element. In *Dilation Theory, Toeplitz Operators, and other Topics*, pp. 345–362. (Ed. by G. Arsene). Basel–Boston–Stuttgart: Birkhäuser Verlag.

ŠVECOV, K. I. (1939). On Hamburger's moment problem with the supplementary requirement that masses are absent on a given interval. *Commun. Soc. Math. Kharkov* **16**, 121–128 (in Russian).

SZAFRANIEC, F. H. (1977). Dilations on involution semigroups. *Proc. Amer. Math. Soc.* **66**, 30–32.

SZANKOWSKI, A. (1974). On Dvoretzky's theorem on almost spherical sections of convex bodies. *Israel J. Math.* **17**, 325–338.

SZ.-NAGY, B. (1960). *Extension of Linear Transformations in Hilbert Space which extend beyond this Space.* New York: Ungar.

TALAGRAND, M. (1976). Quelques examples de représentation intégrale: Valuations, fonctions alternées d'ordre infini. *Bull. Sci. Math.* **100** (2), 321–329.

TONEV, T. W. (1979). Positive-definite functions on discrete commutative semigroups. *Semigroup Forum* **17**, 175–183.

TOPSØE, F. (1970). *Topology and measure.* Lecture Notes in Mathematics 133, Berlin-Heidelberg–New York: Springer-Verlag.

TOPSØE, F. (1978). On construction of measures. In *Topology and Measure* I, pp. 343–381. (Proceedings of a conference in Zinnowitz 1974. Ed. by Flachsmeyer, Frolik and Terpe.) Greifswald: Ernst–Moritz–Arndt Universität.

VARADARAJAN, V. S. (1965). Measures on topological spaces. *Amer. Math. Soc. Transl.*, Ser. II, **48**, 161–228.

WALL, H. S. (1931). On the Padé approximants associated with a positive definite power series. *Trans. Amer. Math. Soc.* **33**, 511–532.

WELLS, J. H. and L. R. WILLIAMS (1975). *Embeddings and Extensions in Analysis.* Berlin–Heidelberg–New York: Springer-Verlag.

WIDDER, D. V. (1941). *The Laplace Transform.* Princeton: Princeton University Press.

WILLIAMSON, J. H. (1967). Harmonic analysis on semigroups. *J. London Math. Soc.* **42**, 1–41.

ZARHINA, R. B. (1959). On the two-dimensional problem of moments. *Dokl. Akad. Nauk SSSR* **124**, 743–746 (in Russian).

# List of Symbols

## 1. Sets of Numbers

| | | |
|---|---|---|
| $\mathbb{N} = \{1, 2, 3, \ldots\}$ | | |
| $\mathbb{N}_0 = \{0, 1, 2, \ldots\}$ | | |
| $\mathbb{Z} = \{0, \pm 1, \pm 2, \ldots\}$ | | |
| $\mathbb{Q}$ | rational numbers | |
| $\mathbb{R}$ | real numbers | |
| $\mathbb{C}$ | complex numbers | |
| $\mathbb{K}$ | common symbol for $\mathbb{R}$ or $\mathbb{C}$ | |
| $\mathbb{Q}_+$ | nonnegative rational numbers | |
| $\mathbb{R}_+$ | nonnegative real numbers | |
| $\underline{\mathbb{R}} = [-\infty, \infty[$ | | 213 |
| $\mathbb{C}_+ = \{z \in \mathbb{C} \mid \operatorname{Re} z \geqq 0\}$ | | |
| $D = \{z \in \mathbb{C} \mid \lvert z \rvert \leqq 1\}$ | | |
| $\mathbb{T} = \{z \in \mathbb{C} \mid \lvert z \rvert = 1\}$ | | |
| $\mathbb{T}_d$ | $\mathbb{T}$ considered as a discrete group | 166 |

## 2. Operations on Sets, Functions and Measures

| | | |
|---|---|---|
| $A^c$ | complement of $A$ | |
| $A \backslash B = A \cap B^c$ | | |
| $\bar{A}$ | closure of $A$ | |
| $\mathring{A}$ | interior of $A$ | |
| $\partial A = \bar{A} \backslash \mathring{A}$ | (topological) boundary of $A$ | |
| $\operatorname{ex}(A)$ | extreme points of $A$ | 55 |
| $\operatorname{conv}(A)$ | convex hull of $A$ | |
| $A^\circ$ | polar of $A$ | 12 |
| $A^\perp$ | dual cone | 13 |
| $f \wedge g$ | minimum of $f$ and $g$ | |
| $f \vee g$ | maximum of $f$ and $g$ | |
| $f^+ = f \vee 0$ | | |

| | | |
|---|---|---|
| $f^- = -(f \wedge 0)$ | | |
| $(f \times g)(x, y) = (f(x), g(y))$ | | |
| $(f \otimes g)(x, y) = f(x)g(y)$ | tensorproduct of $f$ and $g$ | |
| $1_A(x) = \begin{cases} 1 & \text{if } x \in A \\ 0 & \text{if } x \notin A \end{cases}$ | indicator function of $A$ | |
| $\text{sgn} = 1_{]0,\infty[} - 1_{]-\infty,0[}$ | signum function | |
| $\text{supp}(f) = \overline{\{f \neq 0\}}$ | support of $f$ | 39 |
| $f \in o(g)$ | | 42 |
| $f \mid B$ | restriction of the function $f$ to $B$ | 34 |
| $\mu \mid B$ | restriction of the measure $\mu$ to $B$ | 21, 53 |
| $\text{supp}(\mu)$ | support of a Radon measure $\mu$ | 22 |
| $\\|\mu\\|$ | total variation of $\mu$ | 40 |
| $\mu^f$ | image measure | 28 |
| $\mu \otimes \nu$ | product measure of $\mu$ and $\nu$ | 26 |
| $\mu * \nu$ | convolution of $\mu$ and $\nu$ | 29 |
| $\mu^{*n} = \mu * \cdots * \mu$ | ($n$ times) | |
| $f\mu = f\,d\mu$ | measure with density $f$ with respect to $\mu$ | 22 |
| $\mu^+, \mu^-$ | Jordan–Hahn decomposition | 41, 44, 45 |
| $\varepsilon_x(B) = 1_B(x)$ | one-point measure, Dirac measure | 22 |
| $\mathscr{A} \otimes \mathscr{B}$ | product of two $\sigma$-algebras $\mathscr{A}$ and $\mathscr{B}$ | |
| $Y^X$ | set of all mappings from $X$ to $Y$ | |
| $\mathscr{A}_+ := \{f \in \mathscr{A} \mid f(x) \geqq 0 \text{ for all } x \in X\}$ if $\mathscr{A} \subseteq \mathbb{C}^X$ | | |

## 3. Families of Subsets of a Hausdorff Space $X$

| | | |
|---|---|---|
| $\mathscr{G} = \mathscr{G}(X)$ | open subsets of $X$ | 137, 268 |
| $\mathscr{F} = \mathscr{F}(X)$ | closed subsets of $X$ | 136, 268 |
| $\mathscr{F}^B, \mathscr{F}_B$ | | 137 |
| $\mathscr{K} = \mathscr{K}(X)$ | compact subsets of $X$ | 17, 136, 267 |
| $\mathscr{K}^B, \mathscr{K}_B$ | | 268 |
| $\mathscr{B}(X)$ | Borel subsets of $X$ | 17 |
| $\mathscr{U}$ | filter of neighbourhoods of zero | 1 |
| $\sigma(E, F)$ | weak topology | 11 |
| $\tau(E, F)$ | Mackey topology | 12 |
| $\mathscr{V}(K, G) = \{L \in \mathscr{K} \mid K \subseteq L \subseteq G\}$ | | 136, 268 |

## 4. Spaces of Functions and Measures Associated with a Hausdorff Space $X$

| | | |
|---|---|---|
| $C(X)$ | continuous (real-valued) functions on $X$ | 39 |
| $C^b(X)$ | continuous bounded functions | 39 |
| $C^c(X)$ | continuous functions with compact support | 39 |
| $C^0(X)$ | continuous functions tending to zero at infinity | 39 |
| $M_+(X)$ | Radon measures on $X$ | 17 |
| $M_+^b(X)$ | finite (=bounded) Radon measures | 29, 45 |
| $M_+^c(X)$ | Radon measures with compact support | 93 |
| $M_+^1(X)$ | Radon probability measures | 49 |
| $M(X)$ | signed Radon measures | 40 |
| $M^c(X)$ | signed Radon measures with compact support | 96 |
| $\text{Mol}_+(X)$ | molecular measures | 22 |

## 5. Operators, Function Spaces and Spaces of Measures Associated with an Abelian Semigroup $(S, +, *)$

| | | |
|---|---|---|
| $I$ | identity operator | 129 |
| $E_a$ | shift operator | 91, 123 |
| $\Delta_a = E_a - I$ | | 110, 129, 130 |
| $\Gamma_a = \frac{1}{2}(E_a + E_{a*})$ | | 103 |
| $\nabla_a = -\Delta_a$ | | 130 |
| $\nabla_n f(s; a_1, \ldots, a_n)$ | | 130 |
| $S^*$ | dual semigroup | 92 |
| $\hat{S}$ | restricted dual semigroup | 96 |
| $\chi_s$ | evaluation function on $S^*$ | 180 |
| $V(\mathbb{R}), V(\mathbb{C})$ | | 181 |
| $\mathscr{P}(S)$ | positive definite functions on $S$ | 88 |
| $\mathscr{P}_1(S) = \{\varphi \in \mathscr{P}(S) \mid \varphi(0) = 1\}$ | | 88 |
| $\mathscr{P}^\alpha(S)$ | $\alpha$-bounded positive definite functions | 94 |
| $\mathscr{P}_1^\alpha(S) = \{\varphi \in \mathscr{P}^\alpha(S) \mid \varphi(0) = 1\}$ | | 94 |
| $\mathscr{P}^e(S)$ | exponentially bounded positive definite functions | 94 |
| $\mathscr{P}^b(S)$ | bounded positive definite functions | 96 |
| $\mathscr{P}_1^b(S) = \{\varphi \in \mathscr{P}^b(S) \mid \varphi(0) = 1\}$ | | 96 |
| $\mathscr{P}_r(B)$ | | 153 |
| $\mathscr{N}(S)$ | negative definite functions | 89 |
| $\mathscr{N}^l(S)$ | lower bounded negative definite functions | 99 |
| $\mathscr{N}^b(S)$ | bounded negative definite functions | 105 |
| $\mathscr{M}(S)$ | completely monotone functions | 130 |
| $\mathscr{M}^1(S) = \{\varphi \in \mathscr{M}(S) \mid \varphi(0) = 1\}$ | | |
| $\mathscr{A}(S)$ | completely alternating functions | 130 |
| $\mathscr{H}(S)$ | moment functions | 179 |
| $E_+(S^*) = \{\mu \in M_+(S^*) \mid \int \lvert\chi_s\rvert \, d\mu < \infty \text{ for all } s \in S\}$ | | 179 |
| $E_+(S^*, \varphi) = \{\mu \in E_+(S^*) \mid \int \chi_s \, d\mu = \varphi(s) \text{ for all } s \in S\}$ | | 180 |
| $E(S^*) = \{\mu_1 - \mu_2 + i(\mu_3 - \mu_4) \mid \mu_1, \ldots, \mu_4 \in E_+(S^*)\}$ | | 203 |
| $\mathfrak{L}$ | set of Lévy measures | 105 |

## 6. Notations Related to an Abelian Semigroup $(H, +)$ which (Possibly) has no Zero $(S := H \cup \{0\})$

| | | |
|---|---|---|
| $H' = H + H$ | the nonmaximal elements of $H$ | 252 |
| $\check{H} = \hat{S} \setminus \{1_{\{0\}}\}$ | | 255 |
| $\mathscr{P}(H)$ | positive definite functions on $H$ | 253 |
| $\mathscr{P}^q(H)$ | quasibounded positive definite functions | 253 |
| $\mathscr{P}_0^q(H)$ | | 254 |
| $\mathscr{N}(H)$ | negative definite functions on $H$ | 253 |
| $\mathscr{N}^q(H)$ | negative definite functions, quasibounded below | 257 |
| $\mathscr{N}_0^q(H)$ | | 257 |
| $\mathscr{M}(H)$ | completely monotone functions on $H$ | 264 |
| $\mathscr{M}_0(H)$ | | 264 |
| $\mathscr{A}(H)$ | completely alternating functions on $H$ | 264 |
| $\mathscr{A}_0(H)$ | | 265 |

## 7. Miscellaneous Symbols

| | | |
|---|---|---|
| $p_A$ | seminorm determined by $A$ | 3 |
| $\|\cdot\|_\infty$ | supremum norm | |
| $E^*$ | algebraic dual space | 10 |
| $E'$ | topological dual space | 10 |
| alg span($\tau$), alg span$^+(\tau)$ | | 123, 124 |
| $A_\tau^*$ | $\tau$-positive linear functionals | 124 |
| $l^p = \left\{x \in \mathbb{R}^{\mathbb{N}} \,\middle\vert\, \sum_{n=1}^{\infty} \|x(n)\|^p < \infty\right\}, 0 < p < \infty$ | | 148, 153 |
| $\mathscr{L}^p(\mu) = \mathscr{L}^p(X, \mu) = \{f: X \to \mathbb{C} \mid f \text{ measurable}, \int \|f\|^p \, d\mu < \infty\}, 0 < p < \infty$ | | |
| $L^p(\mu)$ | the associated Banach space (if $p \geqq 1$) | |
| $\mathscr{L}$ | Laplace transformation | 75, 114, 115 |
| $\hat{\mu}$ | generalized Laplace transform of $\mu$ | 96, 203 |
| $\hat{\mu}$ | generating function of $\mu \in M_+^1(\mathbb{N}_0)$ | 161 |
| $A^{(k)}$ | polynomials in $k$ variables with real coefficients | 129, 190 |
| $A_{\mathbb{C}}^{(k)}$ | polynomials in $k$ variables with complex coefficients | 197 |
| $A_d^{(k)}$ | polynomials in $A^{(k)}$ of degree at most $d$ | 191 |
| $A_+^{(k)}(F) = \{p \in A^{(k)} \mid p(x) \geqq 0 \text{ for all } x \in F\}$ | | 196 |
| $\mathbb{P}$ | probability | |
| $\mathbb{E}$ | expectation | |
| var | variance | |
| i.i.d. | independent and identically distributed | |
| $B(n, p)$ | binomial distribution | 165 |
| $\pi_b$ | Poisson distribution | 164, 165, 235 |
| $x \prec y$ | order of majorization | 240 |
| $\mu \lhd \nu$ | | 58 |
| $\mathbf{1} = (1, \ldots, 1)$ | | |
| $\Omega_n$ | doubly stochastic $n \times n$ matrices | 60, 240 |
| $\mathscr{S}$ | set of subsemigroups, hereditary on the left | 119, 136, 267 |
| $\mathscr{S}_r = \{J \in \mathscr{S} \mid 1_J \text{ is right continuous}\}$ | | 136, 267 |
| RKHS | reproducing kernel Hilbert space | 81 |

# Index

## Graduate Texts in Mathematics

*continued from page ii*

93 DUBROVIN/FOMENKO/NOVIKOV. Modern Geometry – Methods and Applications Vol. I.
94 WARNER. Foundations of Differentiable Manifolds and Lie Groups.
95 SHIRYAYEV. Probability, Statistics, and Random Processes.
96 ZEIDLER. Nonlinear Functional Ananlysis and its Applications I: Fixed Points Theorem.
97 ZEIDLER. Nonlinear Functional Analysis and its Applications II (in preparation).
98 ZEIDLER. Nonlinear Functional Analysis and its Applications III; Variational Methods and Optimization.
99 ZEIDLER. Nonlinear Functional Analysis and its Applications IV (in preparation).
100 BERG/CHRISTENSEN/RESSEL. Harmonic Analysis on Semigroups: Theory of positive definite and related functions.
101 EDWARDS. Galois Theory.
102 VARADARAJAN. Lie Groups, Lie Algebras and Their Representations (1st Springer Edition).